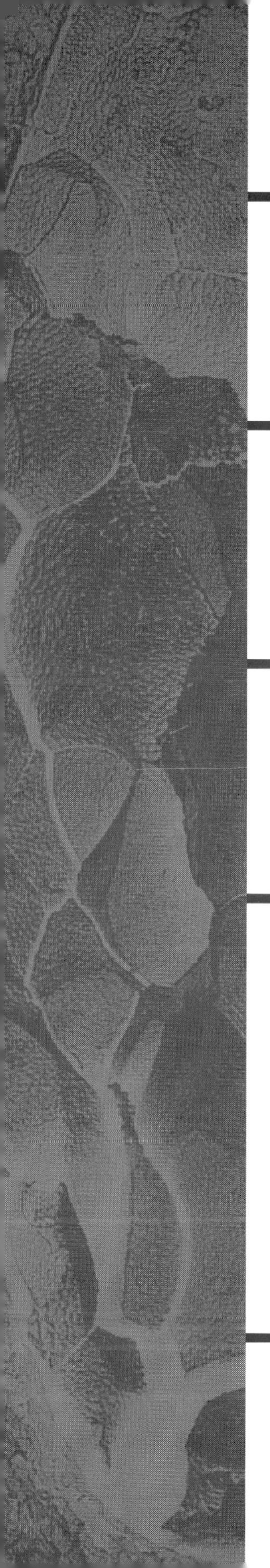

N. EL BASSAM

ENERGY PLANT SPECIES

THEIR USE AND IMPACT ON ENVIRONMENT AND DEVELOPMENT

Published by James & James (Science Publishers) Ltd,
35–37 William Road, London NW1 3ER, UK

A catalogue record for this book is available from the British Library

ISBN 1 873936 75 3

Printed in the UK by MPG Books Limited

Contents

Appendices

Indexes

Foreword

Today, it is estimated that more than two billion people worldwide lack access to modern energy resources. Although we know that energy is absolutely essential for development, relatively little attention has been devoted to this matter at national and international levels.

The magnitude of energy consumed per capita has become one of the indicators of the 'modernization' and progress of a country. Thus energy issues and policies have been strongly concerned with increasing the supply of energy. The strategic and environmental consequences of energy consumption patterns have been neglected for a long time. The world continues to seek energy to satisfy its needs without giving consideration to the social, environmental, economic and security impacts of energy use.

Many of the crises on our planet arise from the desire to secure supplies of raw materials, particularly energy sources, at low prices. The pressure will become even greater as fossil energy feedstocks and uranium are depleted. Although some of these resources might last a little longer than predicted, especially if additional reserves are discovered, the main problem of 'scarcity' will remain, and this represents the greatest challenge to humanity.

It is now clear that current approaches to energy are unsustainable and not renewable. Of all renewable energy sources generally the largest contribution, especially in the short and medium range, is expected to come from biomass. Fuels derived from energy crops are not only potentially renewable, but are also sufficiently similar in origin to the fossil fuels (which also began as biomass) to provide direct substitution. They can be converted into a wide variety of energy carriers using existing and novel conversion technologies, and thus have the potential to be significant new sources of energy into the twenty-first century.

This book deals with various aspects related to the potential of energy plant species that can be grown on plantations for production of fuel feedstocks, and with appropriate upgrading and conversion technologies, along with their environmental, economic and social dimensions.

Most grateful thanks are due to B. Prochnow who did the most arduous and time consuming work of preparing the manuscript. I would also like to thank S.G. Agong, W. Bacher, C. Baldelli, D.G. Christian, L. Dajue, C.D. Dalianis, W. Elbersen,

J. Fernández, A.K. Gupta, K. Jakob, S.F. Khalifa, V. Petriková, A. Riedacker, S. Roy, and M. Satin for providing contributions, as well as S. Klahr and A. Voges for their efforts in providing additional information.

I wish and hope that this book will contribute to the increase in interest and understanding in the vital economic and social roles of biomass to meet the growing demand for energy and to face the future challenges of limited fossil fuel reserves and global warming.

Dr N. El Bassam
Braunschweig, March 1998

Preface

For earth's inhabitants, plants have always served as a vital source of food, raw materials and energy. These roles depend on the ability of green plants to transform light energy into chemical energy. In the process of photosynthesis, carbon dioxide in the atmosphere is accumulated by plants and converted into sugars. This is the basis for the synthesis of a wide variety of natural raw materials, including starch, fibres, vegetable oils, and gums, as well as a vast array of organic compounds. It is the photosynthetic process too that accounts for the world's energy deposits of fossil fuels on which man so heavily relies.

Throughout the history of mankind until the middle of the nineteenth century, biomass, primarily wood, was the most important source of energy and raw materials. The gradual transition to fossil energy sources – coal, mineral oil and natural gas – ushered in the age of industrialization, whereas interest in using biomass for non-nutritional purposes was confined to special areas.

Now that the end of fossil raw materials is in sight, the worldwide search for alternative sources for industrial and energy purposes has intensified. Together with water, wind, solar, tidal and geothermal energy sources, biomass can be considered as an essential key to future renewable energy strategies.

Alexander King, President of the Club of Rome, warned in 1985 about a new global crisis:

> We are approaching the point where the energy consumption for exploration and transportation outside the Middle East is higher than the energy which is extracted from it. Our national economies should be steered by energy balances and not mainly through monetary dimensions. Money is relative and transient, but energy is essential and eternal. We should realize that problems of energy, environment, climate and development are interconnected.

We need to conserve some of our fossil fuel resources for the future and to create adequate substitutes in quantities that could meet the requirements of the people and enable future development. Brundtland (1987) wrote in a Report to the European Parliament: 'every effort should be made to develop the potential for renewable

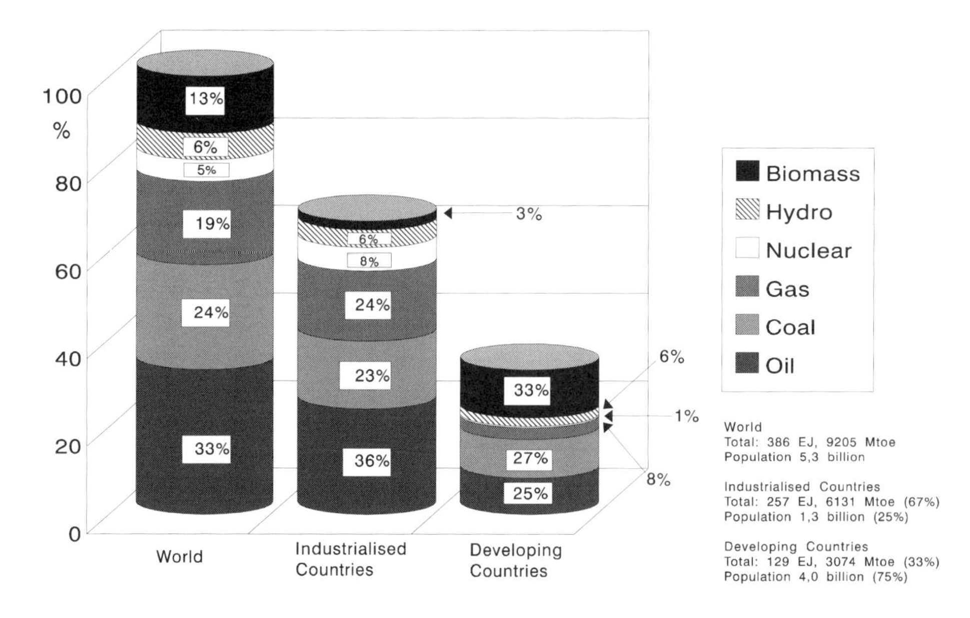

Figure 1. World primary energy consumption, 1990 (Hall and House, 1995).

energy which should form the foundation of the global energy structure during the 21st century'.

Biomass currently contributes almost 15% of global primary energy consumption. Biomass is used mostly in developing countries with much less efficiency than is technically possible and economically feasible. In industrial countries, biomass, mainly in the form of industrial and agricultural residues and municipal wastes, is presently used to produce thermal energy and to generate electricity. Figure 1 provides a look at the distribution of energy consumption in the forms of biomass, hydro, nuclear, gas, coal, and oil worldwide and among industrialized and developing countries.

Crop production strategies need to be developed that are as efficient as possible in capturing sunlight (solar energy) and storing it in plants (the solar battery) (Samson *et al.*, 1994). Desirable characteristics for energy feedstocks include:

- efficient conversion of sunlight into plant material (biofuel);
- efficient water use, because moisture is one of the primary factors limiting biomass production in most parts of the world;
- sunlight interception for as much of the growing season as possible;
- minimal external inputs to the production and harvest cycle (seed, fertilizer, machine operations and crop drying) – i.e. low input plant species;
- a sustainable energy balance – i.e. a positive energy balance;
- high dry matter contents at time of harvest;
- high energy density (expressed in MJ/kg etc.) – i.e. rich in oils, sugars, starches, lignocellulose, etc.;
- their production and use have the lowest possible environmental impacts.

More than 450,000 plant species have been identified worldwide; approximately 3000 of these are used by humans as sources of food, tools and other feedstocks. About 300 plant species have been domesticated as crops for agriculture; of these, 60 species are of major importance. It will be a vital task to increase the number of plant species that could be grown to produce fuel feedstocks. This can be done by identifying, screening, adapting and breeding a part of the large potential of the flora available on our planet. The introduction of new plant species to agriculture would lead to an improvement in the biological and environmental condition of soils, water, vegetation and landscapes. This would be achieved by increasing biodiversity to replace the present monoculture production systems.

Part I

We shall require a substantially new means of thinking if mankind is to survive

Albert Einstein

1 *Basic Elements of Biomass Accumulation*

Nearly all of the energy that we use today has been collected and stored by the process of photosynthesis. The amount of carbon fixed each year by autotrophic plants, in the form of biomass, is roughly 200 billion tonnes, or approximately ten times the energy equivalent used yearly in the world. Of these 200 billion tonnes, 800 million (0.4%) are used to feed the human population.

The fundamental process of biomass accumulation within the context of energy is based on photosynthesis. The green plant is the only organism able to absorb solar energy with the help of the green pigment chlorophyll, converting this energy into the chemical energy of organic compounds with the aid of carbon dioxide and water (Figure 1.1). In addition to carbon, hydrogen and oxygen, plants also incorporate nitrogen and sulphur into organic matter by means of light reactions.

The photosynthetic process also requires other nutrients for optimum functioning, and an adequate temperature at which to take place. The efficiency of photosynthesis, expressed as the ratio between chemical energy fixed by plants and the energy contained in light rays falling on plants, is less than 1%. A small increase in this

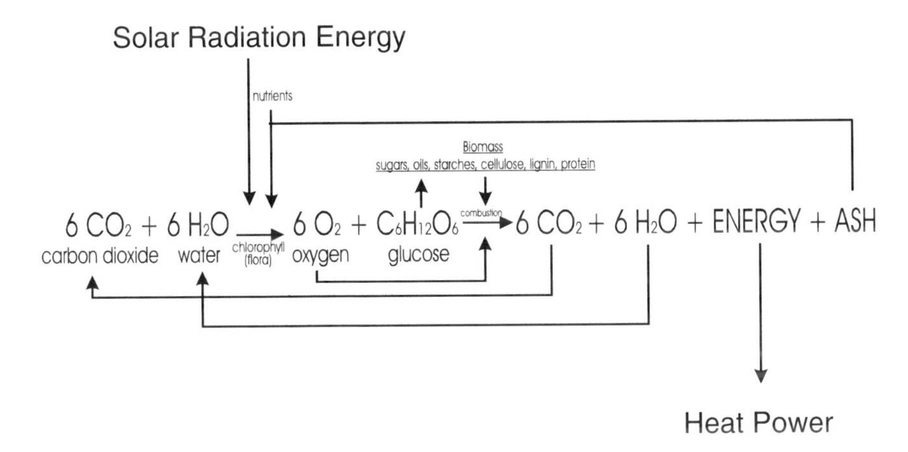

Figure 1.1. The almost closed cycle of energy production from biomass.

efficiency would have spectacular effects in view of the scale of the process involved. However, what has been defined as 'photosynthetic efficiency' is actually the sum of the individual effectiveness of a large number of complex and related processes taking place both within the plants and in their environment.

If we summarize the solar energy that hits every hectare globally, it averages out to be the amount found at 40° latitude where every hectare receives 1.47×10^{13} calories of total energy radiation (TER). If this energy were completely converted into the chemical energy of carbohydrates we would expect a yield of more than 2000 tonnes per hectare. But the photosynthetically active radiation (PAR) amounts to 43% of the total solar radiation on the earth's surface.

In the case of a crop yielding 10 tonnes of dry matter per hectare per year, located at 40° latitude, the photosynthetic efficiency would be 0.27% for TER or 0.63% for PAR. As the PAR is about 43% of TER, the upper limit of efficiency for the absorption of solar radiation is 6.8% of the total radiation, or 15.8% with respect to PAR.

Considering the total amount of radiation incident at certain areas, and keeping in mind that the calorific content of dry matter is about 4000 kilocalories per kilogram (= 4kcal/g), a theoretical upper production limit of about 250 tonnes of dry matter per hectare could be considered for an area having an amount of incident radiation similar to that at 40° latitude. This production value has not yet been attained with any crop, because the vegetation cycle and high growth and solar utilization rates cannot be achieved and maintained continuously throughout the entire year.

From laboratory studies, we know that the photosynthetic apparatus of green plants works with an efficiency of more than 30%. Why this discrepancy? What we really measure when we determine the harvested biomass of our fields is the difference between the efficiencies of photosynthesis and respiration. Furthermore, we have to consider that the growing season for main crops comprises just a fraction of the year, and that mutual shading from the leaves, unfavourable water content of the soil, air humidity and too low or too high temperatures all keep the yield below the theoretical limit. Besides this, the solar spectrum contains broad regions (wavelengths below 400nm and above 700nm) that are photosynthetically almost inactive. Another important factor is the availability of adequate resources of water and nutrients during the growth cycle. There are three main groups of determinants that affect photosynthesis: environmental factors, the genetic structure of genotype and population, and external inputs (Figure 1.2). Their influence on the plant's photosynthesis controls the final biomass yield.

Photosynthetic pathways

The two major pathways of photosynthesis are the C_3 pathway and the C_4 pathway. A third, less common pathway is the Crassulacean acid metabolism (CAM) pathway. In the C_3 pathway, the first product of photosynthesis is a 3-carbon organic acid (3-phosphoglyceric acid), whereas in the C_4 the first products are 4-carbon organic acids (malate and aspartate). In general, the C_3 assimilation pathway is adapted to operate at optimal rates under conditions of low temperature (15–20°C). At a given radiation level, the C_3 species have level relatively lower rates of CO_2 exchange than C_4 species, which are adapted to operate at optimal rates under conditions of higher temperature (30–35°C) and have comparatively higher rates of CO_2 exchange. Furthermore, C_4

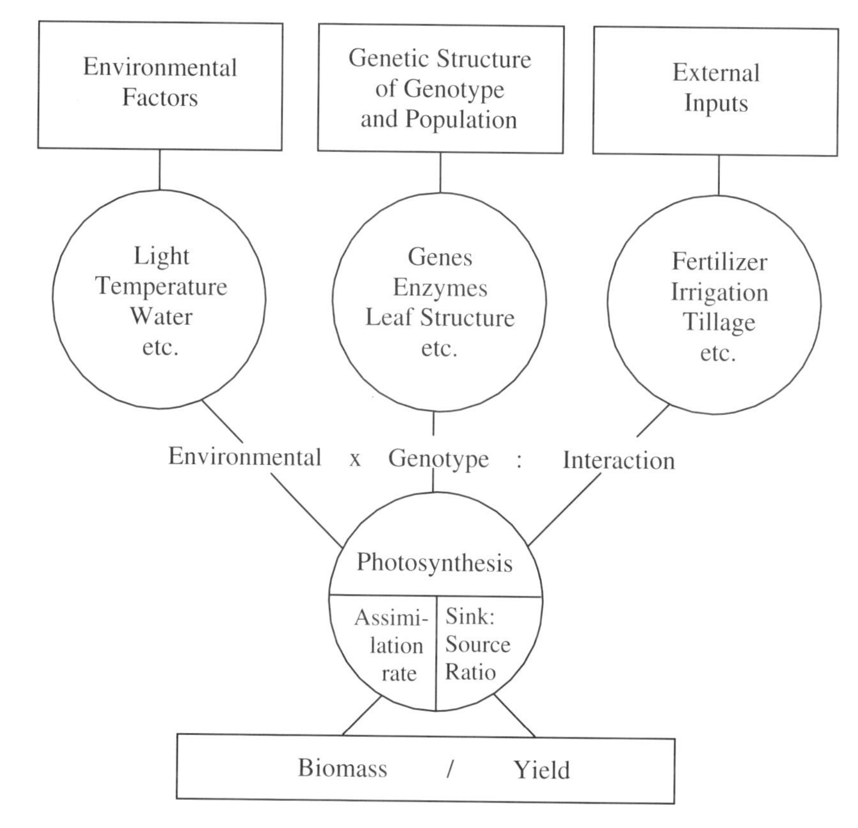

Figure 1.2. Determinants of potential biomass production (El Bassam, 1990).

species have maximum rates of photosynthesis in the range of 70–100mg CO_2/dm²/h with light saturation at 1.0–1.4cal/cm²/min total radiation, while C_3 species have maximum rates of photosynthesis in the range 15–30mg CO_2/dm²/h with light saturation at 0.2–0.6cal/cm²/min.

One additional group of species has evolved and adapted to operate under xerophytic conditions. These species have the CAM. Although the biochemistry of photosynthesis in the CAM species has several features in common with C_4 species, particularly the synthesis of 4-carbon organic acids, CAM species have some unique features that are not observed in C_3 and C_4 species. These include capturing light energy during daytime and fixing CO_2 during the night and, consequently, very high water use efficiencies. There are two CAM species of agricultural importance: pineapple and sisal.

The C_3, C_4 and CAM crops can be roughly classified into five groups (Table 1.1), based on the differences between crop species in their photosynthetic pathways and the response of photosynthesis to radiation and temperature.

C_4 plant species tend to be more efficient in water utilization because they do not have to open their stomata so much for the inward diffusion of CO_2. Thus transpiration is reduced. Figure 1.3 demonstrates that, although irrigation generally leads to higher yields, the C_4 crops *Miscanthus* and maize had higher dry matter yields under

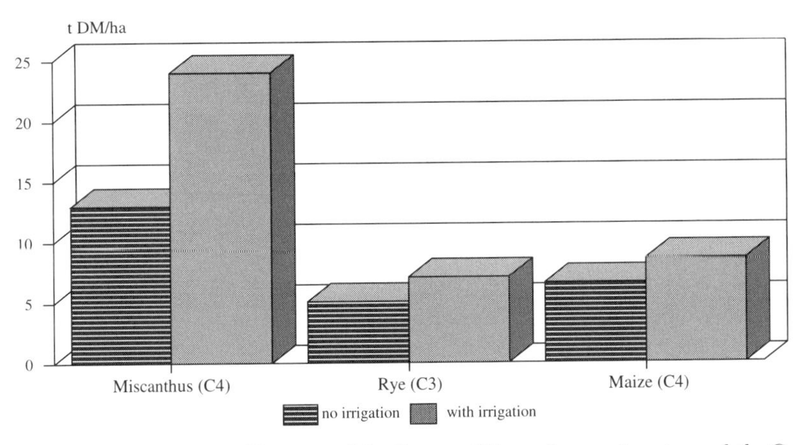

Figure 1.3. Production of biomass of the C_4 crops Miscanthus *and maize and the C_3 crop rye under two water supply conditions.*

Table 1.1. Some physiological characteristics according to photosynthesis type (FAO, 1980; revised El Bassam, 1995).

Photosynthesis characteristics	Crop group 1	2	3	4	5
Photosynthesis pathway	C_3	C_3	C_4	C_4	CAM
Radiation intensity at max. photosynthesis (cal cm^{-2} min^{-1})	0.2–0.6	0.3–0.8	1.0–1.4	1.0–1.4	0.6–1.4
Max. net rate of CO_2 exchange at light saturation (mg dm^{-2} h^{-1})	20–30	40–50	70–100	70–100	20–50
Temperature response of photosynthesis:					
optimum temp.	15–20°C	5–30°C	25–30°C	10–35°C	30–35°C
operative temp.	15–45°C	20–30°C	10–35°C	25–35°C	10–45°C
Max. crop growth rate (g m^{-2} day^{-1})	20–30	30–40	30–60	40–60	20–30
Water use efficiency (g/g)	400–800	300–700	150–300	150–350	50–200
Crop species	field mustard, potato, oat, rye, rape, sugarbeet, wheat, sunflower, olive, barley, lentil, linseed	groundnut, rice, soybean, sesame, tobacco, sunflower, castor, safflower, kenaf, sweet potato, bananas, coconut, cotton, cassava, *Arundo donax*	Japanese barnyard millet, sorghum, maize, sugarcane	millet, sorghum, maize, *Miscanthus*, *Spartina*, *Panicum virgatum*	sisal, pineapple

conditions of non-irrigation than did the C_3 crop rye. The corollary is that C_3 species are usually more efficient net photosynthesizers at lower temperatures than C_4 species. Both metabolic groups are capable of substantial climatic adaptation (Clayton and Renovize, 1986). *Miscanthus* spp. and *Spartina* spp. are both C_4 species and more efficient photosynthesizers than a wide variety of C_3 grasses. This reflects the fact that there are possibilities for the genetic selection of C_4 plants with improved cold tolerance. The C_4 species may utilize nitrogen with greater efficiency.

An estimate of the upper threshold for converting incident solar energy into biomass production for C_3 and C_4 species was made by Hall *et al.* (1993). For C_4 plants (plants where the first stable products of photosynthesis are 4-carbon compounds) it is 100% × 0.50 × 0.80 × 0.28 × 0.60 = 6.7%, where:

- 0.50 represents the 50% of light that is photosynthetically active radiation (PAR), at wavelengths between 400 and 700nm;
- 0.80 represents the roughly 80% of PAR captured by photosynthetically active compounds; the rest is reflected, transmitted, and absorbed by non-photosynthesizing leaves;
- 0.28 represents 28%, the theoretical maximum energy efficiency of converting the effectively absorbed PAR to glucose; and
- 0.60 represents the roughly 60% energy stored in photosynthesis remaining after 40% is consumed during dark respiration to sustain plant metabolic processes.

For C_3 plants the equivalent figure is 6.7% × 0.70 × 0.70 = 3.3%. There are two additional losses for these plants: C_3 plants lose about 30% of the already fixed CO_2 during photorespiration; and C_3 plants become light-saturated at lower light intensities than C_4 plants, so that C_3 plants are unable to utilize perhaps 30% of the light absorbed by photosynthetically active compounds.

These figures can be now used to calculate the potential dry matter yield for each biomass crop. However, the following information is also required:

- length of growing season
- average daily solar radiation in the region
- energy content of the feedstocks
- partitioning of the biomass above and below ground
- availability of water and nutrients.

The final biomass accumulation is then determined by the genetic structure of the genotype and population.

Plants with the C_4 photosynthetic pathway have a maximum possible potential in temperate climates of 55t/ha, compared to 33t/ha for temperate C_3 crops (Bassham, 1980). The potentials could be much higher for warmer and more humid climates. Samson and Chen (1995) estimated that the maximum photosynthetic efficiency of a biomass plantation studied in Canada would be 64.8 ODT/ha for switchgrass (C_4) and 33.6 ODT/ha for willow (C_3), ODT being oven dried tonne.

The growth and productivity of crops depends on their genetic potential. How far this potential is realized is closely related to the environmental factors dominating in the region and to the external inputs. Yields achieved in field experiments do not represent the physiological limits of the present cultivars, but only demonstrate that portion of the genetic potential that is realized by the optimal utilization of present means of cultivation and levels of inputs (Dambroth and El Bassam, 1990). Yield and response of cultivars to environmental conditions and inputs is under genetic control, and therefore an improved response is accessible via screening, selection and breeding.

Climate and local soil properties are the most important constraints influencing the biomass productivity within a specific region. Temperature, precipitation and global radiation are the few classical climatic factors that influence, to a great extent, the regional distribution of various plant species.

Breeding and selection, through both natural and human agencies, have changed the temperature response of photosynthesis in some C_3 and C_4 species. Consequently, there are C_3 species (such as cotton and groundnut) whose optimal temperature is in a medium to high range (25–30°C), and there are C_4 species (such as maize and sorghum), where, for temperate and tropical highland cultivars, the optimal temperature is in a low to medium range (20–30°C).

Algae

Algae too are plants that can absorb solar energy and carbon dioxide from the atmosphere, which they store as biomass. In Europe, research is under way on microalgal cultivation in various laboratories at different scales. This research includes the use of waste waters and mineralized or brackish waters. Productivity seems to be potentially very high (5% photosynthetic efficiency and up to 50 dry t/ha/y) (Palz and Chartier, 1980). Higher rate ponds of up to 25ha are being used in the USA; these are mostly concerned with production of algae of various sorts in association with wastewater treatment.

The production of biodiesel from algae is the target of a large-scale project conducted by several universities and the automobile and energy industries. The project's aims are to produce biomass from solar energy and carbon dioxide from exhaust gases by means of algae, and the first experimental results were very encouraging. The algae growth rate in newly developed reactors was very high (BS, 1996). The goal is to convert the algae biomass into biodiesel.

Because theoretical and claimed yields are at potentially cost effective levels, and there are alternative users for the algal biomass, work is needed on techniques to increase yields and convert feedstocks to usable fuel.

Intensively managed microalgal production facilities are capable of fixing several times more carbon dioxide per unit area than trees or crops. Although CO_2 is still released when fuels derived from algal biomass are burned, integration of microalgal farms for flue gas capture approximately doubles the amount of energy produced per unit of CO_2 released. Materials derived from microalgal biomass can also be utilized for other long term uses, serving to sequester CO_2.

Seeing microalgae farms as a means to reduce the effects of the greenhouse gas CO_2 changes the view of the economics of the process. Instead of requiring that microalgae-derived fuel be cost competitive with fossil fuels, the process economics must be compared with those of other technologies proposed to deal with CO_2 pollution. However, the development of alternative, environmentally safer energy production technologies will benefit society whether or not global climate change actually occurs.

Microalgal biomass production has great potential to contribute to world energy supplies, and to control CO_2 emissions as the demand for energy increases. This technology makes productive use of arid and semi-arid lands and highly saline water, resources that are not suitable for agriculture and other biomass technologies (Chelf *et al.*, 1993).

2 *Harvesting, Transportation and Storage*

The availability of adequate logistic systems, which include harvesting, recovery, compacting, transport, upgrading and storage, represents a basic requirement for the utilization of energy crops as feedstocks for industrial and energy purposes.

Each conversion technology has specific requirements concerning dry matter content, shape, size, and particle consistency of the raw material. The logistics of the raw materials provides the tool for establishing an effective link between agricultural production systems and industrial activities.

Mechanization of harvest, transport and storage is defined by the methods and processing of the primary products and the need for year-round availability. Since semi-finished products can be created with various properties and qualities and can be used for various final products, chain management is needed. This also relates to possible storage techniques and conditions during harvest (in winter and spring), and therefore to the workability and timeliness aspects of harvest and storage.

Harvesting of agricultural products is always associated with the handling, transportation and processing of large volumes. This is particularly true of the harvest of biofuels from annually yielding herbaceous field crops, such as energy cereals, *Miscanthus*, straw and hay. Because the total dry matter contributes to the energetic yield of these crops, the annual turnover can be extremely high. Moreover, cost-effective combustion units have to exceed a certain size; thus the area for biofuel production and supply has to be relatively large and increases the need for long-distance transportation in order to meet the demand for biofuel. This reveals one of the main problems for biofuels: while the mass-related energetic density of solid biomass varies only little, the required volume for a single unit of fuel equivalent can easily vary by a factor of ten, depending on the method of harvesting or processing (Figure 2.1). In the case of the storage of chopped biomass, the volume demand can be reduced tenfold by the application of a high density compression technology. Thus, the development of appropriate harvesting and compacting systems promises to be extremely beneficial in view of logistical improvements. Such systems appear to be essential for a large scale introduction of biofuels to the energy market (Hartmann, 1994).

Harvesting systems and machinery

Biomass can be harvested in different ways. From the harvesting standpoint, the most important characteristics of energy crops are moisture content and hardness of the stem. There are now several existing harvesting systems, and newly developed or modified harvesting machines are being tested. These harvesting systems follow two main procedures:

- Multi-phase procedure (several machines must be used)
  - cutting (mowing)
  - pick-up, compacting and baling
- Single-phase procedure (one machine is used)
  - chopping line
  - baling line
  - bundling line
  - pelleting line

Multi-phase procedure

For the multi-phase procedure, existing harvesting machines such as mowers and balers can be employed. A rotary or double knife mower can be used to cut the whole crop, which must then be put in swaths before baling. After swathing, the following baler picks up almost all of the harvested material.

The numerous types of baling machines produce rectangular bales, round bales or compact rolls. The big round baler (Figure 2.2) and the big rectangular baler have demonstrated good results for compacting: dry matter density in these high pressure bales is about 120kg/m³.

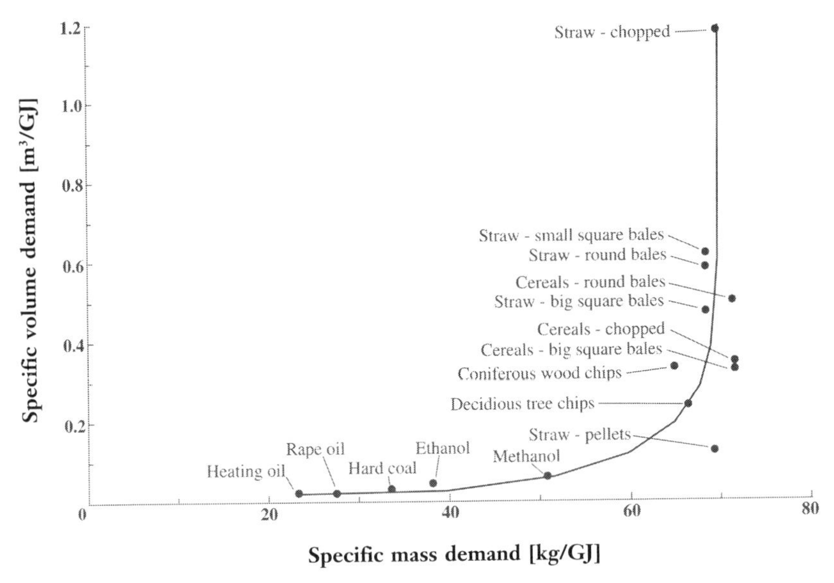

Figure 2.1. Mass and volume related energy density of biofuels (Hartmann, 1994).

Figure 2.2. Big round baler (Welger).

The Welger CRP 400 compact round baler produces very highly compacted bales of 0.40m diameter and varying in length from 0.50m to 2.50m. The dry matter density of the compact roll is about 350kg/m³. This high compaction technology helps to minimize transport and storage: the volume demand of a compact baler is about ten times lower than that of chopped material (Hartmann, 1994). The bulk density can obviously be reduced. Thus, harvesting with a compact roller baler can increase harvesting and transportation efficiency in comparison to other types of balers.

Single-phase procedure

Chopping line

Two different systems can be used for the single-phase chopping line. One is the chopping system normally used for short rotation forestry, which consists of a mowing unit in front of a tractor in combination with a trailer. Depending on the cutting unit, this system can be applied in stands with or without a special row spacing. An alternative approach uses the chop forage harvester designed for harvesting maize. In most cases, a row-independent mowing attachment is necessary, because the planted rows are not distinguishable with increasing age of the stand.

A chopping machine developed by the Claas company has two heavy mowing devices, which also enable cutting and chopping with one machine. A pneumatic conveyor feeds the chopped biomass to a trailer. One disadvantage of this system is that a strong wind can adversely influence the recovery of the biomass.

Depending on the cut length of the material, the densities in dry matter mass of the chopped product are between 70 and 95kg/m³. This low dry matter density is another disadvantage of the chopping system compared to the baling line: the chopped

Figure 2.3. Biomass combine harvester (John Deere).

material needs very high transport and storage capacities; and transportation efficiency, especially over long distances, is extremely low.

Baling line

The single-phase baling line process is based on a type of machine developed by several companies (Figure 2.3). This full plant harvester combines mowing, pick-up, compaction and baling in one pass, and produces a highly compacted square bale. The full harvester combines the advantages of the single-phase procedure with the baling line, and is characterized by very low biomass losses during the whole harvesting process.

Bundling line

A bundling line needs a special machine and should be used only when the whole stem of the plant is necessary for further processing. The harvesting method is based on methods developed for reed culture. One harvesting machine developed by Agostini consists of a mowing unit, binding equipment and a transport/deposit unit attached to the three-point linkage of a tractor. The crop is cut with a cutter bar and transported via the binding unit to the side. The density of the bundles is approximately 140kg/m^3, the weight 9kg and the bundle diameter 0.2m. The machine is a low-cost and comprehensive unit.

Pelleting line

The Haimer company's newly developed 'Biotruck' (Figure 2.4) means that a single-phase pelleting line is now an option. This machine combines mowing, chopping and

Figure 2.4. Biotruck for harvesting and pelleting (Haimer).

pelleting in one procedure on the field. After mowing and chopping, the material is predried using the thermal energy of the engine. The raw material is compacted, pressed and pelleted without additional bonding agents. The result of the process is a corrugated plate with a length of 30–100mm. The single pellet density ranges from 850 to 1000kg/m³, and the bulk density is about 300–500kg/m³. The Biotruck has a capacity of 3–8t/h. Figure 2.5 shows various forms of pellets.

Pelleting offers an excellent opportunity for the reduction of bulk density and hence transportation and storage requirements. In addition, pellets are generally easier to handle than chopped material or bales. If the approach is further utilized, there are good prospects for the automatic handling of such pellets.

Upgrading and storage

For the year-round delivery of biomass, on-farm storage is necessary. Experience with crop harvests has shown that the moisture content at spring harvest is usually between 20 and 40%, so some kind of drying process is required before storage. It is important to preserve the quality of the product. Depending on the harvest method, the material can be stored chopped (at various different lengths), baled, bundled or pelleted.

Chopped material

The maximum moisture content for the storage of chopped material is 25%, though for a safe storage over a longer period (such as one year) the moisture content must be 18% or lower. When harvest conditions are poor, and the material contains more

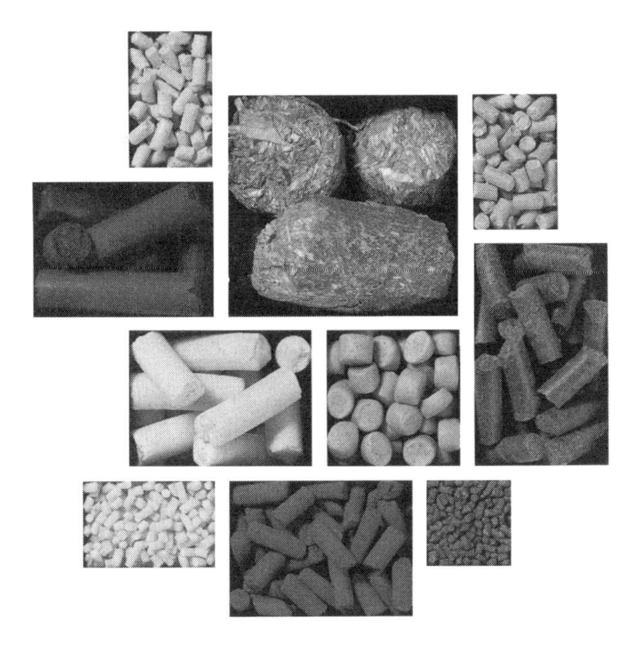

Figure 2.5. Different types of pellets and briquettes (Kahl, 1987).

than 25% moisture, the chopped material can be stored in any storage facility where ventilation from a floor system is possible. Tests show that daily ventilation for 1.5 hours with ambient air reduces the moisture content sufficiently. The cheapest way to store chopped material is in outdoor piles covered with a plastic that allows vapours to pass through.

Experiments with outside storage piles have given different results concerning the moisture content of the plant material. One study compared two piles: one had a ventilation channel and was covered with normal plastic; the other was only covered by a net to prevent the material from blowing away. After five months, the covered pile had a moisture content ranging from 15% in the centre to 10% in the top layer. The open pile showed an increase in moisture content, with contents ranging from 24% in the centre to 64% in the top layer. The results indicate that a covered and non-mechanically ventilated pile is preferable. However, the open pile in this experiment was very small, so storage in a bigger open pile could be possible. The characteristics of the outside layer, such as its thickness and behaviour, are subjects for further research.

Another storage method is to ensilage the fresh biomass, a method commonly used for grass, maize and hemp. The silage could then be pressed so that the solid phase provides a feedstock with more than 50% dry matter content; the liquid would be fermented to produce biogas.

Bales

The compaction rates and density produced by the baler are very important parameters for the storage of bales. Two possible storage methods are field (open) storage, and under-cover storage (with or without a drying system).

The moisture content of the bales at harvest time is an important factor when deciding which type of storage should be implemented in relation to the weather conditions at the site. Big bales can be stored with up to 25% moisture, and are difficult to dry. Bales with a high moisture content (about 40%) at the time of harvest will be prone to mildew, especially under field storage conditions.

Compact rolls with a dry matter density of 350kg/m^3 cannot be further dried, so the moisture content of the material has to be lower than 25%. On the other hand, a higher compaction density will enable storage of the bales under field conditions without large losses: in rainy weather only the first 10cm layer will absorb water, and this layer protects the other, deeper layers. Storage under cover is a favourite method, because the sides are open and allow air to circulate and dry the bales.

Bundles

Bundles can be stored outside. In an experiment to compare two different kinds of bundles, one bundle was tightly bound and the other had a loose binding. There were two further variations for each kind of bundle: covered with plastic, and uncovered. The results showed a great difference in the months of July and September between the covered and uncovered bundles. The difference between loose and tight binding is greater when the bundles are not covered. The water that has infiltrated the tight bundle cannot easily evaporate or trickle through. Thus the outdoor storage of bundles is quite feasible, though a plastic cover against rain prevents fluctuations in moisture content.

Pellets

The advantages of high density compaction systems for biofuel treatment become obvious when the logistical improvements for long-distance transportation to a combustion unit is considered. However, advanced big square bales and small sized compact roller bales (400mm diameter) present an interesting advantage, since their drawbacks compared to pelletized fuels are rather marginal, a position that is confirmed by an examination of the aggregate energy demand and total costs. However, there is a strong indication that pellets are associated with a series of benefits for combustion, such as the reduced risk of breakdown. The interactions of biofuel treatment and combustion performance are still awaiting systematic investigation.

High density biofuel feedstocks (compact bales or pellets) have two important advantages: they need much less space for storage and transport; and their recovery capacity (tonnes of dry matter per manpower hour) is about four times higher than that of chopped feedstocks (Figure 2.6) (Schön and Strehler, 1992).

Chips

Biofuel from short rotation *Salix* is harvested during the winter when the plants are in dormancy and the dry matter content is about 50%. The harvested, chipped material is either burned directly after harvesting or dried in the field for shorter or longer periods before it is chipped and burned. This material is used only at large plants where heat recovery is utilized.

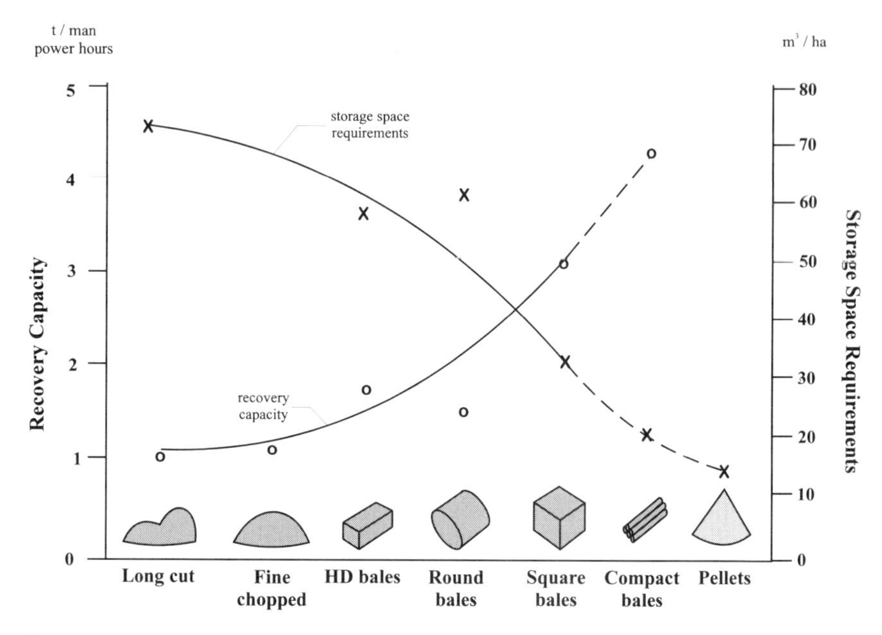

Figure 2.6. The recovery capacity and storage space requirements for different harvest products as biofuel feedstocks (Schön and Strehler, 1992).

Machines for harvesting coppice have been developed along two principal lines (Danfors, 1994):

1 Direct harvest the crop, chop it up and transport it to a heating plant where the material is stored for only a few days before it is burned. There may be

Figure 2.7. Harvesting poplar for fibre and fuel (Photo: Warren Gretz, NREL).

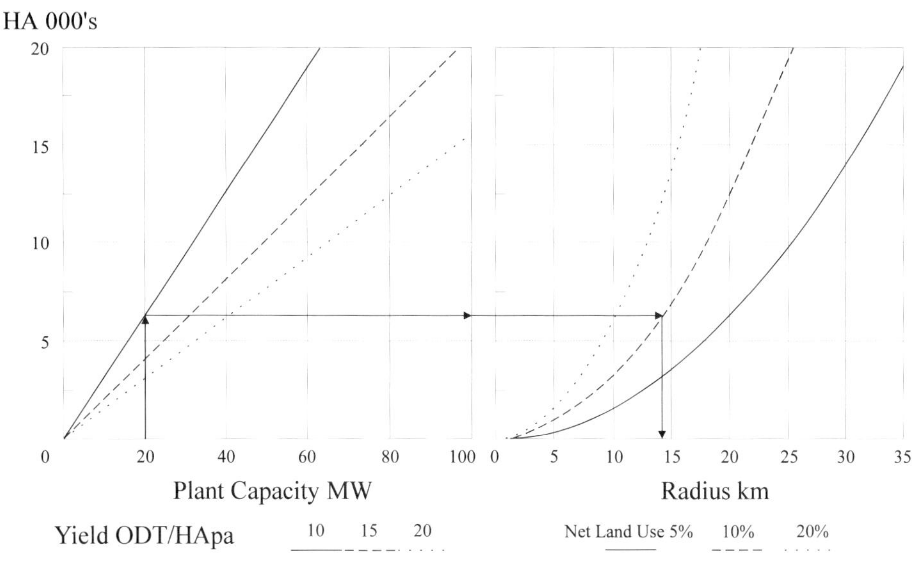

Figure 2.8. Land requirements for biomass power plants (Hall, 1994).

intermediate storage in the field if the transport capacity is too low, or if the fuel cannot be received at the heating plant for some reason.

2 Harvest whole stems with the other type of harvester (Figure 2.7), and store the material in large heaps outside the coppice plantation. The dry matter content decreases gradually during the spring and summer. The material is chipped and delivered at a convenient time to suit demand.

Figure 2.8 shows the land requirements for biomass power plants calculated by taking the power plant's capacity and the yield of the crop to determine the plant's land requirement in hectares. This requirement is correlated with net land use to discover the radius of the land area around the power plant. The example in the figure uses a 20MW power plant with a 10t (oven dry)/ha yield. This plant would require 7000 ha which for 10% land use will require land within a 14km radius of the power plant. This assumes 45% efficiency, 80% load and 18GJ/ODT.

3 Conversion Technologies

Conversion of biomass to fuels

Research and development work in various industrialized countries currently focuses on technologies that are relevant to their needs and future markets. It is important to note that the same technologies can often be employed with minimal modifications in less developed areas of the world. Hence renewable energy will play an increasing role in future cooperation schemes between developed and developing countries and in export markets for several regions.

Grassi and Bridgewater (1992) indicated that the full range of traditional and modern liquid fuels can be manufactured from biomass by thermochemical conversion and synthesis or upgrading of the products. Advanced gasification for the production of electricity through integrated power generation cycles has a significant role to play in the short term, and is being developed in many parts of the world including Europe and North and South America. Longer term objectives are the production of hydrogen, methanol, ammonia and transport fuels through advanced gasification and synthesis.

Biological conversion technologies use microbial and enzymatic processes to produce sugars that can later be developed into alcohol and other solvents of interest to the fuel and chemical industries. Yeast based fermentation, for example, has yielded ethanol from sugar or starch crops. Solid and liquid wastes can be used to produce methane through anaerobic digestion. Figure 3.1 identifies different forms of biomass with high oil, sugar or starch contents, and what technologies are used to convert them into biofuels.

Other conversion technologies concentrate on the conversion of biomass high in lignocellulose. Figure 3.2 demonstrates the variety of conversion technologies that can be used on high lignocellulose biomass to create a range of biofuels.

This chapter provides additional details that demonstrate both up to date and future possibilities for producing biofuels, and their conversion technologies.

Ethanol

Ethanol is ethyl alcohol obtained by sugar fermentation (sugarcane), or by starch hydrolysis or cellulose degradation followed by sugar fermentation, and then by subsequent distillation.

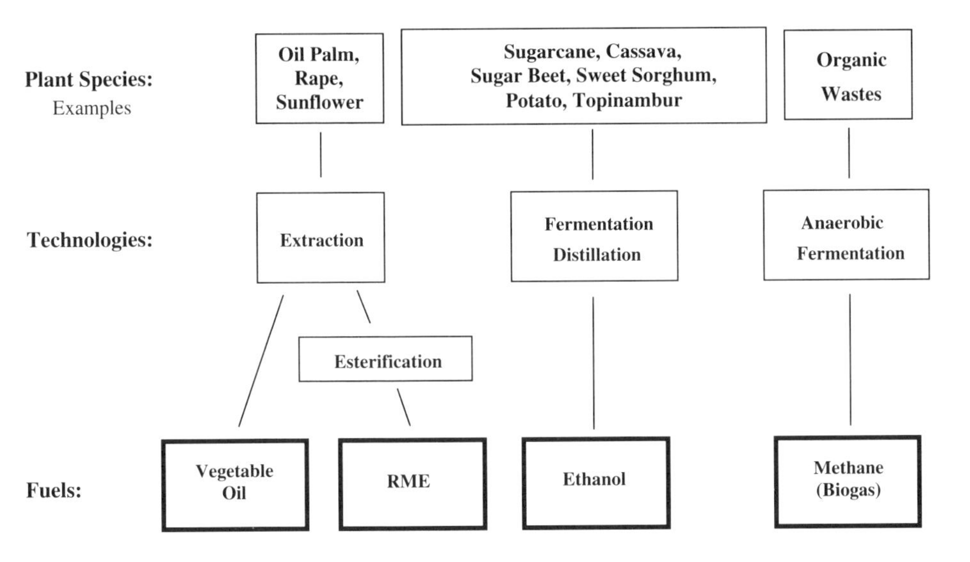

Figure 3.1. Conversion possibilities for biomass rich in oils, sugars and starch (El Bassam, 1993).

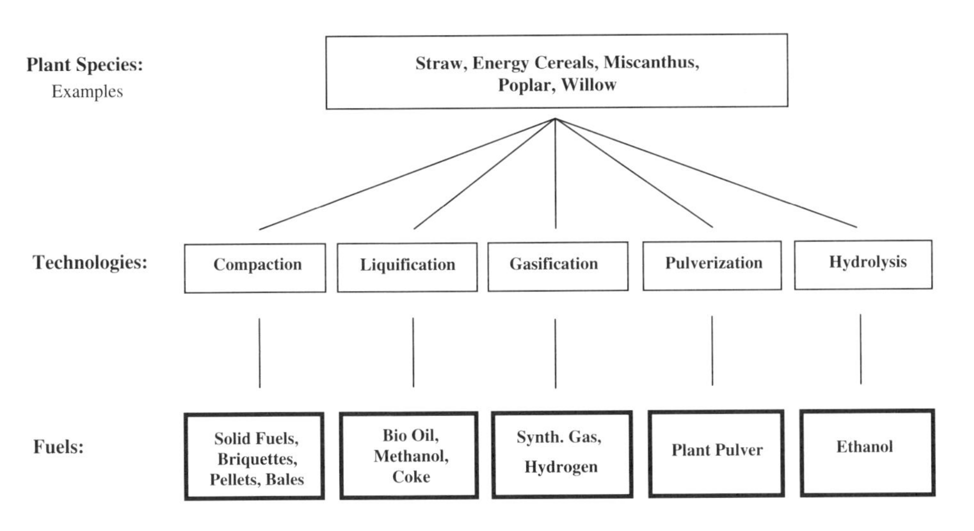

Figure 3.2. Conversion possibilities for lignocellulose biomass (El Bassam, 1993).

Obtaining alcohol from vegetable raw materials has a long tradition in agriculture. The fermentation of sugar derived from agricultural crops using yeast to give alcohol, followed by distillation, is a well established commercial technology. Alcohol can also be produced efficiently from starch crops (wheat, maize, potato, cassava, etc.). The glucose produced by the hydrolysis of starch can also be fermented to alcohol. The goal-directed use of cellulose-containing biomass from agriculturally utilized species for producing alcohol has not yet been practised on a large-scale

Table 3.1. Ethanol yields of some sugar or starch producing plants.

Source	Ethanol yield (litres per tonne biomass)
Molasses	270
Sugarcane	70
Sweet sorghum	80
Corn	370
Sweet potato	125
Cassava	180
Babassu	80
Wood	160

engineering level. Production has been confined to the use of wood, residues and waste materials. Table 3.1 lists the ethanol production of several sugar or starch producing plants.

Technologies that can produce fuel ethanol from lignocellulose are already available (Ingram *et al.*, 1995). The challenge today is to assemble these technologies into a commercial demonstration plant. Many wood wastes and residues are available as feed mat, while high-yielding woody crops will also provide lignocellulose at lower prices than agriculture crops.

Approaches for the improvement of acetone–butanol–ethanol (ABE) fermentation have focused on the development of hyper-amylolytic and hyper-cellulolytic clostridial strains with an improved potential for biomass-to-butanol conversion. The developed strains produce approximately 60% more butanol than the parental wild type strain (Blaschek, 1995). These findings suggest that the developed strains are stable and offer a significant economic advantage for use in commercial fermentation processes for producing butanol.

Bioethanol can be used as a fuel in different ways: directly in special or modified engines as in Brazil or in Sweden, blended with gasoline at various percentages, or after transformation into ethyl tertio butyl ether (ETBE). In the first case it does not need to be dehydrated, whereas blends and ETBE production need anhydrous ethanol. Moreover, ETBE production requires neutral alcohol with a very low level of impurities. ETBE is an ether resulting from a catalytic reaction between bioethanol and isobutylene, just as MTBE is processed from methanol. MTBE is the most commonly known fuel ether, used in European gasoline since 1973 and still generally used throughout the world to improve gasoline. Even the aviation industry has become involved in using ethanol as fuel (Figure 3.3).

Bioethanol can also be the alcohol used for the production of biodiesel, thus giving ethyl esters instead of methyl esters; the other constituent is vegetable oil from rape, sunflower, soybeans or other sources.

European regulations enable direct blending of bioethanol up to 5% and of ETBE up to 15% with other oxygenated products, according to a Directive dated 5 December 1985 concerning crude oil savings by utilization of substitute fuels. On the other hand, the European Union's objective, as expressed in its Altener programme, for example, is to let biofuels take a 5% market share of fuels consumption in the year 2005. In fact, oil companies and car manufacturers in Europe do not like bioethanol

Figure 3.3. An ethanol powered plane (Photo: Aviation Sciences).

blends, primarily because of their lack of water tolerance and their volatility and the need to put a label on the pumps to inform the consumer.

However, oil companies and car manufacturers are interested in ETBE, which is very similar and slightly better than the MTBE they are used to. ETBE has a lower Reid vapour pressure (RVP) than MTBE, and a higher octane value and lower oxygen content. Moreover, most MTBE producers could readily convert their plants to make ETBE after some process and equipment modifications.

ETBE and MTBE blended gasolines have several benefits:

- the oxygen content reduces carbon monoxide emissions;
- the lower RVP lessens the pollution that forms ozone;
- the high octane value reduces the need for hydrocarbon based aromatic octane enhancers like benzene, which is carcinogenic.

This growing interest in one product but not the other brought bioethanol producers into discussions with MTBE producers, and a partnership was initiated. There were only two of them in France: ELF and ARCO. ELF made a first test of ETBE production over five days in 1990 at its refinery in Feyzin near the town of Lyon, and a second test during a one month period in 1992. The result was conclusive, so the company started ETBE production on an industrial scale in 1995.

Oxygenated fuel feedstocks (ETBE and MTBE) are considered as convenient options for reaching the octane value requirements of gasoline fuels. Oxygenated compounds offer considerable benefits if they are mixed with gasoline: reduction in the amount of carbon monoxide emissions and unburnt hydrocarbons. There is a growing demand for oxygenates in Europe, as well as in the USA as a result of the Clean Air Act Amendments of 1990.

Bioethanol can be used in an undiluted form or mixed with motor gasoline. 'Gasohol' is the term used in the USA to describe a maize-based mixture of gasoline (90%) and ethyl alcohol (10%); it should not be confused with gasoil, an oil product used to fuel diesel engines. Although engines need to be adapted to use bioethanol at 100%, unadapted engines can run the mixed fuels. Bioethanol can be used to substitute for MTBE and is added to unleaded fuel to increase octane ratings. In Europe, the preferred percentage – as recommended by the Association of European Automotive Manufacturers (AEAM) – is a 5% ethanol or 15% ETBE mix with gasoline.

Oils

Vegetable oils and fats, in contrast to the simple carbohydrate building blocks glucose and fructose, exhibit a number of modifications to the molecule's structure and are also more valuable regarding energy. From the point of view of plant breeding and plant cultivation of energy crops, it is high oil yields per hectare rather than quality aspects that are of primary importance.

The use of vegetable oils as fuels is not new. Since ancient times, oils have also been used as a material for burning and for lighting (the oil lamp!). The inventor Rudolph Diesel used peanut oil to fuel one of his engines at the Paris Exhibition of 1900, and wrote in 1912 that: 'The use of vegetable oils for engine fuels may seem insignificant today, but such oils may become, in the course of time, as important as petroleum and the coal tar products of the present time' (Nitske and Wilson, 1965). Now, biodiesel is increasingly used as a transportation fuel (Figure 3.4).

Figure 3.4. Biodiesel fuel station (Photo: UFOP).

Worldwide, there are more than 280 plant species with more or less considerable oil contents in their seeds, fruits, tubers and/or roots. The oil can be extracted by compressing the seeds (sunflower, rapeseed, etc.) mechanically with screw presses and expellers, with or without preheating. Pressing with preheating allows for the removal of up to 95% of the oil, whereas without preheating the amount is considerably less. Extraction by the use of solvents is more efficient: it removes nearly 99% of the oil from the seeds, but it has a greater energy consumption than the expeller. Table 3.2 looks at the seed and oil yields of various crops. Table 3.3 shows the oil and fatty acid composition of various crops.

Table 3.2. Oil seed yields and oil contents of various plant species (Peterson, 1985).

Botanical name	Common name	Seed or oil yield (kg/ha)		% oil
Ablemoschus moschatus Medik	Ambrette	836–1693	seed	
Aleurites fordii Hemsl.	Tung-oil tree		seed	
Aleurites montana (Lour.) Wils.	Mu-oil tree	5500	oil	
Anacardium occidentale L.	Cashew	1000	seed	
Arachis hypogaea L.	Peanut	5000	seed	35–55
Brassica juncea (L.) Czern	Mustard, green	1166	seed	30–38
Brassica napus L.	Rape	3000	seed	40–45
Brassica nigra (L.) Koch	Mustard, black	1100	seed	30–35
Brassica rapa L.	Turnip	1000	seed	
Calophyllum inophyllum	Calophyllum			50–73
Camellia japonica	Tsubaki		seed	66
Cannabis sativa L.	Hemp	1500	seed	25–30
Carthamus tinctorius L.	Safflower	4500	seed	35–45
Ceiba pentandra	Kapok			20–25
Cier arietinum L.	Chickpea	2000	seed	
Citrullus colocynthis (L.) Schrad.	Colocynth	6700	seed	47
Cocos nucifera L.	Coconut	1000	copra	62.5
Coronilla varia L.	Crownvetch	500	seed	
Crambe abyssinica Hochst ex R.E. Fries	Crambe	5000	seed	25–33
Croton tiglium L.	Croton, purging	900	seed	50–55
Cucurbita foetidissima HBK	Buffalo gourd	3000	seed	24–34
Cyamopsis tetragonoloba (L.) Taub	Guar	2000	seed	
Elaeis guineensis Jacq.	African oil palm	2200	oil	55 (of kernel)
Glycine max (L.) Merr.	Soybean	3100	seed	13–25
Gossypium hirsutum L.	Cotton	900	seed	16
Guizotia abyssinica (L. f.) Cass.	Niger seed	600	seed	25–35
Helianthus annuus L.	Sunflower	3700	seed	40–50
Juglans nigra L.	Black walnut	7500	seed	60
Juglans regia L.	Persian walnut	7500	seed	63–67
Lens culinaris Medik.	Lentil	1700	seed	
Lesquerella ssp.	Lesquerella	1121	seed	11–39
Limnanthes bakeri J. T. Howell	Meadowfoam, Baker's	400	seed	24–30
Limnanthes douglasii R. Br.	Meadowfoam, Douglas'	1900	seed	24–30
Linum usitatissimum	Flax	650	seed	34
Lotus corniculatus L.	Trefoil, birdsfoot	600	seed	

Table 3.2. (cont.)

Botanical name	Common name	Seed or oil yield (kg/ha)		% oil
Lupinus albus L.	Lupine, white	1000	seed	
Macadamia spp.	Macadamia nut	4000	seed	15–20 (of nut)
Medicago sativa	Alfalfa seed	800	seed	8–11
Mucuna deeringiana (Bort) Merr.	Velvetbean	2000	seed	
Oryza sativa	Rice bran	800	seed	15–20
Pachyrhizus erosus (L.) Urb.	Jicama	600	seed	
Pachyrhizus tuberosus (Lam.) Spreng	Ajipo	600	seed	
Papaver somniferum L.	Poppy, opium	900	seed	45–50
Perilla frutescens (L.) Britt.	Perilla	1500	seed	35–45
Phaseolus acutifolius A. Gray	Tepary bean	1700	seed	
Phaseolus coccineus L.	Scarlet runner bean	1700	seed	
Phaseolus lunatus L.	Lima bean	2500	seed	
Pisum sativum L.	Garden pea	1800	seed	
Prunus dulcis (Mill.) D. A. Webb	Almond	3000	seed	50–55
Ricinus communis L.	Castorbean	5000	seed	35–55
Sambucus canadensis L.	Elderberry			22–28
Sapium sebiferum (L.) Roxb.	Chinese tallow tree	14000	seed	19
Sesamum indicum L.	Sesame	1000	seed	45–50
Simmondsia chinenesis (Link) C. Schneid	Jojoba	2250	seed	43–56
Sinapis alba L.	Mustard, white	8000	seed	25–30
Stokesia laevis (Hill) Green	Stokes aster	1121	seed	27–44
Stylosanthes humilis HBK	Townsville style	1200	seed	
Telfairia pedata (Sm. ex Sims.) Hook	Zansibar olivine	2000	seed	35
Theobroma cacao L.	Cacao	3300	seed	35–50
Trigonella foenum-graecum L.	Fenugreek	3000	seed	
Vigna angularis (Willd.) Ohwi & Ohashi	Adzuki bean	1100	seed	
Vigna radiata (L.) Wilczek var. *radiata*	Mung bean	1100	seed	
Vigna umbellata (Thunb.) Ohwi & Ohashi	Rice bean	200	seed	
Vigna unguiculata (L.) Walp.	Cow pea	2500	seed	
Vitis vinifera	Grape seed oil			6–21

Raw vegetable oils must be refined prior to their utilization in engines or transesterification. The most widespread idea is that fuel must be adapted to present day diesel engines and not vice versa. In order to meet the requirements of diesel engines, vegetable oils can be modified into vegetable oil esters (transesterification). The transesterification procedure includes the production of methylesters – RME (rape methylester) or SME (sunflower methylester) – and glycerol (glycerine) by the processing of plant oils (triglycerides), alcohol (methanol) and catalysts (aqueous sodium hydroxide or potassium hydroxide). Table 3.4 provides a comparison of the fuel properties of three of the most common vegetable oils, their corresponding methylesters, and diesel fuel. Table 3.5 lists the properties of some additional vegetable oils.

Table 3.3. Oil and fatty acid contents of various plant species (Osman and Amahd, 1981).

Species	Family	Oil content (%)	Fatty acid composition (wt %)						
			16:0	18:0	18:1	18:2	18:3	18:4	others
Asphodelus tenuifolius	Liliaceae	22.0	7.6	16.0	12.8	78.0			
Calotropis procera	Asclepidaceae	26.0	14.3	19.3	62.7	3.1			0.6% 16:1
Carthamus oxycantha	Compositae	26.3	8.5	2.4	17.3	71.3	0.5		
Celastrus paniculatus	Celastraceae	50.6	24.1	5.6	16.6	18.3	29.9		2.2% 12:0, 0.9% 10:0
Chrozophora plicata	Euphorbiaceae	26.0	4.1	21.4	13.4	59.3	1.8		
Chrozophora rottleri	Euphorbiaceae	24.0	5.6	18.2	18.9	56.4			0.9% 14:0, 0.2% 12:0
Cisternum divernum	Solanaceae	25.0	28.0	6.2	6.4	40.0			19.3% 14:0
Cynoglossum zeylanicum	Boraginaceae	28.0	11.8	2.8	33.5	20.0	10.5	2.7	0.4% 14:0
Echium italicum	Boraginaceae	24.0	8.4	4.4	36.8	18.6	21.1	5.7	4.9% 20:0
Euphorbia genicelate	Euphorbiaceae	20.6	31.5	8.7	13.9	24.3	21.6		
Evolvulus nummulari	Convolvuaceae	47.0	36.3	14.4	18.4	20.3	7.1		3.5% 14:0
Fimbristylis quinqueangularis	Cyperaceae	23.6	5.5	3.6	21.4	52.9	3.2		3.5% 14:0
Gelonium multiflorum	Euphorbiaceae	21.0	12.8	4.5	17.7	15.2	47.8		1.6% 12:0
Justicia adhatoda	Acanthaceae	26.2	5.5	0.6	80.1	5.2	3.3	5.4	
Lepidum sativum	Cruciferae	29.0	8.9	0.2	33.7	29.1	10.1		0.4% 14:0, 16.3% 20:0
Martynia diandra	Pedialiaceae	48.0	48.2	0.5	37.0	0.8	9.2		1.7% 14:0, 2.5% 12:0
Ochna squarrosa	Ochnaceae	23.4	73.5	1.5	14.1	10.9			
Passiflora foetida	Passifloraceae	21.2	13.3	4.1	17.2	65.4			
Pogestemon plactranthoides	Labiatae	32.0	22.8	trace	3.2	70.0			
Ruellia tuberosa	Acanthaceae	24.0	30.3	7.2	17.5	32.5			3.8% 14:0, 3.6% 12:0, 5.1% 10:0
Salvia plebeia	Labiatae	23.0	7.8	3.3	13.4	43.9	28.5		3.1% 14:0
Sesamum indicum	Pedaliaceae	28.6	10.5	2.3	42.3	41.8	2.3		0.8% 12:0
Solanum sisymbrifolium	Solanaceae	20.1	14.0	2.9	22.7	54.9	3.8		1.7% 14:0
Tabernaemontana coronaria	Apocyanaceae	42.6	21.3	0.5	50.5	23.1	4.6		
Tragia involucrata	Euphorbiaceae	27.0	10.6	2.4	15.8	61.7	7.3		2.4% 14:0
Xantrium strumarium	Compositae	36.0	20.9	2.6	54.3	22.2			

Table 3.4. Fuel properties of vegetable oils and methylesters in comparison with diesel fuel (Ortiz-Cañavate, 1994).

Fuel property	Diesel fuel	Sunflower oil	Sunflower methylester	Rape oil	Rape methylester	Linseed oil	Linseed methylester
Specific gravity (kg/dm^3)	0.835	0.924	0.88	0.916	0.88	0.932	0.896
Viscosity (cSt):							
at 20°C	5.1	65.8		77.8	7.5	50.5	8.4
at 50°C	2.6	34.9	4.22	25.7	3.8		
Heat of combustion:							
Gross (MJ/litre)	38.4	36.5	35.3	37.2		36.9	35.6
Net (MJ/litre)	35.4	34.1	33.0	34.3	33.1		
Cetane number	>45	33	45–51	44–51	52–56		
Carbon residue (%)	0.15	0.42	0.05	0.25	0.02		0.22
Sulphur (%)	0.29	0.01	0.01	0.002	0.002	0.35	0.24

Table 3.5. Physical-chemical properties of some additional vegetable oils (Guibet, 1988).

	Groundnut	Copra	Cotton	Palm oil	Soya
Density (kg/litre) (20°C)	0.914	0.915 (30°C)	0.915		0.916
Viscosity (cSt):					
20.0°C	88.5		69.9		
50.0°C	29.0		24.8	28.6	
100.0°C		6.1	8.4	8.3	7.6
Melting point	(−3)–0	20–28	(−4)–0	23–27	(−29)–(−12)
Composition (weight %):					
C	77.35	73.4	77.7	76.4	78.3
H	11.8	11.9	11.7	11.7	11.3
O	10.9	14.7	10.6	11.5	10.3
Formula	$CH_{1.83}O_{0.11}$	$CH_{1.95}O_{0.15}$	$CH_{1.81}O_{0.10}$	$CH_{1.84}O_{0.11}$	$CH_{1.73}O_{0.10}$
Cetane number	39–41	40–2	35–40	38–40	36–9
Heat of combustion (MJ/kg)	36.7	37.4	36.8	36.5	36.8
Heat of combustion (MJ/litre)	33.5	34.2	33.7		33.7

Cracking is another option for modifying the triglyceride molecule of the oil (Pernkopf, 1984). However, the cracking products are very irregular and more suitable for gasoline substitution; the process has to be conducted on a large scale; costs are considerable; and conversion losses are also significant. These negative aspects together with the much lower efficiency of gasoline engines make the cracking process of minor interest.

The Veba process is yet another procedure for converting the triglyceride molecule. During the refining of mineral oil to form the different conventional fuels (gasoline, diesel, propane, butane, etc.), up to 20% rapeseed oil is added to the vacuum distillate. The molecules are cracked and the mixture is treated with hydrogen. The generated fuel molecules are not different from conventional fuel molecules. Advantages of the Veba process are that no glycerine is produced as a by-product, and that the generated fuel is not different from standardized fuels so there is no need for a special distribution system and handling. But there are also disadvantages: the high consumption of valuable hydrogen, and reductions in biodegradability (Vellguth, 1991).

Pure plant oils, particularly when refined and deslimed, can be used in pre-chamber (indirect injected, like the Deutz engine described below) and swirl-chamber (Elsbett) diesel engines as pure plant oil or in a mixture with diesel (Schrottmaier *et al.*, 1988). Pure plant oil cannot be used in direct injection diesel engines, which are used in standard tractors and cars, because engine coking occurs after some hours of operation. The addition of a small proportion of plant oils to diesel fuel is possible for all engine types, but will also lead to increased deposits in the engines in the long term (Ruiz-Altisent, 1994).

The 'Elsbett' engine is a recently developed diesel engine variant with a 'duothermic combustion system' that uses a special turbulence swirl chamber and runs on pure vegetable oils. Another engine, developed by Deutz, uses the principle of turbulence with indirect injection and can run on purified plant oils. Its consumption is about 6% higher than that of other diesel engines, but it has proved to be robust and reliable.

Other sources (Löhner, 1963; van Basshuysen, 1989) indicate that the consumption level of indirect injection swirl chamber diesel engines is 10–20% higher in relation to comparable diesel engines with direct injection. Another engine, still in development, is the John Deere 'Wankelmotor', which works with multi-type fuels, including vegetable oils.

The transesterification of vegetable oils permits the wide utilization of these oils in existing engines, either as a 100% substitute or in blends with mineral diesel oil. Although vegetable oil esters have good potential, are well suited for mixture with or replacement of diesel fuels, and are effective in eliminating injection problems in direct injection diesel engines, their use still creates problems.

Solid biofuels

There are four basic groups of plant species rich in lignin and cellulose that are suitable for conversion into solid biofuels (as bales, briquettes, pellets, chips, powder, etc.):

- annual plant species such as cereals, pseudocereals, hemp, kenaf, maize, rapeseed, mustard, sunflower, and reed canary grass (whole plants);
- perennial species harvested annually, such as *Miscanthus* and other reeds;
- fast-growing tree varieties like poplar, aspen or willow with a perennial harvest rhythm (short rotation or cutting cycle), short rotation coppice (SRC);
- tree species with a long rotation cycle.

The raw materials can be used directly after mechanical treatment and compaction, or are converted to other types of biofuels (Figure 3.5). The lignocellulose plant species offer the greatest potential within the array of worldwide biomass feedstocks. The following basic processes are of major importance for the conversion of lignocellulose biomass into fuel suitable for electricity production: direct combustion of biomass to produce high grade heat; advanced gasification to produce fuel gas of medium heating value; and flash pyrolysis to produce bio-oil, with the possibility of upgrading to give hydrocarbons similar to those in mineral crude oils.

The production of methanol or hydrogen from woody biomass feedstocks (such as lignocellulose biomass woodchips from fast growing trees, *Miscanthus* spp. or *Arundo donax*) via processes that begin with thermochemical gasification can provide consid-

Figure 3.5. The 'bio-pipelines' of the future (Photo: El Bassam).

erably more useful energy per hectare than the production of ethanol from starch or sugar crops and vegetable oils like RME. However, methanol and hydrogen derived from biomass are likely to be much more costly than conventional hydrocarbon fuels, unless oil prices rise to a level that is far higher than expected prices in coming decades.

The most promising thermochemical conversion technology of lignocellulose raw materials currently available seems to be the production of pyrolytic oil or 'bio-oil' (sometimes called bio-crude oil). This liquid can be produced by flash or fast pyrolysis processes at up to 80% weight yield. It has a heating value of about half that of conventional fossil fuels, but can be stored, transported and utilized in many circumstances where conventional liquid fuels are used, such as boilers, kilns and possibly turbines. It can be readily upgraded by hydrotreating or zeolite cracking into hydrocarbons for more demanding combustion applications such as gas turbines, and further refined into gasoline and diesel for use as transport fuels. In addition, there is considerable unexploited potential for the extraction and recovery of specialized chemicals.

The technologies for producing bio-oil are evolving rapidly with improving process performance, larger yields and better quality products. Catalytic upgrading and extraction also show considerable promise and potential for transport fuels and chemicals; this is at a much earlier stage of development with more fundamental research underway at several laboratories. The utilization of the products is of major importance for the industrial realization of these technologies, and this is being investigated by several laboratories and companies. The economic viability of these processes is very promising in the medium term, and their integration into conventional energy systems presents no major problems.

Charcoal is manufactured by traditional slow pyrolysis processes, and is also produced as a by-product from flash pyrolysis. It can be used industrially as a solid speciality fuel/reductant, for liquid slurry fuels, or for the manufacture of activated charcoal.

To provide a better understanding of the fossil fuels (gasoline and diesel) and the biofuels (ethanol and methanol), Table 3.6 makes a closer comparison of these four important fuels. Some physical and chemical properties of biofuel feedstocks are provided in the subsequent tables. Table 3.7 provides a comparison of the energy contents of the most common biofuels and fossil fuels. Table 3.8 (overleaf) primarily compares the elemental constituents of several fossil fuels and a variety of crops . The biofuels have an average energy content that is approximately equal to that of brown coal (20MJ/kg).

There is no expected environmental impact or any disadvantage created by mixing some percentage of alcohol with diesel or gasoline fuel. There are no legal constraints at present.

Table 3.6. Some properties of ethanol, methanol, gasoline and diesel fuel.

Property	Ethanol	Methanol	Gasoline	Diesel
Boiling temperature(°C)	78.30	64.5	99.2	140.0
Density (15°C)	0.79	0.77	0.70–0.78	0.83–0.88
Heating value (MJ/kg)	26.90	21.3	43.7	42.7
Vaporization heat (kJ/kg)	842.00	1100.0	300.0	
Stoichiometric ratio	9.00	6.5	15.1	14.5
Octane number (MON)	106.00	105.0	79.0–98	

Table 3.7. Heating values of selected fuels.

Fuel	Heating value (MJ/kg)
Biogas	61.0
Brown coal	20.0
Cellulose	15.0
Diesel oil	42.0
Energy cereals	15.0
Ethanol	26.9
Gasoline	46.0
Hydrogen	144.0
Lignin	28.0
Methanol	19.5
Miscanthus	17.0
Pit coal	32.0
Rape oil	40.0
Straw	14.5
Wood	17.0

Wet biofuels

Biomass which can be harvested only in wet conditions and cannot be dried directly in the field can be upgraded as a fuelstock through the technique of ensiling (Figure 3.6): Wet biomass → Ensiling → Pressing (mechanical water reduction) → Fuel combustion (Scheffer, 1997). After the reduction of water content in the ensiled biomass by mechanical pressing, the biomass can be used as fuel for thermal power generation.

Table 3.8. Energy (MJ/kg) and chemical content (%) of different biofuel feedstocks.

Feedstock	Heating value (HHV, MJ/kg)	Ash	Volatile constituents	Water	C	H	N	S	O	P	K	Mg
Bamboo	15.85	3.98	67.89	10.4								
Tree bark	19.5	3.2	76		52	5	0.4	0.05	39	2.3	6.8	2.8
Barley, full plant	17.6	3.7	77.8	14.1	46.1	6.63	1.24	0.11	42	7.6	15.4	2.5
Brown coal	27	7.6	55		68.4	5.5	1.8	1.3	15			
Coal	27–30	5	25–33	8	65–80	5	1.5–2	0.5–1.5	2			
Eucalyptus	19.35	0.52	82.55		48.33	5.89	0.15	0.01	45			
Heating oil	41	0.03	99		85	11–13	0.5	0.5–1.0				
Maize silage	17.1	5.5	75.7		47.3	7.54	1.85	0.43	39			
Maize straw	16.8	5.3	75–81		45.6	5.4	0.3	0.04	43	2.2	21.8	4.3
Meadow hay	16.9	5.8	80.5		45	6.2	1	0.08	44	1.7	8	1.8
Miscanthus spp.	18	3.7	71.2	12.5	45–50	5.7	0.4	0.3	42	3	23.7	3.3
Pseudo acacia	19.5		80									
Poplar	18.7	0.3	84		50.9	6.6	0.2	0.02	42	3.5	10.4	3.3
Rape straw	17	6.5	78.7	48.3	6.3	0.7	0.2	38				
Reed	16.3	3–5	70–80		45–50	5.5–6.5	1.0	0.1–0.3				
Sorghum, full plant	16.9	6.3	28–40	6.2	44	5.6	2	0.05	38			
Triticale straw	17.4	4.9	80		47	6.1	0.4	0.1	41	2.9	16.3	1.5
Turf, dry	14.7	0.4–9.0	60–65									
Wheat, full plant	16.99	3.6	78.9	16.5	46.5	6.84	1.71	0.13	41	5.8	14.5	2
Wheat straw	17.1	5.3	79.6		46.7	6.3	0.4	0.01	41	3.1	17	1.5

Figure 3.6. Ensilaging of biomass as a fuel feedstock (Photo: G.M. Prine).

Harvested crops are preserved in a clamp silo, which has a solid ground plate with drainage furrows and a collector for water.

Water reduction by mechanical pressing processes silage into fuel with a minimum dry matter content of 60%. The addition of straw, crops and hay leads to a further increase of dry matter content which, with less pressing, results in a dry matter content of only 50%.

For the same dry matter content, fuels with a water content of 40% have a 7% lower heating value than dry straw of 15% water content. This energy difference results from the evaporation of the higher water content during combustion, though most of the energy evaporation can be reclaimed by steam condensation.

Besides a reduction of water content, pressing produces an additional qualitative improvement in the fuel. Up to 50% of the plant's nitrogen and 40–80% of the other minerals are transported in the effluent by pressing. This reduces NO_x emission caused by nitrogen, the formation of dioxin by chlorine and damage from corrosion, whereas the reduction of the potassium content increases the melting point of the ash. Table 3.9 shows as an example the result of a pressing experiment with rye silage. The proportions of the elements that remain in the output are far below the maximum limits set by the technical staff of energy plants (Scheffer, 1997).

Apart from nitrogen, all the elements important for crop nutrition are contained in the effluent and in the ashes produced by combustion. A partly closed nutrient cycle can be achieved when the effluent and the ashes are utilized in the field.

Biogas

Biogas consists of similar proportions of methane and carbon dioxide and is produced by the anaerobic fermentation of wet organic feedstocks in a process called

Table 3.9. Content of dry matter and nutritive material (in dry matter) before and after pressing of winter rye silage, harvested during the milky stage (%) (Scheffer, 1997).

	Input	Output	Relative content in fuels
Dry matter	28	55	
Nitrogen	1.25	0.59	47
Phosphorus	0.28	0.08	29
Potassium	1.68	0.53	32
Magnesium	0.10	0.03	30
Sodium	0.01	< 0.01	
Chloride	0.57	0.07	12

'biomethanization'. Biomethanization has a certain significance for the disposal of organic residues and waste products in the processing of agricultural products and in animal husbandry. Here, environmentally relevant aspects are of primary concern.

The cultivation of plants with the goal of producing biogas is hardly practised, though there are a number of green plants that are suitable in their fresh or in their ensiled form for biogas production. Through biomethanization, a broad spectrum of green plants could be used for energy because, in contrast to combustion, the raw material could be used in its natural moist state. This cannot be put into practice, however, because a number of fundamental conditions are not met. These will not be considered here.

The plant species most suitable for the production of biogas are those that are rich in easily degradable carbohydrates, such as sugar and protein matter. According to investigations by Zanuer and Küntzel (1986), the methane yields from maize, reed canary grass and perennial rye grass after silation and fermentation were identical.

Raw materials from lignocellulose-containing plant species are hardly suitable for the production of biogas. Biogas production from green plants has turned out to be very complicated, and the control of the fermentation process is very expensive and awkward. Besides this, for an acceptable yield of biogas from vegetable raw materials, production on a continuous, long term basis and a homogeneous substrate would be indispensable. The biogas yields from several plant species after 20 days retention time are given in Table 3.10.

Table 3.10. Biogas yield from fresh green plant materials, 20 days retention time (Weiland, 1997).

Plant material	Biogas yield (m^3/t VS)*
Alfalfa	440–630
Clove	430–450
Grass	520–640
Jerusalem artichoke	480–590
Maize plant	530–750
Sugarbeet leaves	490–510
Sweet sorghum	640–670

* Volatile Solids

Conversion of biofuels to heat, power and electricity

Heat

Direct combustion of biomass is the established technology for converting biomass to heat at commercial scales. Hot combustion gases are produced when the solid biomass is burned under controlled conditions. The gases are often used directly for product drying, but more commonly the hot gases are channelled through a heat exchanger to produce hot air, hot water or steam.

The type of combustor most often implemented uses a grate to support a bed of fuel and to mix a controlled amount of combustion air. Usually, the grates move in such a way that the biomass can be added at one end and burned in a fuel bed which moves progressively down the grate to an ash removal system at the other end. More sophisticated designs permit the overall combustion process to be divided into its three main stages – drying, ignition and combustion of volatile constituents, and burn-out of char – with separate control of conditions for each activity being possible. A fixed grate may be used for low-ash fuels, in which case the fuel charge is input by a spreader–stoker that maintains an even bed and fuel distribution and hence optimum combustion conditions.

Grates are well proven and reliable, and can tolerate a varied range in fuel quality (moisture content and particle size). They have also been shown to be controllable and efficient. The goal of reducing emissions is one of the driving forces behind current developments. This goal has also driven the development of fluidized bed technology, which is the main alternative to grate based systems.

A fluidized bed involves the fuel being burned in a bed of inert material. Air blown through the fluidized bed suspends the inert material, and is used to combust or partially combust the biomass fuel. In the case of partial combustion, additional secondary air is added downstream of the bed to allow total combustion. At any given time, there is a large ratio of inert material to fuel in the fluidized bed itself and the bed is in continuous motion because of its suspended state. The process thus promotes rapid fuel–air mixing and rapid heat transfer between bed and fuel during ignition and combustion, minimizing the impact of variations in fuel quality. The bed and overbed temperatures can be controlled to maintain optimum combustion conditions, maximizing combustion efficiency and reducing emissions. Powerful fans are therefore needed to maintain fluidization of the bed, which increases the auxiliary power required compared to grate systems and may make it impossible to realize all of the potentially available efficiency improvements (Dumbleton, 1997).

Electricity

Steam technology

At present, the generation of electricity from biomass uses one of the combustion systems discussed above. This process consists of creating steam and then using the steam to power an engine or turbine for generating electricity. Even though the production of steam by combustion of biomass is efficient, the conversion of steam to electricity is much less efficient. Where the production of electricity is to be maximized, the steam engine or turbine will exhaust into a vacuum condenser and

Figure 3.7. The wood fired McNeil electric generating facility in Burlington, Vermont, USA (Photo: Warren Gretz, NREL).

conversion efficiencies are likely to be in the 5–10% range for plants of less than 1MWe, 10–20% for plants of 1 to 5MWe and 15–30% for plants of 5 to 25MWe. Low-temperature heat (< 50°C) is usually available from the condenser, though this is insufficient for most applications so it is normally wasted by dispersal into the atmosphere or a local waterway. The average conversion efficiency of steam plants in the USA is approximately 18%. The USA has installed 7000MWe of wood fired plants since 1979. Figure 3.7 shows the 50MWe McNeil power plant in Burlington, Vermont.

Where there is a need for heat as well as electricity – for example, for the processing of biomass products such as the kilning of wood, or the processing of sugar, palm oil, etc. – the plant can be arranged to provide high-temperature steam. This is achieved by taking some steam directly from the boiler, by extracting partially expanded steam from a turbine designed for this purpose, or by arranging for the steam engine or turbine to produce exhaust steam at the required temperature.

All three options significantly reduce the amount of electricity available from the plant, though the overall energy efficiency may be much higher – 50–80% being common (Dumbleton, 1997). See Table 3.11 at the end of this chapter for a summary of power ranges and efficiencies of different technologies.

Bioelectricity, particularly in the form of combined heat and power (CHP) is in the mainstream of current technological trends. Aside from the large number of small individual heating systems, more than 1000MW of bioelectricity are currently produced annually by European utilities through conventional or advanced combustion (Palz, 1995). Biomass, mainly in the form of industrial and agricultural residues and municipal solid waste, is presently used to generate electricity with conventional steam turbine power systems. The USA has an installed biomass (not just wood) electricity generating capacity of more than 8000MW (Williams, 1995). Although the power plants are small – typically 20MW or less – and relatively capital intensive and energy inefficient, they can provide cost competitive power where low cost biomass is available, especially in combined heat and power applications.

However, the use of this technology will not expand considerably in the future because low cost biomass supplies are limited. Less capital intensive and more energy efficient technologies are needed to make the more abundant, but more costly, biomass sources usable. This involves the production of biomass grown on plantations dedicated to energy crops, which will be competitive particularly for power-only applications (Williams, 1995).

Higher efficiency and lower unit capital costs can be realized with cycles involving gas turbines. Present developmental efforts are focused on biomass integrated gasifier/gas turbine (BIG/GT) cycles (Williams and Larson, 1993).

In all of the cases so far described, steam technology can be considered as sound and well proven, though expensive. Electricity and CHP systems using biomass and steam technology can be competitive with electricity produced from fossil fuels in places where biomass residues are available at low or no cost, but electricity prices will not be competitive if the biomass fuel has to be purchased at market prices. In this case, biomass to electricity systems need other reasons for their utilization, and electricity price structures that recognize these conditions. Steam technology will continue to be an acceptable alternative for biomass to electricity plants where these price structures exist. However, the environment and other benefits of using biomass are not maximized and continuous price support is needed, which are unacceptable factors in many instances.

The cost-effectiveness of electricity generation from biomass can be improved if conversion efficiencies increase and capital costs decrease. Raising conversion efficiencies also helps to maximize environmental benefits and associated environmental tax credits by decreasing the general dependency on fossil fuels. Because steam technology is in essence fully developed, there is unfortunately limited scope for finding improvements. New conversion technologies are therefore important: gasification and pyrolysis are two relatively new technologies that are close to commercialization (Dumbleton, 1997).

Gasification

Gasification is a thermochemical process in which carbonaceous feedstocks are partially oxidized by heating at temperatures as high as 1200°C to produce a stable fuel gas (Dumbleton, 1997). By an exothermic chemical reaction with oxygen, a proportion of the carbon in the biomass fuel is converted to gas. The normal producer gas chemical reactions occur when the remainder of the biomass fuel is subjected to high temperatures in an oxygen-depleted atmosphere. The resulting fuel gas is mainly carbon monoxide, hydrogen and methane, with small amounts of higher hydrocarbons such as ethane and ethylene. These combustible gases are unfortunately reduced in quantity by carbon dioxide and nitrogen if air is used as the source of oxygen. Because carbon dioxide and nitrogen have no heating value, the heating value of the final fuel gas mixture is low: 4–6MJ/Nm^3. This is equivalent to only 10–40% of the value for natural gas, usually 32MJ/Nm^3, for which the current engines and turbines were created. The low heating value also makes pipeline transportation of the gas inappropriate. Unwanted by-products such as char particles (unconverted carbon), tars, oils and ash are also present in small amounts. These could damage the engines and turbines and therefore must be removed or processed into additional fuel gas.

If oxygen is incorporated instead of air, the heating value of the gas is improved to 10–15MJ/Nm^3. This improvement would permit unmodified engines and turbines to be used to generate electricity. However, the cost of oxygen production and the potential dangers associated with its utilization mean that the use of oxygen-blown gasifiers is not a favoured option (Dumbleton, 1997).

Biomass may also be gasified under pressure, though it is not known if this will prove to be more cost-effective. Pressurized gasifiers are more expensive but will be smaller than the present normal gasifiers. The pressurized gasifier system would be more efficient overall because tars will be more completely converted, sensible heat will be retained, and the need for fuel gas compression will be eliminated. However, pressure seals and the need to purge the feed system with inert gas will make the fuel feed system for pressurized gasifiers much more complicated. Such factors increase the capital and operating costs of pressurized gasifier plant, though a demonstration plant at Vernamo in Sweden uses this technology so it will be possible to evaluate these factors in the near future.

Direct combustion of the hot fuel gas from the gasifier in a boiler or furnace may only be possible for heating applications. The condensation of the tars that are cracked and burned in the combustor is prevented by maintaining high temperatures. If the biomass fuel or gasification process leads to the production of dust (ash), it may well be possible to remove this simply, using a hot gas cyclone. Some combustion systems will be able to withstand a limited dust loading with no gas clean up necessary. Biomass gasification for combustion is often found in the pulp and paper industry, where waste products are utilized as fuel. There are also a number of Bioneer gasifiers on district heating schemes in Finland, though it is not clear whether this approach is more cost-effective than combustion.

The quality of the gas must be improved before it can be used in combustion engines or turbines, and the gas may have to be cooled to intermediate, if not low, temperatures because of temperature limitations in the engine's or turbine's fuel control systems. Reducing the gas temperature will increase the volumetric heating value of the gas, but it will also increase the condensation of tars making the gas even less suitable for engine and turbines. A gas cleaning system will be essential in this case, and may consist of cyclones, filters or wet scrubbers. Wet scrubbers are particularly effective: in a single operation, they reduce the gas temperature and also capture tars (which are water soluble) and inert dust in the form of ash and mineral contaminants. However, a contaminated liquid waste stream is produced. This casts a shadow on the idea of biomass fuels and the concept of sustainability being clean alternatives.

Fuels with ashes that have low ash softening and melting temperatures are inappropriate because of the high temperatures used during the gasification process. Numerous annual crops and their residues fall into this category (Dumbleton, 1997).

Advanced gasification for the production of electricity through integrated power generation cycles has a significant short term potential, and is being developed in many parts of the world including Europe and North and South America. Both atmospheric and pressurized air gasification technologies are being promoted that are close coupled to a gas turbine. There are still problems to be resolved in interfacing the gasifier and turbine and meeting the gas quality requirements of the turbine fuel gas, but both areas are receiving substantial support. The production of hydrogen, methanol, ammonia and transport fuels through advanced gasification and synthesis are longer term objectives that rely on larger capacity plants that can take advantage of economies of scale.

PYROLYSIS

Pyrolysis is the thermal degradation of carbonaceous material in the absence of air or oxygen. Temperatures in the 350–800°C range are most often used. Gas, liquid and solid products (char/coke) are always produced in pyrolysis reactions, but the amounts of each can be influenced by controlling the reaction temperatures and residence time. The output of the desired product can be maximized by careful control of the reaction conditions. Because of this, heat is commonly added to the reaction indirectly.

Charcoals for barbecues and industrial processes are the most common products of present pyrolysis. An overall efficiency of 35% by weight can be achieved by maximizing the output of the solid product through the implementation of long reaction times (hours to days) and low temperatures (350°C). Heat for the process in traditional kilns can be produced by burning the gas and liquid by-products.

Higher temperatures and shorter residence times are used to produce pyrolysis oils. Optimum conditions are approximately 500°C with reaction times of 0.5 to 2 seconds. The necessary short reaction time is achieved by rapid quenching of the fuel, which prevents further chemical reactions from taking place. Preventing additional chemical reactions also allows higher molecular weight molecules to survive. The fast heating and cooling requirements of the raw material creates process control problems, which are most often dealt with by fine milling the feedstock to less than a few millimetres in size.

Conversion efficiency to liquid can be up to 85% by weight. Some gas is produced, and is commonly utilized in the fuel heating process. Solid char is also produced. The char remains in the pyrolysis oil, and must be removed before the oil can be used in combustion engines or turbines; it too can be used in the fuel heating process. The liquid product is a highly oxygenated hydrocarbon with a high water content, which is chemically and physically unstable. This instability may create difficulties in all aspects having to do with the storage and use of this product. Detrimental chemical reactions within the liquid could also limit the maximum practical fuel storage time.

Figure 3.8 shows an example of the technology of flash or fast pyrolysis being utilized in a double screw reactor. This form of low temperature pyrolysis (450–550°C) is characterized by a reaction time of under one second. The main products produced are gas, bio-oil and coke. Figure 3.9 demonstrates the mass balance for this technology with respect to the initial weight of the biomass, using *Miscanthus* as an example. The energy balance can also be determined with respect to the heating value of the fraction: Figure 3.10 shows the energy balance obtained if the drying step is regarded as a separate process (Shakir, 1996).

The production of pyrolysis oil from biomass fuels has a number of potential advantages over gasification and combustion. The pyrolysis process can be separated from the final energy conversion process by both space and time. Pyrolysis oil could also be transported from a central production site, close to the biomass source, to one or more energy conversion plants situated in the most appropriate location(s). It is

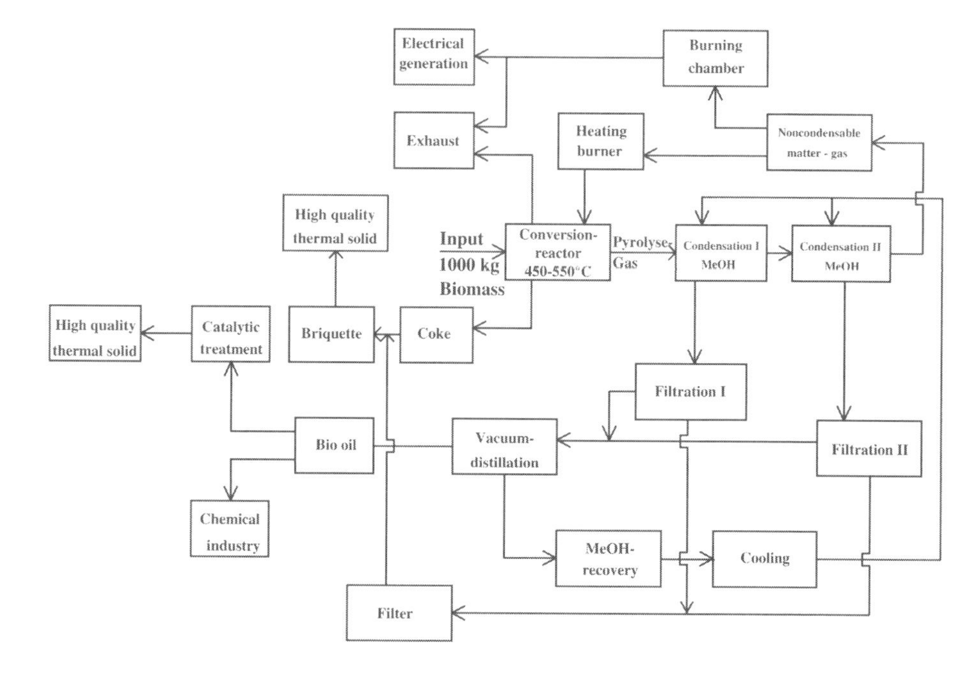

Figure 3.8. Flow diagram of flash or fast pyrolysis biomass conversion (Shakir, 1996).

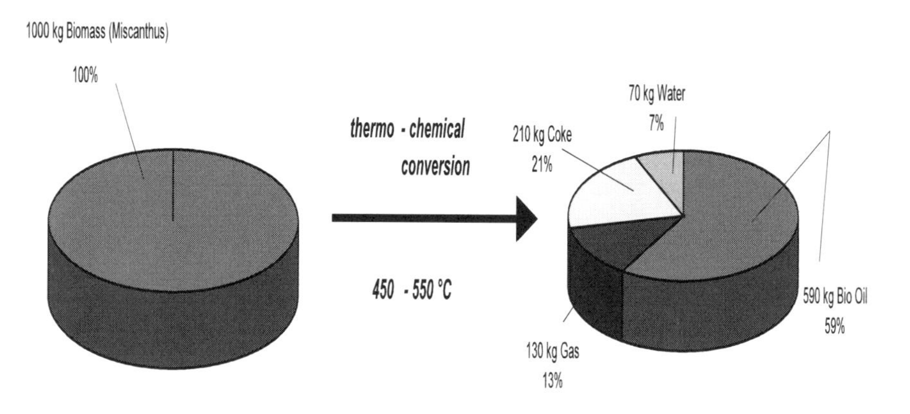

Figure 3.9. Mass balance of Miscanthus *biomass conversion using flash or fast pyrolysis (Shakir, 1996).*

also theoretically possible to upgrade the crude pyrolysis oil into more familiar petroleum products, though the technology for this is yet to be used commercially. Because the moderate process temperatures are not likely to create ash melting or softening, and alkali metals are likely to be retained in the char, pyrolysis is the favoured technology for annual crops and their residues.

Successful trials have shown that pyrolysis oil can be used as a substitute for heavy fuel oil in boilers. Some modification of burners is required, and care must be taken

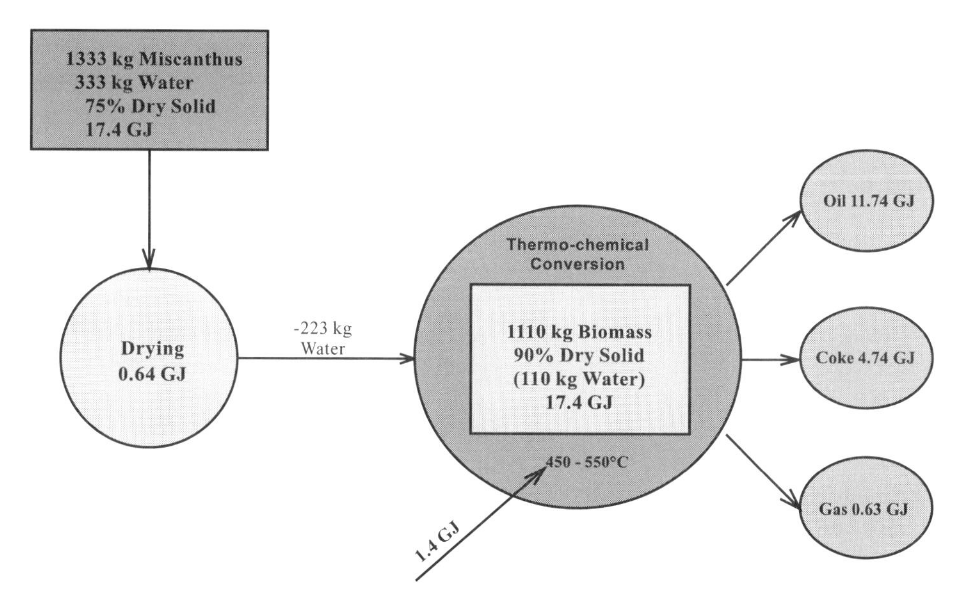

Figure 3.10. Energy balance of Miscanthus *biomass conversion using flash or fast pyrolysis with preconversion drying (Shakir, 1996).*

to limit pyrolysis oil temperature in storage tanks and heated fuel feed lines because of the oil's instability. There have also been limited trials of pyrolysis oil as a substitute for diesel fuel in diesel engines, and the results have been sufficiently encouraging for full scale trials to be planned. The diesel engine needs to be adapted to include a pilot injector ignition system for pyrolysis oil because it will not self-ignite.

Pyrolysis reactions produce a much greater proportion of gas, with correspondingly less liquid and solid products, at temperatures as high as 700–800°C. Gas yields can be greater than 80% by weight. The gas has a moderate heating value, 15–20 MJ/Nm^3, which is enough to allow various types of combustion engines and turbines to use the fuel gas without modification. Liquids, in the form of tars, usually make up less than 5% of the fuel by weight, but as with gasification, the tars must be removed or converted before the gas fuel can be used in engines and turbines. The solid residue consists of highly active dry char that tends to attract any contaminants, such as alkali metals, which may have been present in the biomass feedstock. The solid fraction is usually less than 10% by weight of the fuel (Dumbleton, 1997).

Methanol

Biomethanol is a fuelwood based methyl alcohol obtained by the destructive distillation of wood or by gas production through gasification. Biomethanol is a substitute product for synthetic methanol and is extensively used in the chemical industry (methanol is generally manufactured from natural gas and, to a lesser extent, from coal).

Bio-oil

Bio-oil is directly produced from biomass by fast pyrolysis. The oil has some particular characteristics but has been successfully combusted for both heat and electricity generation. The largest plant built today produces 2 tonnes per hour, but plants for 4 and 6t/h (equivalent to 6–10MWe) are at an advanced stage of planning. In the long term, pyrolysis oil could be used as a transportation fuel.

Fuel cells

The different fuel cell types can be divided into two groups: low temperature, or first generation (alkaline fuel cell, solid polymer fuel cell and phosphoric acid fuel cell); and high temperature, or second generation (molten carbonate fuel cell and solid oxide fuel cell). Low temperature cells have been commercially demonstrated, but are restricted in their fuel supply and are not readily integrated into combined heat and power applications. The high temperature cells can use a wide variety of fuels through internal reforming techniques, have a high efficiency, and can be integrated into combined heat and power systems. Although fuel cell capital costs are falling, increasing environmental restrictions are required before fuel cells will begin to replace conventional technologies in either the power generation or the transport sectors (Williams and Campbell, 1994).

Methanol (MeOH) and hydrogen (H_2) derived from biomass have the potential to make major contributions to transport fuel requirements by competitively addressing all of these challenges, especially when used in fuel cell vehicles (FCVs).

In a fuel cell, the chemical energy of fuel is converted directly into electricity without first burning the fuel to generate heat to run a heat engine. The fuel cell offers a quantum leap in energy efficiency and the virtual elimination of air pollution without the use of emission control devices. Dramatic technological advances, particularly for the proton exchange membrane (PEM) fuel cell, focused attention on this technology for motor vehicles. The FCV has the potential to compete with the petroleum fuelled ICEV (internal combustion engine vehicle) on both cost and performance grounds (AGTD, 1994; Williams, 1993, 1994), while effectively addressing air quality, energy security and global warming concerns.

Ballard Power Systems, Inc., of Vancouver, Canada, introduced a prototype PEM fuel cell bus in 1993 and plans to introduce PEM fuel cell buses on a commercial basis beginning in 1998. In April 1994, Germany's Daimler Benz introduced a prototype PEM fuel cell light duty vehicle (a van) and announced plans to develop the technology for commercial automotive applications. In the USA, the FCV is a leading candidate technology for accelerated development under the 'Partnership for a New Generation of Vehicles', a joint venture launched in September 1993 between the Clinton Administration and the US automobile industry. The partnership's goal is to develop in a decade production-ready prototypes of advanced, low polluting, safe cars that could be run on secure energy sources, especially renewable sources, that would have up to three times the fuel economy of today's gasoline ICEVs of comparable performance, and that cost no more to own and operate (Williams *et al.*, 1995).

The Stirling engine

The Stirling engine's name comes from its implementation of the Stirling cycle. Nitrogen or helium gas (Weber, 1987) is shuttled back and forth between the hot and

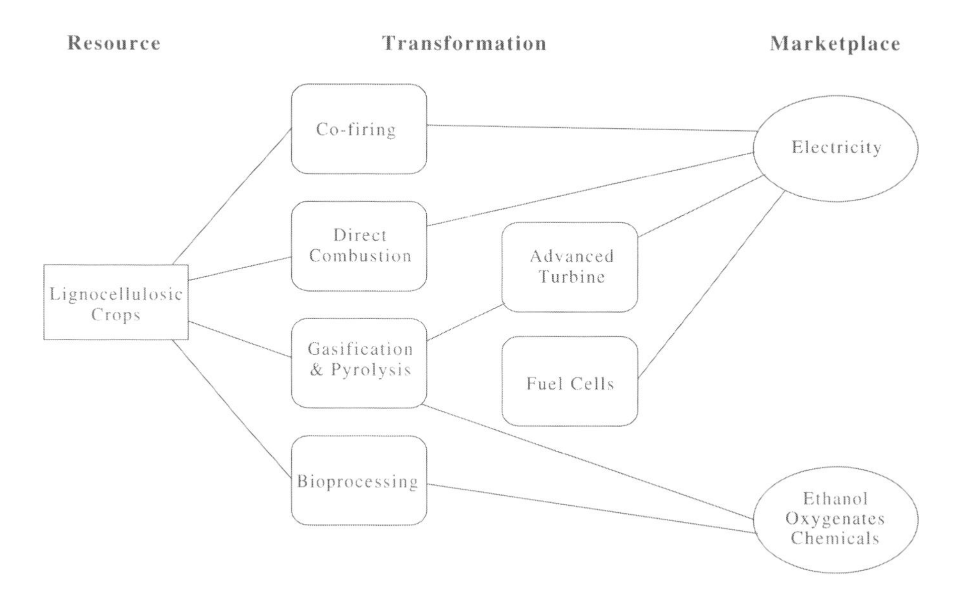

Figure 3.11. Bioenergy pathways (Overend, 1995).

cold ends of the machine by the displacer piston. The power piston, with attached permanent magnets, oscillates within the linear alternator to generate electricity. New developments (Sunpower Inc., 1997) indicate that the Stirling engine's generator efficiently converts rough biomass into electricity with intrinsic load matching capacity. A single engine produces 2.5kW of 60Hz, 120V alternating current power, and up to four engines may be used with a single burner. It is proposed for use at domestic and light industrial sites, and may be used for cogeneration, delivering both electricity and heat.

Positive features of the engine include the direct conversion of biomass heat into electricity, and an integrated alternator with greater than 90% efficiency. Further development of these types of engines is necessary and will open significant future opportunities to produce electricity directly from biomass.

Table 3.11 summarizes the power ranges and efficiencies of the biomass conversion technologies mentioned in this chapter.

Table 3.11. Power range and reported efficiencies of different technologies (Grimm, 1996).

Technology	Power range (MWe)	Typical overall efficiency (%)
Steam engine	0.025 – 2.0	16
Steam turbine (back pressure)	1–150	25
Steam turbine (extraction–condensation cycle)	5–800	35
Steam turbine (condensation cycle)	1–800	40
Gas engine	0.025–1.5	27
Gas turbine	1.0–200	35
Integrated gas and steam combined cycle	5–450	55
Stirling engine	0.0003–?	40
Fuel cell	0.005–?	70

Biofuel properties

Certain properties of biomass detrimentally affect its efficient conversion as a biofuel. The utilization of biomass for high efficiency thermal electricity generation has been shown to be limited not by the properties of the organic component of biomass, but by the characteristics of the associated mineral matter at high temperatures (Overend, 1995). On a moisture and ash free basis, biomass has a relatively low heating value of 18.6GJ/t. This, however, would not limit its use in high efficiency combustion systems since sufficiently high temperatures could be reached to achieve high Carnot cycle efficiencies. The high temperatures cannot be reached because of the fouling and slagging properties of the minerals in the biomass. Mainly responsible is the potassium component alone or in combination with silica. Under combustion conditions the potassium is mobilized at relatively low temperatures and can then foul heating transfer surfaces and corrode high performance metals used in the high temperature sections of burners and gas turbines (Overend, 1995).

Techniques that separate the mineral matter from the fuel components (carbon and hydrogen) at low temperatures can result in very high efficiency combustion applications in combustors, gas turbines and diesel engines. Gasification and various types of clean up systems, as well as pyrolysis techniques are able to separate the minerals from the fuel component and provide the organic part of the biomass as a suitable fuel for high efficiency energy conversion in gas turbines, fuel cells and other engines (Overend, 1995). The conversion pathways are shown in Figure 3.11 (previous page).

4 *Environmental Impact*

There is an increasing awareness that excessive use of fossil fuels may be causing serious damage to the environment and climate, contributing to global warming and possible negative effects on vegetation through acid rain. There is also major concern about the damage that could be caused by transportation accidents and the risks from nuclear power plants.

The potential contamination by mineral oils of the soil and water and air is considerable: one litre of mineral diesel contaminates one million litres of water. In cases of leakage and transport accidents, oils of fossil origin create extensive damage to flora and fauna in both terrestrial and aquatic environments.

Nuclear fuels are limited, but their residues are nearly unlimited in their radioactivity. In addition, the transportation, handling, installation, deposition and storage of radioactive materials need long term monitoring and high inputs. Accidents in nuclear power plants like Chernobyl can be the cause of extensive disasters.

Wide utilization of plant raw materials for energy offers the chance to reorganize agricultural production towards an environmentally consistent system through increasing the number of plant species, reintroducing traditional crops and introducing new alternative crops. This will lead to the production of different energy feedstocks with greater outputs and lower environmental inputs. It will also lead to diversification, improving the appearance of the landscape, reducing the inputs, such as fertilizers, herbicides, fungicides, fuels, etc., in crop management, and improving the microclimate through water use and recycling mechanisms. The introduction of energy crops allows a significant quantity of renewable fuels to be consumed and energy to be produced without markedly increasing the CO_2 content of the atmosphere. Crops can absorb CO_2 released from the combustion of biofuels and produce oxygen from water during photosynthesis, thus reducing the depletion of oxygen in the atmosphere.

Biofuels make a beneficial contribution to the problems of the 'greenhouse effect' and global climatic change in both industrialized and developing countries. There is no doubt that energy crops can be grown in ways that are environmentally desirable. It is possible to improve the land environmentally relative to present use through the production of biomass for energy. The environmental outcome closely depends on how the biomass is produced (Larson and Williams, 1995). Table 4.1 takes a closer look at the environmental benefits and constraints that are present when solid energy crops are produced and harvested.

Thermochemical conversion of vegetative feedstocks leads to the recovery of all mineral nutrients, except nitrogen, in the form of ash at the biomass conversion facility. The ash can then be returned to the fields as fertilizers. Fixed nitrogen can be restored in environmentally consistent procedures – that is, by improving the crop rotation systems through selection of plant species that fix nitrogen (legumes) to eliminate or to reduce the levels of the necessary mineral nitrogen fertilizers. Perennial lignocellulose energy crops need less nitrogen than annual food crops. Energy crops also offer flexibility in dealing with erosion and chemical pollution from herbicide use. The erosion and herbicide pollution problems would be similar to those for annual food or feed crops. But the growing of cereals (whole crop) as biofuel feedstocks needs much less nitrogen – about one-third of that needed to produce grain cereals, in which most of the nitrogen is used to improve the nutritional value. In the case of whole cereal crops for energy, the food quality aspects are not valid.

Some perennial grasses (*Miscanthus* and *Arundo donax*) can be harvested annually for decades after planting, and fast growing trees (woods) that are harvested only every 2 to 5 years are replanted perhaps every 10 to 20 years. Erosion would be sharply reduced in both cases as compared to annual crops such as corn and soybeans. There would also be a reduction in the need for fertilizers and herbicides (Table 4.2) (Larson and Williams, 1995).

Careful selection of crop species and good plantation design and management can be helpful in controlling pests and disease, and could thereby minimize or even eliminate the use of chemical pesticides.

Table 4.1. Environmental benefits and constraints through production and harvest of some energy crops as solid feedstocks.

	Wood		Straw	Hay	*Miscanthus*
	Short rotation	Forest residues			
Environmental benefits	Little soil erosion risk; little soil compaction risk; relaxation of areas with little forest	Possible reduction of risk of disease and/or pest infestation	No need for additional production means	Very little soil erosion risk; very little soil compaction risk; little risk of nitrogen removal; very little risk of pesticide removal	Little soil erosion risk; little risk of nitrogen removal; very little risk of pesticide removal
Environmental constraints	Higher Nutrient and pesticide demand compared to conventional forest; monoculture plantation	Nutrient extraction; reduction of humus supply	Reduction of humus supply		Medium soil compaction risk; monoculture plantation; water demand for irrigation possible

Table 4.2. Fertilizer and herbicide application and soil erosion rates in the USA (Hohenstein and Wright, 1994).

Cropping system	N-P-K application rate (kg/ha/year)	Herbicide application rate (kg/ha/year)	Soil erosion rate (tonne/ha/year)
Annual crops			
Corn	135-60-80	3.06	21.8[a]
Soybeans	20[b]-45-70	1.83	40.9[a]
Perennial energy crops			
Herbaceous	50[c]-60-60	0.25	0.2
Short-rotation woods	60[c]-15-15	0.39	2.0

[a] Based on data collected in the early 1980s. New tillage practices used today may lower these values.
[b] The nitrogen input is inherently low for soybeans, a nitrogen fixing crop.
[c] Not including nitrogen fixing species.

Energy crops could replace monoculture food crops, and in many cases the shift would be to an ecologically more diversified landscape. Improving biodiversity on a regional basis might enable many species to migrate from one habitat to another and could help to address the aesthetic concerns sometimes expressed about intensive monocultures.

The use of energy crops on available land to substitute for fossil fuels would be an effective means of decreasing atmospheric carbon dioxide pollution, because these crops can absorb CO_2 released from the combustion of biofuels.

The biodegradability of various biodiesel fuels in an aquatic environment, including rape methyl ester (RME) and soyate methyl ester (SME) and blends of biodiesel and mineral diesel at different ratios, was determined by Zhang *et al.* (1995). All of the biodiesel fuels are 'readily biodegraded' compounds according to Environmental Protection Agency (EPA) standards, and have a relatively high biodegradation rate in an aquatic environment. Biodiesel can promote and speed up the biodegradation of mineral diesel. The more biodiesel present in a biodiesel/mineral diesel mixture, the faster the degradation rate. The biodegradation pattern in a biodiesel/diesel mixture is that microorganisms metabolize both biodiesel and diesel at the same time and at almost the same rate.

Measurements were carried out on buses and private vehicles to compare CO, CO_2, SO_x and NO_x emissions with respect to mineral diesel, biodiesel and a 30% biodiesel blend. The results are given in Figure 4.1. CO_2 and SO_x emissions were eight times higher when mineral diesel was used. The emission of NO_x was 10% higher from biodiesel (RME), but the NO_x emission was similar when a blend of 30% biodiesel was used (ADEME, 1996).

The energy output/input ratio, a characteristic value for the energetic efficiency of production processes, for the main products was 2.14 for RME production. The ratio for RME was more than twice as high as the ratio for bioalcohol production from sugarbeet, which was 1.04. Taking into account the energy of the by-products, the output/input ratio as a measure for the energy balance of all the involved production processes was about 1.63 for bioalcohol and 3.18 for RME production (Graef *et al.*, 1994).

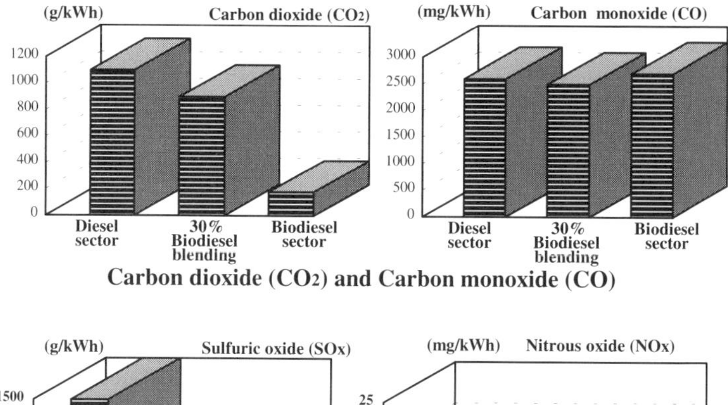

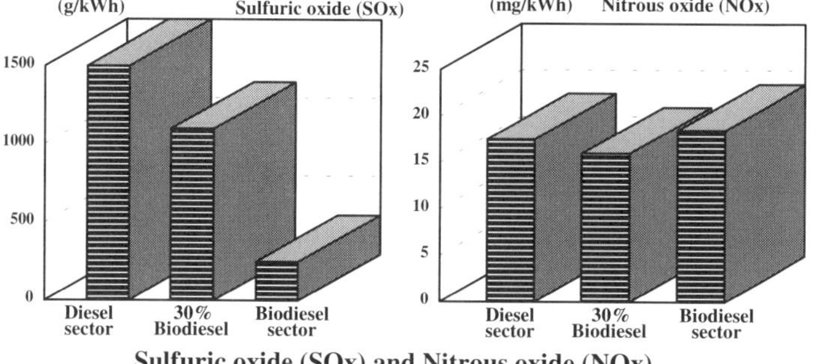

Figure 4.1. Comparison of oxide emissions for diesel, biodiesel and blended fuels (ADEME, 1996).

If bioalcohol is used instead of gasoline in Otto cycle engines there is an improvement in knock resistance, and a reduction of carbon monoxide (CO) and hydrocarbons (HC) in the exhaust gas. Running diesel engines on bioalcohol leads to reduction of smoke, nitrogen oxides (NO_x) and HC emissions. The use of rape seed oil and RME in diesel engines instead of diesel fuel reduces the emission of carcinogens and polycyclic aromatic hydrocarbons (PAH) in the exhaust gas (Krahl, 1993).

Biomass integrated gasifier/gas turbine (BIG/GT) power plants would be characterized by low levels of sulphur and particulate emissions. The low sulphur content of biomass makes it possible to avoid the costly capital equipment and operating cost penalties associated with sulphur removal. However, NO_x emissions could arise from nitrogen in the biofuels. These could be kept at low levels by growing biomass feedstocks with low nitrogen contents and/or by selectively harvesting the portions of the biomass having high C/N ratios. Also, feedstocks from old cereal varieties with a low harvest index contain considerably less nitrogen than new high yielding varieties.

The combustion of biomass (*Miscanthus*) reduced the carbon dioxide (CO_2) emission up to 90% in a comparison with the combustion of fossil feedstocks (hard coal) (Lewandowski and Kicherer, 1995). Investigations have been carried out into the potential reductions of different gases emitted into the atmosphere where biofuels are used instead of fossil fuel in a thermal energy plant. Total emissions were determined using a base of 12.7MWh net heat capacity; the biofuel was straw and the fossil fuel brown coal. The results are summarized in Figure 4.2, and show that the bioenergy heat power plant in Schkölen, Germany, emits 28% of the CO_2-equivalent gases emitted by the comparable brown coal plant.

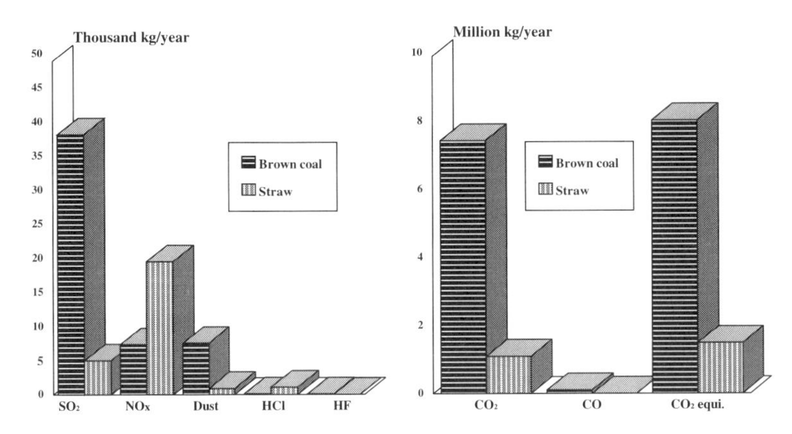

Figure 4.2. Emission of different gases at the energy plant in Schkölen (Garbe, 1995).

A significant amount of research is presently under way to develop and refine the use of fuel cell vehicles (FCVs). Williams *et al.* (1995) explain that the fuel cell converts its fuel (presently hydrogen and methanol) directly into electricity, without burning. Through the use of indirectly heated gasifiers, biomass can be used to produce methanol and hydrogen for FCV fuels. When the biomass feedstock is grown sustainably, the use of these fuels in FCVs would result in extremely low levels of CO_2 emissions and no or very little air pollution, and would make a significant contribution to road transportation energy needs. Table 4.3 compares current and future energy yields from alternative feedstocks and conversion technologies for internal combustion

Table 4.3. Energy yield for alternative feedstock/conversion technologies (Williams et al., *1995).*

Option	Feedstock yield (dry tonnes/ha/year)	Transport fuel yield (GJ/ha/year)	Transport services yield[a] (10^3 v-km/ha/year)
Rape methyl ester (Netherlands, year 2000)	3.7 (rape seed)	47	21 (ICEV)
EthOH from maize (USA)	7.2 (maize)	76	27 (ICEV)
EthOH from wheat (Netherlands, year 2000)	6.6 (wheat)	72	26 (ICEV)
EthOH from sugarbeet (Netherlands, year 2000)	15.1 (sugarbeet)	132	48 (ICEV)
EthOH from sugarcane (Brazil)	38.5 (cane stems)	111	40 (ICEV)
EthOH, enzymatic hydrolysis of wood (present technology)	15 (wood[b])	122	44 (ICEV)
EthOH, enzymatic hydrolysis of wood (improved technology)	15 (wood[b])	179	64 (ICEV)
MeOH, thermochemical gasification of wood	15 (wood[b])	177	64/133 (ICEV/FCV)
H_2, thermochemical gasification of wood	15 (wood[b])	213	84/189 (ICEV/FCV)

[a] Fuel economy assumed (in litres of gasoline-equivalent per 100km): 6.30 for rape methyl ester, 7.97 for ethanol, 7.90 for methanol, 7.31 for hydrogen used in ICEVs and 3.81 for methanol and 3.24 for hydrogen used in FCVs

[b] Including high yielding lignocellulose crops and short rotation coppice.

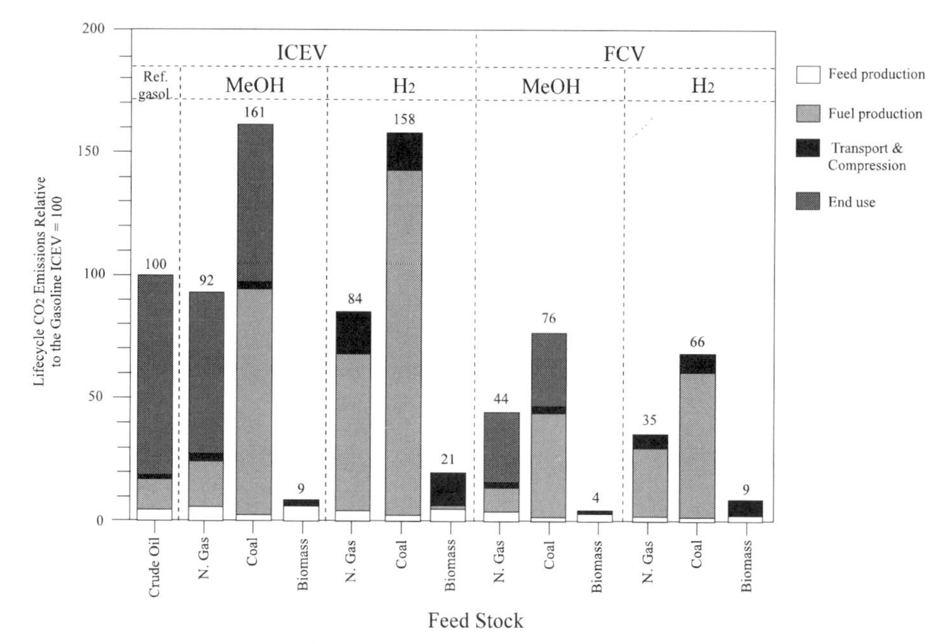

Figure 4.3. Estimated life cycle of carbon dioxide emissions from methanol and hydrogen ICEVs and FCVs, with comparisons to emissions from a gasoline ICEV (Williams et al., *1995).*

engine vehicles (ICEVs) and FCVs. In the cases of MeOH and H_2, the FCVs offer more than double the transport services' yield.

Figure 4.3 continues the comparison of alternative fuel feedstocks and traditional fuel feedstocks (natural gas and coal) with respect to carbon dioxide emissions. The factors taken into consideration are feed production, fuel production, transport and compression, and end use. The biomass feedstocks, which in each case have considerably lower life-cycle carbon dioxide emissions than the traditional feedstocks, have most of their emissions coming from feed production and transport and compression. There are no end-use emissions, and fuel production emissions are virtually non-existent.

Production of biomass biofuel for transportation, along with the subsequent improvement in land usage, could lead to less dependence on oil for traditional transportation fuels. Biomass derived methanol and hydrogen fuels would offer the additional advantages of providing a sustainable source of income for the rural areas of developing countries, and improving competition and price stability in world transportation fuel markets. These fuels bring increased land use efficiency and decreased adverse environmental impacts when compared with conventional biofuels such as ethanol from maize. They can also be used in environmentally friendly and efficient FCVs.

It is important that biomass producers are aware of the variety of ways in which their biomass can be utilized and of the increasing need for methanol and hydrogen fuels to be used in transportation vehicles such as FCVs. In turn, FCV developers should be aware that the production of these fuels through sustainable biomass utilization makes their innovation and research efforts worthwhile.

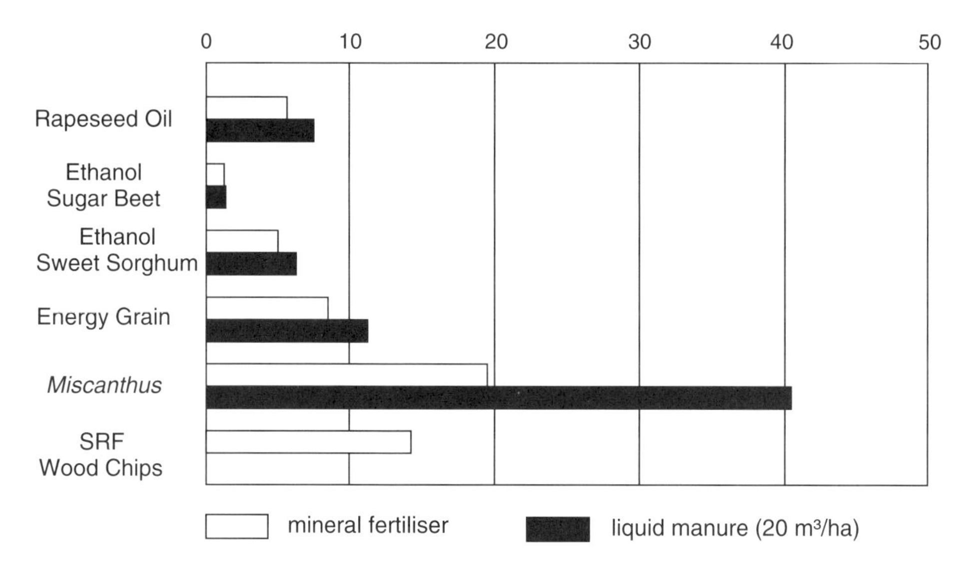

Figure 4.4. Energy balance, output/input of different energy crops as a result of mineral or organic fertilizer application (Hartmann, 1994).

The environmental advantages of biomass from energy plant species can be summarized as follows:

- They are a renewable resource that can be used to generate heat, electricity and transport fuels.
- The energy balance is positive – that is, more output energy than input energy, up to 25:1 (Figure 4.4).
- The ashes are recyclable.
- If they are cultivated and harvested properly, they make no or very little net contribution to global carbon dioxide.
- Other important environmental benefits, especially of the perennial plant species, are the protection and improvement of local watersheds and surrounding wildlife habitats, and reduction of the use of chemical pesticides and fertilizers.
- They create no major environmental risks through transportation, storage, processing and conversion.

5 *Economic and Social Dimensions*

Energy and development

Fossil fuel utilization is highest in developed countries, where it is usually used for transportation and for the production of heat and electricity. Fossil fuels have a detrimental impact on the environment and our planet, and supplies are limited, so it is now time to look for a new source of energy that is environmentally friendly and that can be sustainably produced for future generations. The sustainable production of biomass is one of the most promising possibilities for moving the energy production trend away from fossil fuels.

Biomass utilization is most commonly found in developing countries, where it is usually the primary source of energy (making up approximately one-third of the total energy production). In addition, in some countries (Rwanda, Nepal and Tanzania) biomass utilization accounts for over 90% of their energy production.

Economic development is closely correlated with the availability and utilization of modern energy sources. Also, the production and consumption of food is linked to the amount of energy used (Figure 5.1). In 1990, the per capita consumption of modern energy for 21 African countries was less than 100 kgoe (World Bank, 1992). In many of these countries, daily per capita caloric supply is below 2000 calories. Food production is unlikely to increase without greater access to modern energy (FAO, 1995b).

According to Hohmeyer and Ottinger (1994), one of the main obstacles to the expansion and acceptance of renewable fuels and biomass energy technologies into world markets is that the markets do not acknowledge the adverse social and environmental costs and risks connected with the usage of fossil and nuclear fuels. Oil shipping accidents, such as the Exxon Valdez disaster (US$2.2 billion clean-up costs), cause devastating and long lasting effects on the environment and wildlife, yet these disasters are becoming increasingly common occurrences. Nuclear accidents such as Three Mile Island (US$1 billion medical and clean-up costs) and Chernobyl still leave questions as to the effects on future generations (Miller, 1992). The costs of maintaining channels to fossil fuel sources through military means, along with the non-renewable depletion of these limited resources, should also to be taken into consideration when promoting increased use of biomass fuels. Table 5.1 takes a comparative

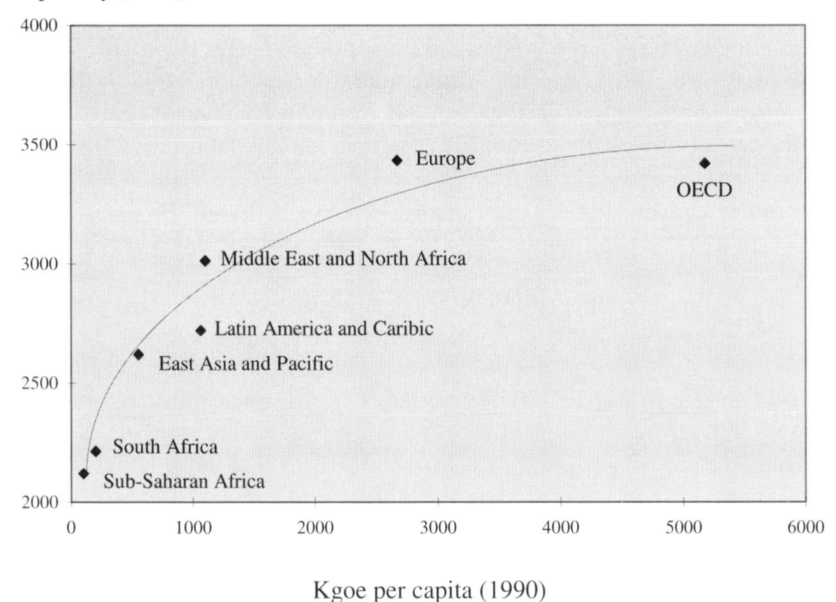

Figure 5.1. Energy consumption and calorie supply in different regions of the world (FAO, 1995b).

Table 5.1. Social, economic and environmental aspects of utilization of different energy sources (Scheer, 1993; El Bassam, 1996).

	Energy fuels		
	Fossil	Nuclear	Biomass
Renewable	no	no	yes
CO_2 reduction	no	yes	yes
Reduction of heat emission	no	no	yes
Landscape	no	no	yes
Avoidance of large accident risks	no	no	yes
Excessive costs of environmental repair	yes	yes	no
Reduction of administrative costs	no	no	yes
Innovation	no	yes	yes
Creating new industrial jobs	no	yes	yes
Promoting decentralization of economic structure	no	no	yes
Promoting export	no	yes	yes
Increasing autonomous energy supply (industrialized countries)	no	yes	yes
Increasing autonomous energy supply (developing countries)	no	no	yes
Improving farmers' incomes	no	no	yes
Significant time of waste decay	yes	yes	no
Migration to urban areas	yes	yes	no
Favourable public opinion	no	no	yes
Avoidance of international conflicts and wars	no	no	yes
Gene deformation	no	yes	no

look at the social, economic and environmental aspects of the utilization of fossil, nuclear and biomass fuels.

The comparatively large amount of subsidies and support provided to conventional energy sources is another problem preventing biomass derived fuels from playing a more substantial role in global energy supply (Hubbard, 1991). The amount of money spent by developed countries of the International Energy Authority (IEA) on research and development from 1988 to 1990 was US$73 billion for nuclear, $12 billion for coal, $11 billion for all renewables and $1 billion for biomass (OECD, 1991). Energy prices in numerous developing countries are subsidized by as much as 30–50%.

Although biomass energy has become competitive with conventional energy sources in many areas, in order for biomass derived fuels to compete more evenly it is necessary that subsidies and support be more equally distributed and guaranteed for longer time periods, and that environmental effects are given the recognition that they deserve.

A study was undertaken by Environmental Resources Management (ERM, 1995) to look at the economics of bioethanol, along with the effects of subsidies for helping increase the income of wheat and sugarbeet farmers. The study divided the farmers into 'base case' and 'favoured region' categories. The base case considered small farms that were not in traditional sugarbeet farming territories, making up approximately 70% of EU farmers. Here, yields were estimated at 49t/ha for sugarbeet and 6t/ha for wheat; the set-aside subsidy was 270Ecu/ha. The favoured region took larger farms in traditional sugarbeet farming territories, with bioethanol processing facilities in the region. The crop yields were estimated at 63t/ha for sugarbeet and 7t/ha for wheat; the set-aside subsidy was 315Ecu/ha.

Figure 5.2 shows the results of the study in terms of farmers' net income in both the base case and favoured region for sugarbeet and wheat production when the crops are grown as a food crop, as a bioethanol crop on set-aside land, and when the land is left fallow and the subsidy payment is collected. Only the wheat farmer is presently able to collect an additional set-aside subsidy for growing wheat for a non-food purpose.

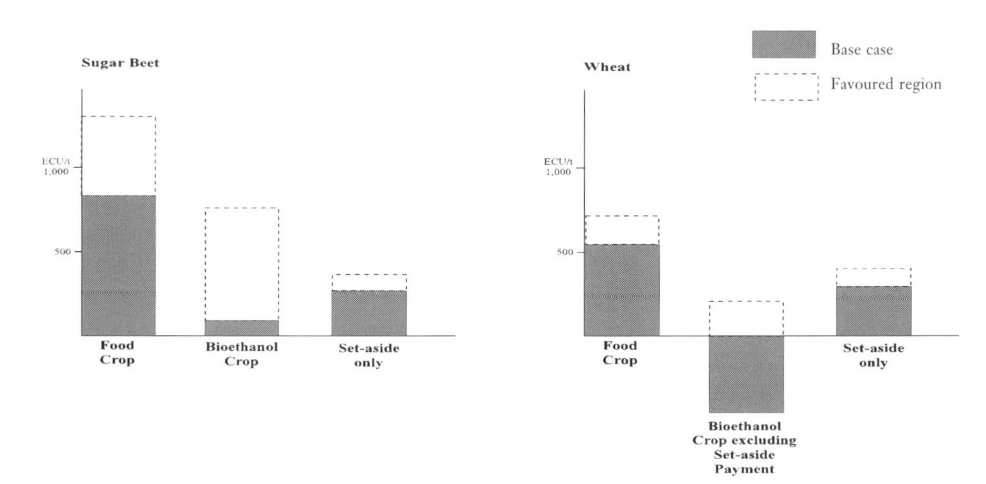

Figure 5.2. Farmers' net incomes from food crops, biofuel crops and set-aside subsidies (ERM, 1995).

The results of the study also showed the money that would be returned to the base case farmers because of a tax subsidy paid at the gasoline blending stage. For sugarbeet, a 40Ecu/t tax would be forgone by the EU. At a crop price to the farmer of 23Ecu/t, this would contribute 2Ecu/t to the farmer's income. In the case of wheat, a 151Ecu/t tax would be forgone by the EU. At a crop price to the farmer of 28Ecu/t, along with a 45Ecu/t set-aside payment, this would contribute 2.5Ecu/t to the farmer's income.

The study concluded that it would not be very feasible for farmers in the EU to grow sugarbeet or wheat crops for bioethanol production. In addition, no significant amount of money would be returned to the farmer because of a tax reduction on bioethanol blended fuels. The tax subsidy for bioethanol in motor fuels would not be sufficient to help support rural communities.

Medium and long term projects show an unavoidable increase in oil prices, which will make biofuels more competitive. Energy crops and biomass in general seem to have, in many instances, much lower external costs than fossil fuels, provided they are grown and processed in an environmentally consistent way. It is unhelpful to argue the case for bioenergy with yesterday's technology without taking into account the innovation possibilities and the prices of today, along with the scarcity, the growing future demand for energy and the environmental damage and risks associated with fossil and nuclear energy feedstocks.

There are various concerns regarding the production of biomass as a fuel source, especially the establishment of large scale crop production systems. These concerns include environmental impact, land availability, and possible conflicts with food production. The penetration of bioenergy into the market will require overcoming preconceived ideas of bioenergy within governments, the market and by the general public (Hall and House, 1995).

On the other hand, more energy will be available from a specific unit area as a result of:

- the introduction of naturally high yielding plant species, such as *Miscanthus* spp., *Arundo donax*, etc.;
- the enhancement of the acreage yield of traditional crops through further developments in plant breeding, gene technology and biotechnology;
- improvements in the efficiency of engines and conversion technologies: in 1995, the major car companies in Europe agreed to develop 3 litre per 100km diesel car engines by the year 2000, which means more mileage per hectare!

Biotechnological advances in biomass conversion may initially alter the prevailing technical and economic conditions in comparison with the production of non-biomass fuels. Assessments of emerging technologies (using the biomass integrated gasifier as an example) suggest that power plants of medium scale (20–50MW) could achieve thermal efficiencies in excess of 40% within a few years (eventually reaching 50% or more) combined with capital costs well below those of comparable conventional biomass plants utilizing boiler/steam technology (Elliott and Booth, 1993).

The competitiveness of biomass utilization for biofuels will thus be considerably improved. Energy crops, as converters of solar energy and grown with environmentally sustainable methods, fulfil the sustainability criterion because they avoid the depletion of non-renewable fossil resources. This will impose positive impacts on soils (less erosion), landscapes, ecosystems and biodiversity, which should be of economic importance.

Although biomass feedstock prices are presently high, it would be possible for biomass methanol and hydrogen fuel prices to be competitive with those of coal. If the cost of fuel production from natural gas rises, as expected, it is estimated that by about the year 2010 biomass methanol and hydrogen fuels will be nearly competitive with natural gas. At that time, a tax of less than 2% for using natural gas derived fuels in fuel cell vehicles (FCVs) would be enough to make biomass derived fuels more economical than natural gas.

In industrialized countries, the contribution of biofuels to energy supply varies. In Germany the percentage of biomass to total primary energy supply is less than 2%, in Denmark it is 7%, in the USA 4.5%, in Austria 13% and in Sweden almost 17%. Sweden has developed a very effective taxation system for energy sources, and has implemented three environmental taxes: a sulphur tax, a CO_2 tax and an energy tax. All these led to increases in the prices of fossil energy resources in comparison to biofuels (Figure 5.3).

Jobs and employment

A study by Grassi (1997) looked at the impact of bioenergy development on job creation in the European Union (EU). The study came to the following conclusions:

- The renewable energy sector is labour intensive, and is thus a significant source of employment.
- Bioenergy, especially given its ability to penetrate all energy markets in the future (power, heat, transport, chemicals), can offer the largest contribution to the EU, with an envisaged 12% renewable energy target for the year 2010.
- At the EU level, the foreseen bioenergy contribution of about 113mtoe/y by the year 2020 should open up opportunities for about 1,500,000 new jobs.
- The total average investment cost for about 1 million new direct jobs (the cost of the indirect jobs will derive automatically from private investment) will be

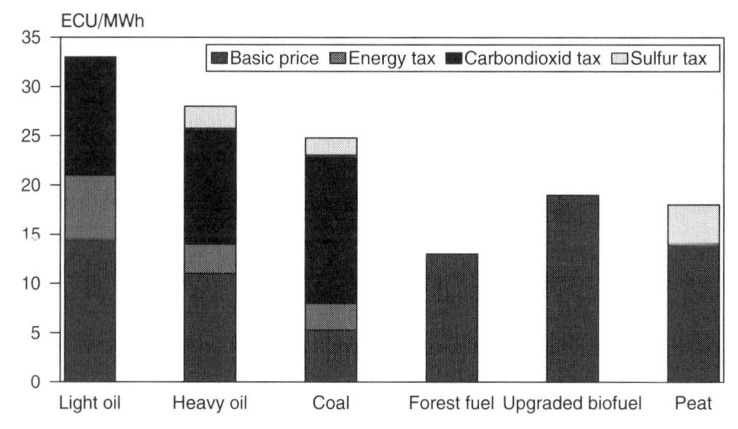

Figure 5.3. Energy prices for heating plants in Sweden, 1996 (Burvall, 1996).

> around 250 billion Ecu less than the estimated 345 billion ECU of total subsidies paid to 1.5 million non-employed people in our socially protected EU society during the 20 year minimum life of the investments (plants).

As reported by the US Department of Energy (DOE, 1997), in the USA economic activity associated with biomass currently supports about 66,000 jobs, most of which are in rural regions. It is predicted that by the year 2020, over 30,000MW of biomass power could be installed, with about 60% of the fuel supplied from over 10 million acres of energy crops and the remainder from biomass residues. This would support over 260,000 US jobs and would substantially revitalize rural economies (Figure 5.4).

For farmers, biomass energy crops can be a profitable alternative that will complement, not compete with, existing crops and thus provide an additional source of income for the agricultural industry. It is envisioned that biomass energy crops will be grown on currently underutilized agricultural land.

In addition to rural jobs, expanded biomass power will create high skill, high value job opportunities for utility and power equipment vendors, power plant owners and operators, and agricultural equipment vendors.

Biomass stoves

The most important energy service today in many developing countries is cooking. Traditional fuels – fuelwood, crop residues and dung – are the main cooking fuels in rural areas of these countries. In many urban areas, charcoal and coal are also used. More than half of the world's 2 billion poor people depend on these crude, polluting fuels for their cooking and other heating needs. Figure 5.5 shows the large amount of

Figure 5.4. Projected annual US employment impacts from biomass power (DOE, 1997).

Figure 5.5. Extremely inefficient, traditional type of cooking stove (stove is on the right side of the photograph), northwest China (Photo: El Bassam).

firewood, collected over great distances, necessary for fuelling the small and inefficient stoves (right side of photo). Figure 5.6 shows a slightly more modern and efficient stove, but there is still a large demand for significant improvement and distribution.

Higher incomes, and reliable access to fuel supplies, enable people to switch to modern stoves and cleaner fuels such as kerosene, LPG and electricity. This transition can be observed worldwide in various cultural traditions. These technologies are

Figure 5.6. Improved type of biomass stove, northwest China (Photo: El Bassam).

Figure 5.7. Biofuel marketplace in Ethiopia (Photo: Liese).

preferred for their convenience, comfort, cleanliness, ease of operation, speed, efficiency and other attributes. The efficiency cost and performance of stoves generally increase as consumers shift progressively from wood stoves to charcoal, kerosene, LPG or gas, and electric stoves.

There can be a substantial reduction in both operating costs and energy use in going from traditional stoves using commercially purchased fuelwood to improved biomass, gas or kerosene stoves. There are also opportunities to substitute high performance biomass stoves for traditional ones, or to substitute liquid or gas (fossil or biomass based) stoves for biomass stoves. Local variations in stove and fuel costs, availability, convenience and other attributes, and in consumer perceptions of stove performance, will then determine consumer choice.

In rural areas, biomass is likely to be the cooking fuel for many years to come. This is exemplified by the common establishment of biofuel marketplaces on which many towns and villages are dependent throughout the world (Figure 5.7). Alternatively, particularly in urban areas, liquid or gas fuelled stoves offer the consumer greater convenience and performance at a reasonable cost.

From a national perspective, public policy can help to shift consumers towards the more economically and environmentally promising cooking technologies. In particular, improved biomass stoves are probably the most cost-effective option in the short to medium term, but require significant additional work to improve their performance.

In the long term, the transition to high quality liquid and gas fuels for cooking is inevitable. With this transition, substantial amounts of labour now expended to gather biomass fuels in rural areas could be freed; the time and attention needed to cook using biomass fuels could be substantially reduced; and household, local and regional air pollution from smoky biomass (or coal) fires could be largely eliminated. The use of biomass derived liquid or gaseous fuels (such as ethanol, biogas and producer gas) for cooking and other advanced options is particularly relevant (UNDP, 1997).

The development, improvement and manufacture of biomass stoves and cookers will not only lead to improved efficiency of the energy conversion outputs and

Figure 5.8. Women carrying firewood in Ethiopia (Photo: Liese).

improve the environment, but will also create large opportunities for job creation both directly and indirectly throughout the world, especially in rural areas. In some cases, women in Africa and Asia need up to 8 hours and must go as far as 20km to collect firewood in order to cook one meal for the family (Figure 5.8). The growing of energy crops and utilization of plant residues, the development and production of efficient cookers, and the development of technologies for pelleting and briquetting of biomass will have great positive impacts on rural markets. The money that has to be spent on fossil energy fuels will stay in the region instead.

Investment in the energy supply sector

The present level of worldwide investment in the energy supply sector, $450 billion per year, is projected to increase to perhaps $750 billion per year by 2020, about half of which would be for the power sector. Such investment levels cannot be sustained by traditional sources of energy financing.

Also, the multilateral financing agencies are able to provide only a small fraction of the capital needed. All of this undermines self-reliance, and leads to deals that reflect the high price of capital arising from the high financial risks involved.

The dependence on fossil fuels has created a wide variety of problems for non-oil-producing developing countries, as well as for some industrialized countries and economies in transition. In over 30 countries, energy imports exceed 10% of the value of all exports, a heavy burden on their balance of trade that often leads to debt problems. In about 20 developing countries, payments for oil imports exceed payments for external debt servicing. This is an important aspect of the energy—foreign-exchange nexus.

Advanced technology in bioenergy would facilitate decentralized rural electrification, and thereby promote rural development as well as considerably reduce the costs of imported fuels (UNDP, 1997).

6 Prospects for Renewable Energy

We are consuming and exhausting energy feedstocks (coal, oil, gas and peat) that took nature millions of years to develop. We extract these resources from deep layers in our earth and then burn them, causing pollution of the environment and the atmosphere, with partly irreversible adverse effects. In the future, energy resources will be obtained from the surface and not from inside the earth.

Energy cannot be created; energy can only be converted from one form to another. Every energy source, except radioactive materials, arises from solar energy directly or indirectly. Today's primary sources of energy are mainly non-renewable: fossil sources and conventional nuclear power. Of these two, predicted fossil sources have only a short supply remaining (Figure 6.1 shows the position for oil). Future development depends on the long term availability of energy in increasing quantities from sources that are dependable, safe and environment friendly. At present, no single source or combination of sources can meet this future need.

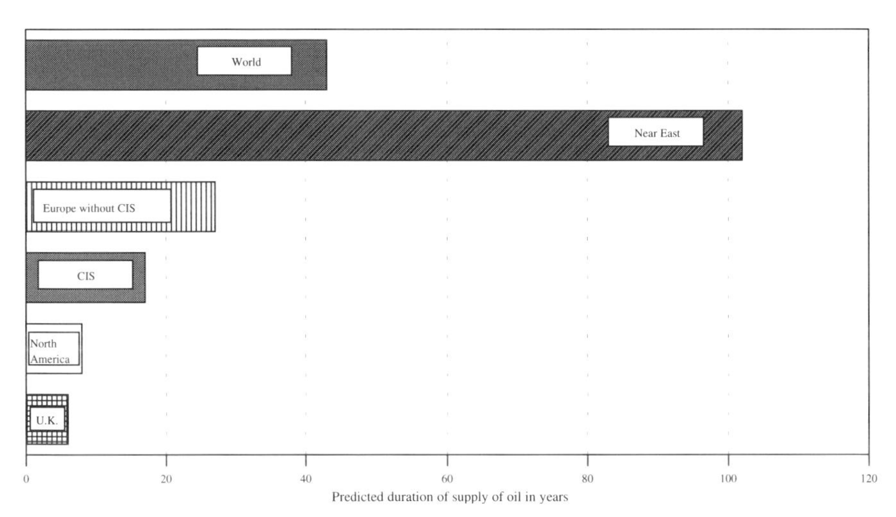

CIS = Commonwealth of Independent States (Former Soviet Union)

Figure 6.1. Predicted length of supply of oil in selected regions, based on current utilization rates (BMWI, 1994).

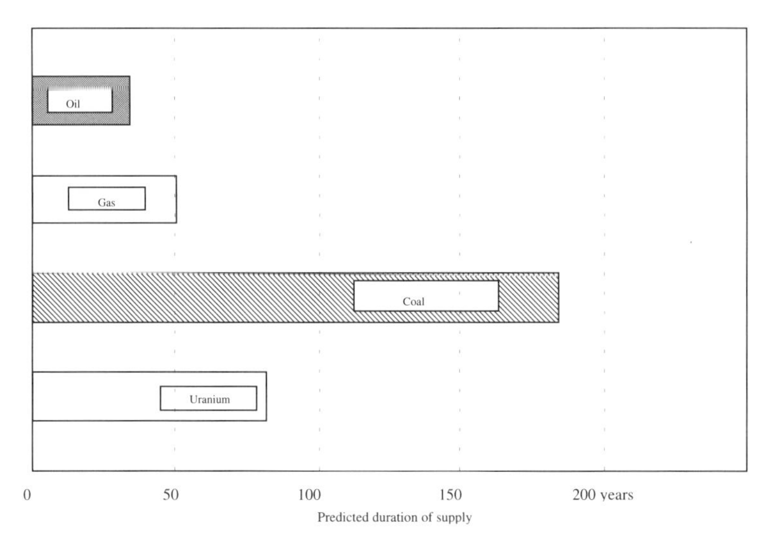

Figure 6.2. Predicted worldwide availability of primary fuels (BMWI, 1994).

The key elements of sustainability in the energy supply are:

- The conscientious development of renewable energy sources, as well as their subsequent utilization.
- Improved energy use efficiency and conservation measures to minimize the loss of primary resources.
- The protection of the biosphere, and prevention of more localized forms of pollution.
- The development of adequate political, administrative and financial support for the implementation and utilization of renewable energy sources.

According to the German Federal Ministry of Economy, the predicted supply durations of the non-renewable energy sources, based on current available rates, are oil 45 years, gas 55 years, coal 150 years, and uranium 80 years (Figure 6.2) (BMWI, 1994).

Although some of these resources might last a little longer, especially if more reserves were to be discovered, the main problem of 'scarcity' will remain, and this represents the greatest economic and social challenge to humanity.

Because fuel is relatively cheap now, this is an optimum time to build a new industry throughout the world. In addition to creating new jobs, we will be conforming to the United Nations Clean Air Act and reducing our dump sites.

We have the key in our hands; we need only to open the right door or to press the right button. And the earlier the better: we don't have much time to lose. Biomass energy plant species will light the world's future, and keep people mobile (Figure 6.3).

Integrated energy farms

The implementation of integrated energy farms to produce food and energy for all represents a realistic vision that could be established in different regions of the world. The concept can be summarized as follows.

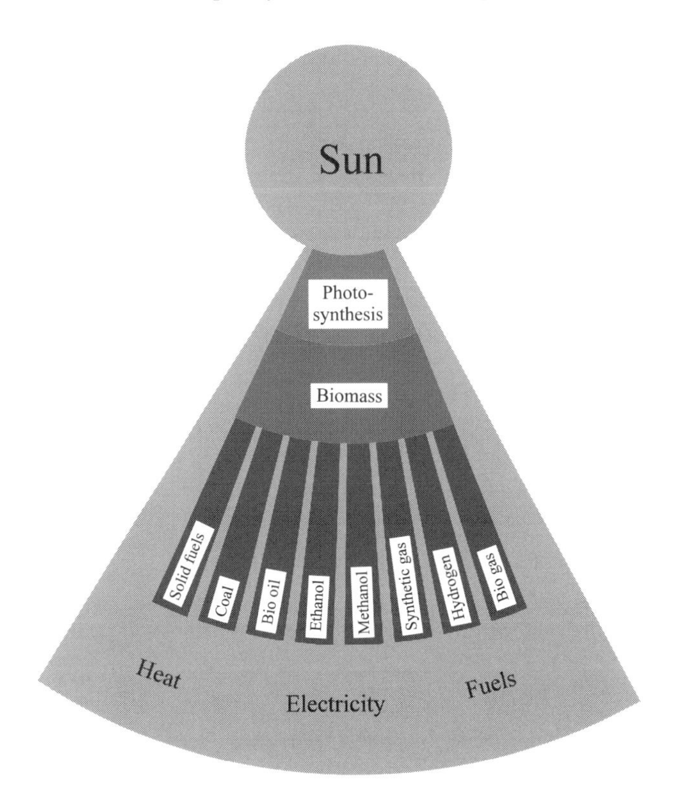

Figure 6.3. Energy plant species – the key to lighting the world's future and keeping people mobile.

The integrated energy farm is centralized around a main building or farmhouse (Figure 6.4, overleaf). The land is then divided up into consecutive rings, expanding away from the main building. The land immediately surrounding the main building is used for growing spices, herbs and vegetable crops. The next ring is used for growing food crops, followed by a ring of land for growing oil, ethanol and annual energy crops. This is surrounded by a ring used for growing fruit trees, and for raising poultry and other animals, as well as providing pasture for grazing and energy. The next consecutive ring carries perennial energy crops and short rotation forests, while the outermost ring of land is used for permanent trees that would provide shelter for the inner crops. The energy farm would also have a combined fish lake and water reservoir, along with windmills and solar collectors dispersed around the farm.

At present, the optimum size of the farm and the individual crop plots, and the appropriate on-site facilities and equipment required for maximum production, still await calculation. Equipment and facilities would probably include:

- windmills
- solar collectors
- photovoltaic (PV) units
- briquetting and pelleting machines
- biogas production units

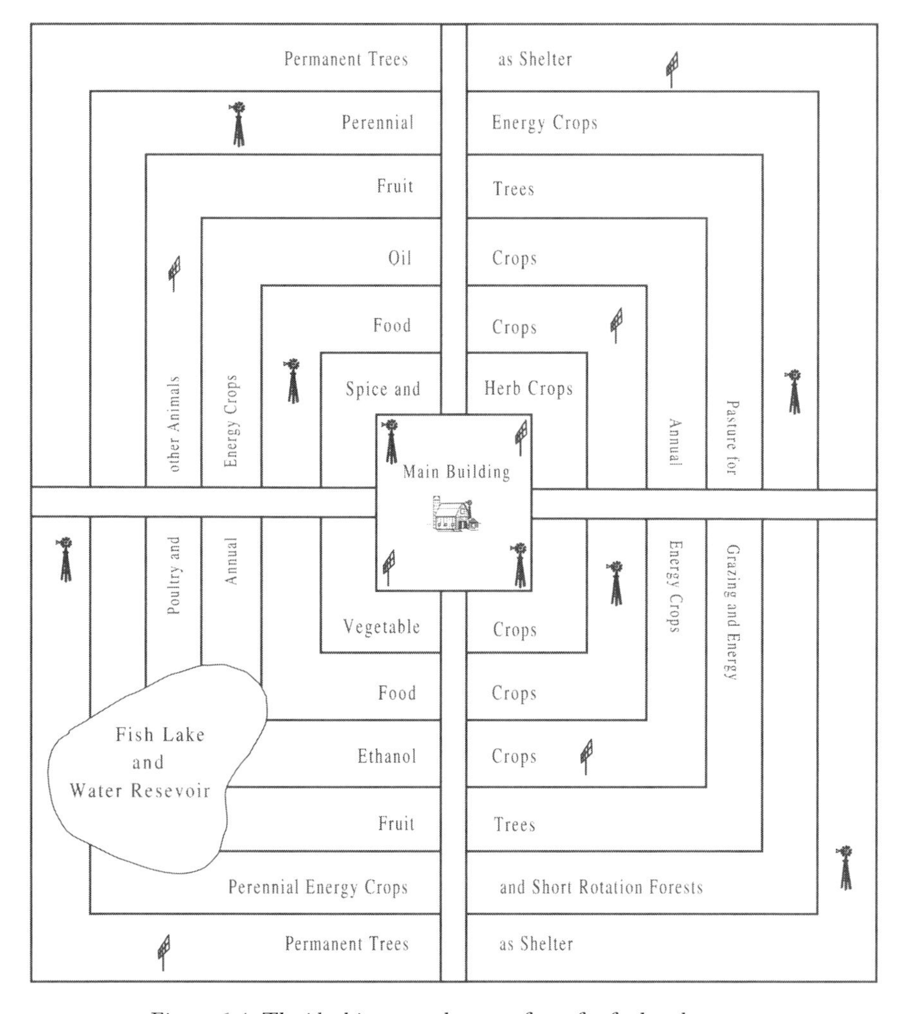

Figure 6.4. The ideal integrated energy farm for food and energy.

- power generators (vegetable oil, wind and PV operated)
- vegetable oil extraction and purifying equipment
- fermentation and distillation facilities
- water pumps (solar, wind or vegetable oil operated)
- vehicles (solar or biofuel operated)
- monitoring systems.

Current energy patterns contribute to unsustainability, and are directly related to the most pressing issues that affect sustainable development: poverty, jobs and income levels, access to social services, gender disparity, population growth, agricultural production and food scarcity, health, land degradation, climate change and environmental quality, and economic and security issues. Without adequate attention to the critical importance of energy to all these issues, global social and economic goals cannot be achieved. Indeed, the magnitude of change needed is large, fundamental, and directly related to the energy produced and consumed internationally.

Energy can play a major role in stemming and reversing the problem of land degradation. Specifically, the introduction of modern biomass energy systems would put a high enough market price on biomass to make it profitable to restore many of the potentially productive degraded lands to 'energy farm quality' so as to serve lucrative biomass energy markets (UNDP, 1997).

Growing crops for energy will bring industries into rural areas. New energy industries will create jobs and return money into rural regions. Energy systems based on various energy plant species will change the farming and energy businesses for ever (DOE, 1994). Farmers will harness the sun to feed and fuel the world in a way that will be integrated and sustainable to ensure the supply for generations to come.

Part 1 – References

ADEME (1996) *Rape Methyl Ester – Energy, Ecological and Economic Assessments.* TELLES Report, EUREC Network on Biomass (Bio-electricity).

AFB-NETT Final Report (1995) *Environment, Legislation and Information Availability*, pp. 50–59.

AGTD (Allison Gas Turbine Division) of General Motors Corporation (1994) Research and development of proton-exchange membrane (PEM) fuel cell system for transportation applications. Initial Conceptual Design Report prepared for the Chemical Energy Division of Argonne National Laboratory, US Department of Energy.

Andreuccetti, P., Bacchiet, P., Barbucci, P., Belletti, A. and Frati, G. (1991) Energy crops: Farming tests in two different Italian sites and preliminary assessment of technologies to use them for electricity production. In: *Proceedings of 1st European Forum on Electricity Production from Biomass and Solid Wastes by Advanced Technologies*, Italy, pp. 209–13.

Bassham, J.A. (1980) Energy crops (energy farming). In: *Biochemical and Photosynthetic Aspects of Energy Production*, 6, Academic Press, New York, pp. 147–73.

Blaschek, H.P. (1995) Recent developments in the ABE fermentation. *Workshop on Energy from Biomass and Wastes*, Dublin Castle, Ireland, p. 28.

BMWI (1994) *Energie Daten '94. Nationale und internationale Entwicklung*, Bundesministerium für Wirtschaft. Union-Druckerei, Frankfurt.

Breittschuh, G. and Eckert, H. (1995) Landwirtschaft und Umwelt aus gegenwärtiger Sicht. *Mitteilungen der Deutschen Bodenkundlichen Gesellschaft*, 78: 231–46.

Brundtland, G.H. (1987) *Our Common Future*, World Commission on Environment and Development, and Oxford University Press.

BS (1996) *Braunschweiger Zeitung*, 10 February 1996.

Bullard, M.J., Heath, M.C., Nixon, P.M.I., Speller, C.S. and Kilptrick, J.B. (1994) The comparative physiology of *Miscanthus sinensis*, *Triticum aestivum* and *Zea mays* grown under UK conditions. In: *Proceedings of the Workshop on Alternative Oilseed and Fibre Crops for Cool and Wet Regions of Europe*, Wageningen, The Netherlands, pp. 176–80.

Bundesministerium für Ernährung, Landwirtschaft und Forsten (1995) *Bericht des Bundes und der Länder über Nachwachsende Rohstoffe*, Landwirtschaftsverlag GmbH, 48165 Münster-Hiltrup, 340 pp.

Burvall, J. (1996) Personal communication.

Chelf, P., Brown, L.M. and Wyman, C.E. (1993) Aquatic biomass resources and carbon dioxide trapping. *Biomass and Bioenergy* 4(3): 175–84.

Clayton, W.D. and Renovize, S.A. (1986) Genera Graminum, Grasses of the World. *Kew Bulletin*, Additional Series XIII.

Crop Development for Cool and Wet Regions of Europe (1994) Alternative oilseed and fibre crops for cool and wet regions of Europe. Workshop, Wageningen, The Netherlands.
Dambroth, M. (1992) Nachwachsende Rohstoffe. *Wissenschaft und Fortschritt* 42: 249–53.
Dambroth, M. and El Bassam, N. (1990) Genotypic variation in plant productivity and consequences for breeding of 'low-input cultivars'. *Genetic Aspects of Plant Mineral Nutrition, Developments in Plant and Soil Sciences*, Kluwer, Dordrecht, pp. 1–7.
Danfors, B. (1994) Harvesting *Salix*: Technical data on harvesting machines, harvesting capacity and further development. In: *Harvesting Techniques for Energy Forestry*, Swedish University of Agricultural Sciences, pp. 46–9.
Dierke (1981) *Weltwirtschaftsatlas* 1: *Rohstoffe, Agrarprodukte*, Deutscher Taschenbuch Verlag und Westermann.
DOE (US Department of Energy) (1994) *The American Farm*. Document prepared by the National Renewable Energy Laboratory, NREL/SP-420-5877. 24 pp.
DOE (1997) *Biomass Power Program*. Document prepared by the National Renewable Energy Laboratory, DOE/GO-10097-412.
Dumbleton, F. (1997) Biomass conversion technologies: An overview. *Aspects of Applied Biology* 49, Biomass and energy crops, 341–7.
El Bassam, N. (1990) Requirements for biomass production with higher plants under artificial ecosystems. *Proceedings* of 'Artificial Ecological Systems' workshop, Marseille, pp. 59–65.
El Bassam, N. (1993) Möglichkeiten und Grenzen der Bereitstellung von Energie aus Biomasse. *Landbauforschung Völkenrode* 43(2/3): 101–11.
El Bassam, N. (1995) Auswirkungen einer Klimaveränderung auf den Anbau von C4-Pflanzenarten zur Erzeugung von Energieträgern. *Mitteilungen der Gesellschaft für Pflanzenbauwissenschaften* 8: 169–72.
El Bassam, N. (1996) *Renewable Energy: Potential Energy Crops for Europe and the Mediterranean Region*, REU Technical Series, 46, FAO, Rome, 200pp.
El Bassam, N., Dambroth, M. and Jacks, I. (1992) Die Nutzung von *Miscanthus sinensis* (Chinaschilf) als Energie- und Industriegrundstoff. *Landbauforschung Völkenrode* 42(3): 199–205.
Elliott, P. and Booth, R. (1990) *Sustainable Biomass Energy*, Shell Selected Papers, E. Hallett & Co, England, 11 pp.
Elliott, P. and Booth, R. (1993) *Brazilian Biomass Power Demonstration Project*, Special Brief, Shell Centre, London, 10 pp.
Elsbett, K. (1991) *Die Elsbett-Motortechnologie*. Symposium Biokraftstoffe für Dieselmotoren, Esslingen.
ERM (1995) Effectiveness of Tax Incentives: Bioethanol (as ETBE) for Gasoline. Environmental Resources Management, Research project summary.
European Commission, Agro-Industrial Research Division (1994) *Biofuels, Application of Biologically Derived Products as Fuels Additives in Combustion Engines*, Directorate-General XII, Science, Research and Development, 185pp.
European Farmer Association (1992) *Proceedings of the Arable Crops for Fuel Workshop*, Solihull, European Farmer Association, 188 pp.
FAO (1980) *Report on the Agro-ecological Zones Project*, vol. 1, pp. 24–48, FAO, Rome.
FAO (1995a) *Production Year Book 1994*, FAO, Rome, pp. 15–87.
FAO (1995b) *Future Energy Requirements for Agrica's Agriculture*, FAO, Rome, 96pp.
Franke, W. (1985) *Nutzpflanzenkunde: Nutzbare Gewächse der gemäßigten Breiten, Subtropen und Tropen*, Georg Thieme Verlag, Stuttgart and New York, 470 pp.
Garbe, C. (1995) Strohheizwerk Schkölen – nur ein Strohfeuer in Deutschland? *Agronomical* 1/95: 10–13.
Gesellschaft für Pflanzenbauwissenschaften (1995) Kurzfassungen der Vorträge und Poster, 39, Jahrestagung in Zürich, *Mitteilungen der Gesellschaft für Pflanzenbau-wissenschaften*, vol. 8, Wissenschaftlicher Fachverlag, Giessen, 1995, 442 pp.

Goering, C.E. and Daugherty, M.J. (1982) Energy accounting for eleven vegetable oil fuels. *Transactions of the ASAE*: 1209–1215.
Graef, M., Vellguth, G., Krahl, J. and Munack, A. (1994) Fuel from sugar beet and rape seed oil – mass and energy balances for evaluation. In: *Proceedings of the 8th European Conference on Biomass for Energy, Environment, Agriculture and Industry*, Vienna, pp.1–4.
Grassi, G. (1997) Operational employment in the energy sector. Presentation at the Tagung des SPD – Umweltforums, 5 July.
Grassi, G. and Bridgewater, T. (1992) *Biomass for Energy and Environment, Agriculture and Industry in Europe: A Strategy for the Future*, Commission of the European Communities, Directorate General for Science Research and Development, Luxembourg, 58pp.
Grimm, H.P. (1996) Personal communication.
Guibet, J. C. (1988) L'utilisation des huiles vegetales et de derivés comme carburants diesel. In: *Liquid Fuels from Biomass*, Report and Proceedings of CNRE Technical Consultation Saint-Rémy-lés-Chevreuse (France), 2–4 December 1988, *Food and Agriculture Organization of the United Nations CNRE Bulletin*, No. 20.
Hall, D.O. (1994) Biomass energy in industrialized countries – a view from Europe. In: *Agroforestry and Land Use Change in Industrialized Nations*, 7th International Symposium of CIEC, Berlin, pp. 287–329.
Hall, D.O. and House, J. (1995) Biomass: an environmentally acceptable fuel for the future. *Proceedings of the Institution of Mechanical Engineers*, 209: 203–13.
Hall, D.O., Rosillo-Calle, F., Williams, R.W. and Woods, J. (1993) Biomass for energy: supply prospects. In: T.B. Johannsson et al. (eds), *Renewable Energy: Sources of Fuels and Electricity*, Island Press, Washington DC, pp. 593–651.
Hartmann, H. (1994) Systems for harvesting and compaction of solid herbacious biofuels. In: *Environmental Aspects of Production and Conversion of Biomass for Energy*, REUR Technical Series 38, FAO, Rome, pp. 86–93.
Hartmann, H. and Strehler, A. (1995) *Die Stellung der Biomasse, im Vergleich zu anderen erneuerbaren Energieträgern aus ökologischer, ökonomischer und technischer Sicht*, Landwirtschaftsverlag GmbH, 48165 Münster, 396pp.
Hennink, S., van Soest, L.J.M., Pithan, K. and Hof, L. (1994) *Alternative Oilseed and Fibre Crops for Cool and Wet Regions of Europe*, European Commission, 230pp.
Hohenstein, W.G. and Wright, L.L. (1994) Biomass energy production in the United States: An overview. *Biomass and Bioenergy* 6 (3). 161–73.
Hohmeyer, O. and Ottinger, R.L. (eds) (1994) *Social Costs of Energy: Present Status and Future Trends*, Springer-Verlag, Berlin, pp. 61–8.
Hovi, M. (1994) Perspektiven von Energiegräsern in der Republik Estland. In: *Environmental Aspects of Production and Conversion of Biomass for Energy*, REUR Technical Series 38, FAO, Rome, pp. 59–61.
Hubbard, H.M. (1991) The real cost of energy. *Scientific American* 264: 18–23.
Ingram, L.O. (1995) Personal communication.
Ingram, L.O., Bothast, R.J., Doran, J.B., Beall, D.S., Brooks, T.A., Wood, B.E., Lai, X., Asghari, K. and Yomano, L.P. (1995) Genetic engineering of bacteria for the conversion of lignocellulose to ethanol. *Proceedings, Workshop on Energy from Biomass and Wastes*, Dublin Castle, Ireland, p. 17.
Kahl, A. (1987) *Pelleting Plants for All Grades of Powdered or Granular Raw Materials Either Dry or in Paste Form*, Tech. Report 1326, Amandus Kahl, Hamburg. 8pp.
Kahnt, G., Lewandowski, I. and Eusterschulte, B. (1995) Umweltgerechte Produktion und Bereitstellung von Ölen, Ligno-Cellulose und Fasern als nachwachsende Rohstoffe. *Landinfo* 3/95, Hochschultag, pp. 11–19.
Kalbhenn, M. (1995) Zur Wassernutzungseffizienz der Pseudocerealien Amarant, Reismelde und Buchweizen. *Mitteilungen der Gesellschaft für Pflanzenbauwissenschaften.* 8: 181–2.
Kassler, P. (1994) *Energy for Development.* Shell Selected Paper, Well Hall Press, England, 11pp.

Kechagia, U. (1994) Research on euphorbia and other plant species for energy production. In: *Environmental Aspects of Production and Conversion of Biomass for Energy*, REUR Technical Series 38, FAO, Rome, pp. 55–8.
Kilpatrick, J.B., Heath, M.C., Speller, C.S., Nixon, P.M.I., Bullard, M.J., Spink, J.G. and Cromack, H.T.H. (1994) Establishment, growth and dry weather yield of *Miscanthus sacchariflorus* over two years under UK conditions. In: *Proceedings of the Workshop on Oilseed and Fibre Crops for Cool and Wet Regions of Europe*, Wageningen, The Netherlands, pp. 181–5.
Krahl, J. (1993) Bestimmung der Schadstoffemissionen von landwirtschaftlichen Schleppern beim Betrieb mit Rapsölmethylester in Vergleich zu Dieselkraftstoff. Dissertation TU Braunschweig.
Kurse, M. (1995) Zur N-Aufnahme und N-Verwertung der Pseudocerealien Amarant, Reismelde und Buchweizen. *Mitteilungen der Gesellschaft für Pflanzenbauwissenschaften* 8: 179–80.
Landinfo (1995/3) *Neue Chancen für die Pflanzenproduktion – Nachwachsende Rohstoffe*, Baden-Württemberg, 66 pp.
Larson, E.D. (1993) Technology for electricity and fuels from biomass. *Annual Review of Energy and the Environment* 18: 567–630.
Larson, E.D. and Williams, R.H. (1995) Biomass plantation energy systems and sustainable development. In: Goldemberg, J. and Johansson, T.B. (eds), *Energy as an Instrument for Socio-Economic Development*, United Nations Development Program, New York, pp. 91–106.
Lehmann, H. and Reetz, T. (1995) *Zukunftsenergien, Strategien einer neuen Energiepolitik*, Birkhäuser Verlag, Berlin, Basel and Boston, 282 pp.
Leroudier, J.P. (1995) The outlook for bioethanol/ETBE in the EU. In: *Proceedings, Workshop on Energy from Biomass and Wastes*, Dublin, p. 6.
Lewandowski, I. and Kicherer, A. (1995) CO_2-balance for the cultivation and combustion of Miscanthus, *Biomass and Bioenergy* 8(2): 81–90.
Liquid Biofuels Newsletter (1995) Information from the IEA-BA 'Liquid Biofuels Activity', Wieselburg, 2: 1–20.
Löhner, K. (1963) *Die Brennkraftmaschine*. VDI Verlag, Düsseldorf.
Miller, G.T. (1992) Living in the Environment, 7th ed. Wadsworth, Belmont, CA.
Münzer, W. (1993) Züchtungsbedarf bei nachwachsenden Rohstoffen. *Inform* 13–16.
National Academy of Sciences (1975) *Underexploited Tropical Plants with Promising Economic Value*, Washington, DC, 188 pp.
Nitske, W.R. and Wilson, C.M. (1965) *Rudolph Diesel, Pioneer of the Age of Power*. University of Oklahoma Press, pp. 122–3.
OECD (1991) Energy policies of IEA countries, 1990 review, OECD, Paris.
Ortiz-Cañavate (1994) Technical applications of existing biofuels. In: *Application of Biologically Derived Products as Fuels or Additives in Combustion Engines*, Directorate-General XII, Science, Research and Development, EUR 15647 EN, pp. 1–20, 52–67.
Osman, S.M. and Ahmad, F. (1981) Forest oilseeds. In: Pryde, E.H., Princen L.H. and Mukherjee K.D. (eds), *New Sources of Fats and Oils*, American Oil Chemist's Society, Indianapolis. IN.
Overend, R.P. (1995) Production of electricity from biomass crops – US perspective. In: *Proceedings, Workshop on Energy from Biomass and Wastes*, Dublin Castle, Ireland, p. 5.
Palz, W. (1995) Future options for biomass in Europe. In: *Proceedings*, Workshop on Energy from Biomass and Wastes, Dublin Castle, Ireland, p. 2.
Palz, W. and Chartier, P. (1980) *Energy from Biomass in Europe*, Applied Science Publishers, London, 234 pp.
Pernkopf, J. (1984) The commercial and practical aspects of utilizing vegetable oils as diesel fuel substitute. Bio-Energy 84 World Conference, 18–21 June, Gothenburg, Sweden.

Peterson, C.L. (1985) Vegetable oil as a diesel fuel – status and research priorities. American Society of Agricultural Engineers. Summer Meeting, 1985, Michigan State University, East Lansing, 23–26 June, Paper No. 85-3069.

Petrini, C. Bazzocchi, R. and Belletti, A. (1995) Kenaf for paper production: varieties, yield potential and adaptability. In: *Proceedings of European Seminar on Kenaf (Hibiscus cannabinus L.)*, Bologna, pp. 13–21.

Petrini, C., Bazzocchi, R. and Montalti, P. (1994) Yield potential and adaptation of Kenaf (*Hibiscus cannabinus* L.) in north-central Italy. *Industrial Crops and Products* 3: 11–15.

Petrini, C., Belletti, A., Scozzoli, A. and Bazzicchi, R. (1992) Sweet sorghum breeding for energy production. In: *Proceedings, Biomass for Energy and Industry*, 7th EC Conference, Italy, pp. 447–51.

Petrini, C., Scozzoli, A., Bazzocchi, R. and Montalti, P. (1993) Sweet sorghum breeding for energy production. In: *Proceedings, Breeding and Molecular Biology: Accomplishments and Future Promises. Proceedings of the XVIth Conference of the Eucarpia, Maiz and Sorghum Section*. Bergamo, Italy, pp. 332–7.

Rehm, S. (1989) *Spezieller Pflanzenbau in den Tropen und Subtropen*, Ulmer, Stuttgart, p. 653.

Rehm, S. and Espig, G. (1991) *The Cultivated Plants of the Tropics and Subtropics*, Verlag Josef Margraf, Weikersheim, 552pp.

Rice, B. and Bulfin, M. (1995) Biomass-energy systems in Ireland: Opportunities and experience to date. In: *Proceedings, Workshop on Energy from Biomass and Wastes*, Dublin Castle, Ireland, pp. 3–4.

Riddell-Black, D. and Hall, J.E. (1994) Organic wastes as fertilisers for biomass energy crops. In: *Environmental Aspects of Production and Conversion of Biomass for Energy*, REUR Technical Series 38, FAO, Rome, pp. 62–74.

Ruiz-Altisent, M. (ed.) (1994) *Biofuels – Application of Biologically Derived Products As Fuels Or Additives in Combustion Engines*, Directorate-General XII, Science, Research and Development, EUR 15647 EN, 185 pp.

Samson, R. and Chen, Y. (1995) Short-rotation forestry and water problem. In: *Proceedings, Canadian Energy Plantation Workshop*, pp. 43–9.

Samson, R., Girouard, P. Omielan, J., and Quinn, J. (1994) *Technology Evaluation and Development of Short Rotation Forestry for Energy Production*, Annual Report 1993–94, Resource Efficient Agricultural Production (REAP) Canada. Report written under the Federal Panel on Energy R&D (PERD) for Natural Resources Canada and Agriculture Canada. DSS contract 23440-2-9493/01-SQ, 107pp.

Scheer, H. (1993) *Sonnenstrategie – Politik ohne Alternative*. Piper Verlag, München.

Scheffer, K. (1997) Usability of wet biomass for thermal power generation, oil and fibre production. In preparation.

Schön, H. and Strehler, A. (1992) Wirtschaftlich sinnvolle Erzeugung von Biomasse – Kostentrends, Umweltentlastung. In: *Nachwachsende Rohstoffe*, Proceedings of the International CLAAS Symposium, Harsewinkel, pp. 1–16.

Schröder, R. (1963) *Öl- und Faserpflanzen, Wirtschaftspflanzen der warmen Zonen*, Part II, Stuttgart, Kosmos, Gesellschaft der Naturfreunde, Franckh'sche Verlagshandlung, p. 63.

Schrottmaier, J., Pernkopf, J. and Wörgetter, M. (1988) Plant oil as fuel. A preliminary evaluation. *Proceedings, ÖKL Colloqium*, pp. 44–8.

Scozzoli, A., Bazzocchi, R., Montalti, P. and Petrini, C. (1995) Field performance and adaptability of sorghum hybrids for energy production in the North Centre of Italy. ESNA XXV Annual Meeting, Italy, Book of Abstracts, pp. 1–6.

Shakir, S. (1996) Flash pyrolysis: a technology to produce biofuels from Miscanthus. *RENTEC – Technische Mitteilung*, pp. 2–14.

Sneyd, J. (1995) *Alternative Nutzpflanzen*, Ulmer, Stuttgart, 143pp.

Speller, C.S. (1993) The potential for growing biomass crops for fuel on surplus land in the UK. *Outlook on Agriculture* 22(1): 23–9.

Speller, C.S., Bullard, M.J., Heath, M.C. and Kilpatrick, J.B. (1994) Establishment of *Miscanthus* for biomass production under UK conditions. In: *Proceedings of the Workshop on Alternative Oilseed and Fibre Crops for Cool and Wet Regions of Europe*, Wageningen, The Netherlands, p. 186.
Strehler, A. (1994) *Environmental Aspects of Production and Conversion of Biomass for Energy*. REUR Technical Series 38, FAO, Rome, p. 243.
Sunpower Inc. (1997) *Status of Sunpower Technology*, Information brochure, April.
UNDP (United Nations Development Programme) (1997) *Energy after Rio: Prospects and Challenges*, Executive Summary, 35 pp.
van Basshuysen, R. (1989) Audi Turbodieselmotor mit Direkteinspritzung. *Motortechnische Zeitschrift* 50(10): 458–65.
Vellguth, G. (1991) Energetische Nutzung von Rapsöl und Rapsölmethylester, *Dokumentation Nachwachsende Rohstoff*, FAL, pp. 17–21.
Vetter, A. and Wurl, G. (1994) Möglichkeiten der Nutzung von Topinambur und weiteren Pflanzenarten als Festbrennstoff. In: *Environmental Aspects of Production and Conversion of Biomass for Energy*, REUR Technical Series 38, FAO, Rome, pp. 43–53.
World Energy Council (1993) *Energy for Tomorrow's World*, St. Martin's Press, New York.
Weber, R. (1987) Stirling-motor treibt Wärmepumpe an. *VDI-Nachrichtung* Nr. 19.
Weiland, P. (1997) Personal communication.
Williams, B.C. and Campbell, P.E. (1994) Application of fuel cells in 'clean' energy systems. *Transactions of the Institution of Chemical Engineers B. Process Safety and Environmental Protection Transactions*, 72(B): 252–6.
Williams, R.H. (1993) Fuel cells, their fuels and the US automobile. Prepared for the First Annual World Car 2001 Conference, University of California, Riverside, CA, 20–24 June 1993.
Williams, R.H. (1994) The clean machine: Fuel cells and their fuels for cars. *Technology Review*, April, pp20–30.
Williams, R.H. (1995) The prospects for renewable energy. *Siemens Review*: 1–16.
Williams, R.H. and Larson, E.D. (1993) Advanced gasification-based biomass power generation. In: T.B. Johansson, H. Kelly, A.K.N. Reddy and R.H. Williams (eds), *Renewable Energy: Sources for Fuels and Electricity*, Island Press, Washington D.C. pp. 729–785.
Williams, R.H., Larson, E.D., Katofsky, R.E. and Chen, J. (1995) *Methanol and Hydrogen from Biomass for Transportation, with Comparisons to Methanol and Hydrogen from Natural Gas and Coal*, The Centre for Energy and Environmental Studies, Princeton University, 47 pp.
Wintzer, D., Fürniß, B., Klein-Vielhauer, S., Leibe, L., Nieke, E., Rösch, Ch. and Tangen, H. (1993) *Technikfolgenabschätzung zum Thema Nachwachsende Rohstoffe*, Landwirtschafts-Verlag GmbH, 48165 Münster, pp. 14–26.
World Bank (1992) *World Development Report 1992*, Washington DC, World Bank.
Zauner, E. and Küntzel, U. (1986) Methane production from ensiled plant material. *Biomass* 10: 207–23.
Zhang, X., Peterson, C.L., Reece, D., Möller, G. and Haws, R. (1995) Biodegradability of biodiesel in the aquatic environment. Paper presented to ASAE meeting, p. 14.

Part II

The greatest service which can be rendered any country is to add a useful plant to its culture

Thomas Jefferson, 1821

7 Energy plant species

Definition and scope

Vegetative biomass consists of living plant species all around us. As they grow, plants store the sun's energy in their leaves, stems, bark, fruits, seeds and roots. Bioenergy plant species are so diverse that they grow in virtually every part of the world.

Energy plant species are understood to mean those annual and perennial species that can be cultivated to produce solid, liquid or gaseous energy feedstocks. The organic residues and wastes from the most widely diverse types of plant production, also used for producing energy, do not fall under this term but nevertheless represent a large potential.

All plant species that store primarily carbohydrates or oils are suitable for producing liquid energy sources. Cellulose, starch, sugar and inulin can be used to produce ethanol. Vegetable oils can be used as fuels. Parts of plants containing lignocellulose can provide energy directly as solid fuels or indirectly after conversion.

Obtaining alcohol from vegetable raw materials has a long tradition in agriculture. From an agricultural point of view, the species available for ethanol production are starch plant species, and cereals including maize and grain sorghum, as well as potato, topinambur, and the sugar crops (sugarbeet, root chicory, sweet sorghum and sugarcane). Most of the crop species presently used to produce ethanol are given in Table 7.1. The goal directed use of cellulose-containing biomass from agricultural crops for producing alcohol has not been practised on a large scale engineering level, but the potential for the future is very large. According to a communication from Ingram (1995), the entire gasoline needs of the USA could be substituted in this way if systems developed at the laboratory scale could be implemented at large technical production scales.

Oil crops are well distributed throughout the world from north to south, but only a few of them have a high oil yield per unit area (tonnes per hectare) and this is a disadvantage in comparison to other fuel feedstocks like ethanol or solid biofuel. The main oil yielding crops are rape, sunflower and soybean. Table 7.2 lists several oil crops and their potentials. Most of them need further improvements in breeding and agricultural practices to improve the yield.

Table 7.1. Productivity of ethanol crops.

Plant species		Sugar/starch		Ethanol
	yield (t/ha)	contents (% FS)	yield (t/ha)	efficiency (l/ha)
Barley (*Hordeum vulgare*)	5.8	58.0	3.36	2150
Cassava, manioc (*Manihot esculenta*)	9.0	35.0	3.15	2900
Fodder beet (*Beta vulgaris* var. *rapacea*)	98.5	8.2	8.08	4923
Maize (*Zea mays*)	6.9	65.0	4.49	2874
Potato (*Solanum tuberosum*)	32.4	17.8	5.77	3693
Root chicory (*Cichorium intybus*)	35.0	16.0	5.60	3248
Sugarbeet (*Beta vulgaris* var. *altissima*)	57.4	16.0	9.18	5600
Sugarcane (*Saccharum officinarum*)	80.0	10.0	8.00	5400
Sweet potato (*Ipomoea batatas*)	12.0	25.0	3.00	2400
Sweet sorghum (*Sorghum bicolor*)	90.0	10.0	9.00	5400
Topinambur (*Helianthus tuberosus*)	30.0	15.0	4.50	2610
Wheat (*Triticum aestivum*)	7.2	62.0	4.46	2854

The number of plant species that can be utilized as solid biofuels is much higher than those usable for ethanol and oil production. The level of production of these crops is, besides the genetic potential, largely influenced by the availability of water and other external inputs, but the competition for available water between food crops and energy crops increases from northern to southern regions.

Potential energy plant species

Depending on the time and methods of harvest and utilization, and the prevailing economics, energy plant species could include roots, tubers, stems, branches, leaves, fruits and seeds, or even whole plants.

The main goals of growing biomass on plantations dedicated to energy crops can be summarized as follows:

- The growing of starch and sugar plant species to produce ethanol.
- The cultivation of oil crops as sources for biodiesel.
- The production of solid biomass to obtain heat and electricity, either directly through combustion or indirectly through conversion for use as fuels. Ligno-cellulose-rich raw materials can be used to produce fuels like methanol, biodiesel, synthetic gas and hydrogen (using thermal and thermochemical processes, by direct or indirect liquefaction or gasification) and ethanol (through hydrolysis and subsequent fermentation to produce).
- The cultivation of biomass to produce biogas.

The plant species discussed in this publication were not in all cases described in a uniform structure. This is due to the varied and unbalanced level of available information pertaining to the various crops.

Special emphasis will be placed on aspects related to identification of high yielding energy plant species in which the production system should be sustainable and environmentally acceptable.

Table 7.2. Productivity of oil crops.

Common name	Botanical name	Seed yield (t/ha)	Oil content (%)	Oil yield (t/ha)
Abyssinian kale	*Crambe abyssinica*	2.0–3.5	30–45	0.74
Abyssinian mustard	*Brassica carinata*	1.1–3.0	23–40	0.26–1.2
Bird rape	*B. rapa* ssp. *oleifera*	1.0–2.5	38–48	0.38–1.2
Black mustard	*B. nigra*	0.5–2.0	24–38	0.12–0.76
Brown (Indian) mustard	*B. juncea*	1.5–3.3	30–40	0.72
Castor	*Ricinus communis*	1.2	50	0.60
Coconut palm	*Cocos nucifera*	4.17	36	1.5
Coriander	*Coriandrum sativum*	2.0–3.0	18–22	0.4–0.7
Cotton	*Gossypium* spp.	1.2	15–25	0.29
Dill	*Anethum graveolens*	1.0–1.5	16–20	0.16–0.30
Fennel	*Foeniculum vulgare*	1.0–2.0	10–12	0.08
Flax, linseed	*Linum utitatissimum* spp. *utitatissimum*	1.8	30–48	0.70
Gold of pleasure	*Camelina sativa*	2.25	33–42	0.88
Groundnut	*Arachis hypogaea*	2.0	45–53	1.00
Hemp	*Cannabis sativa*	0.5–2.0	28–35	0.14–0.7
Jojoba	*Simmondsia chinensis*	2.1	48–56	1.01–1.18
Meadowfoam	*Limnanthes alba*	1.2	20–30	0.34
Nigerseed, ramtil	*Guizotia abyssinica*		35–45	
Oil palm	*Elaeis guineensis*	30	26	7.8
Oil radish	*Raphanus sativus* var. *oleiformis*	0.7–1.1	38–50	0.27–0.55
Oil squash	*Cucurbita pepo*	0.8–1.6	40–58	0.32–0.93
Olive	*Olea* spp.	1–12.5	40	0.4–5.0
Opium poppy	*Papaver somniferum*	1.0–1.8	40–55	0.4–1.0
Penny flower	*Lunaria* spp.		30–40	
Pot marigold	*Calendula officinalis*	1.5	18–20	0.29
Rape	*Brassica napus* spp. *oleifera*	2–3.5	40–50	1.26
Rocket	*Eruca sativa*	0.9	24–35	0.27
Safflower	*Carthamus tinctorius*	1.8	18–50	0.63
Sesame	*Sesamum indicum*	0.5	50–60	0.25
Soybean	*Glycine max*	2.1	18–24	0.38
Spurge	*Euphorbia latyris*	1.5	48	0.72
Spurge	*Euphorbia lagascae*	0.6	46	0.28
Sunflower	*Helianthus annuus*	2.5–3.2	35–52	0.88–1.67
White mustard	*Sinapis alba* ssp. *alba*	1.5–2.5	22–42	0.64

ALEMAN GRASS (CARIB GRASS) (*Echinochloa polystachya* (H.B.K.) Hitchc.)

Description

Echinochloa is a C_4 plant that occurs from Mexico to Argentina. It forms large monotypic stands on the fertile flood plains along the white waters of the Amazon region. Each plant has a single unbranched stem, which varies little in diameter or bulk density along its length (Figure 7.1). The stem consists of internodes, which at maturity are of roughly equal length. It is a perennial grass that forms large monotypic

Figure 7.1. Aleman grass, Germany.

stands on the fertile flood plains of white water rivers and lakes (white water because of the light coloured suspended sediment). In the wet tropics the C_3 plants normally dominate, but not in some communities on river and lake margins. There monotypic stands of C_4 species are frequent, like *Cyperus papyrus* in Africa, and *Paspalum regens*, *Paspalum fasciculatum* and *Echinochloa polystachya* in South America.

Ecological requirements

The lifecycle of plants that grow on the flood plains is regulated by the annual oscillation of the water level. The variation in water level shows a mean amplitude of 10m, but this may range from 6m in some years to 14m in others. *Echinochloa* is no floating plant. The roots are fixed in the soil and the leaves reach out of the water. The stem has to be rigid and long enough to withstand the flowing and rising water. The C_4 plants benefit from their higher productivity. They can invest more assimilates into the stem. From November onwards the new plants grow steadily in length at approximately 1.1m per month. The crown reaches 1–2m outside of the water if the stems stand vertical. The increase of stem length is achieved by successive addition of internodes at a rate of 7 per month.

When the water level drops, exposing the sediment surface in October, new shoots form at the nodes of the old stems, and root in the sediment. The old stems die and rot away, and each of the shoots becomes an individual plant. Each new plant normally remains as a single unbranched shoot. As the water level rises to cover the sediment in late November or December, the stems grow upwards, keeping pace with the steady rise in water level. As each node becomes submerged its leaves die and adventitious roots form. If the water is highly turbid, submerged leaves cannot get enough light for photosynthesis. Photosynthesis is only possible if the leaves form a canopy above the water surface. The longevity of the leaves is only 34 ±4 days. The leaves die and decay rapidly on submersion by the rising water.

New leaves are formed at the top of the stems with sufficient rapidity to maintain a dense canopy. As the water level drops the stems become increasingly exposed and bent, eventually collapsing onto the re-exposed sediment surface in October.

Propagation

The stands are perpetuated by vegetative propagation, but micropropagation is also feasible.

Production

The maximum standing of the crop at 80 Mg/ha was determined at a site in central Amazon. The mean rate of dry matter production per unit of solar radiation intercepted by the stand was 2.3g/MJ, which is close to the considered maximum for C_4 plants.

From November on, the biomass increases steadily to peak in September with 6880g/m^2, or 8000g/m^2 if attached dead material is included. In the 9 month period from the December harvest until the September harvest, the total dry mass increases by 7000±35g/m^2 with a daily growth rate of 25.9±0.1g/m^2. The peak dry mass of 8000g/m^2 does not represent the total net primary production. If the dead leaves are taken into account 9930±670g/m^2 are obtained, of which 9420±660g/m^2 are constituted by the shoots.

The accumulated net primary production rose steadily in each month from November, suggesting a complete absence of seasonal limitation of productivity.

The cultivation of this plant species for several years on dry land (in Germany as a perennial crop) showed that a biomass yield of up to 20t dry matter per ha is possible (El Bassam, 1998).

Processing and utilization

Aleman grass is capable of very high productivity levels, up to 100t dry matter per hectare, which exceeds the productivity of almost all other plant species. The biomass is currently being used as an energy source for cooking. There are a wide range of possibilities to upgrade the biomass to produce solid biofuels and to convert them to ethanol, coal or bio-oil through hydrolysis, fermentation or pyrolysis.

Selected references

El Bassam, N. (1998) The productivity of *Echinochloa* as a potential energy crop. (In preparation.)

Piedade, M.T.F., Junk, W.J. and Long, S.P. (1991) The productivity of the C_4 grass *Echinochloa polystachya* on the Amazon Floodplain. *Ecology* 72(4): 1456–63.

ALFALFA *(Medicago sativa L.)*

Contributed by: H.W. Elbersen

Description

Alfalfa, which originated near Iran, also has related forms and species found wild growing scattered over central Asia and into Siberia. Nowadays, alfalfa is grown worldwide (IFFS, 1995), and is one of the most important fodder crops. Alfalfa is a perennial herbaceous legume, reaching a height of 30–90cm. The plant is upright or ascending and highly branched. The flowers are bluish-purple to yellow in colour. They are a source for honey production (FAO, 1996). Normally the leaves are trifoliate, but leaves with more than three leaflets are not uncommon. The content of proteins and the production of protein for a given area is higher than for non-legumes (Franke, 1985).

Ecological requirements

A wide range of soil and climatic conditions are suitable for alfalfa, but for good production some conditions have to be fulfilled. The optimum temperature for growing is about 25°C. However, alfalfa has survived summer temperatures of 50°C in California. In the winter, alfalfa can resist temperatures down to –25°C: although the top growth may be killed by the frost, the plant will regenerate from the roots. Alfalfa needs an annual precipitation of 600 to 1200mm for optimal growth; 350mm is the minimum. The plant can be found at elevations of 4000m in Bolivia, but normally it is grown up to 2400m. The pH value of the soil should be nearly neutral, in the range 6.5 to 7.5. Alfalfa does not tolerate an acid soil, especially in the seedling state. A well drained, deep and light soil is preferred. Because of the high production of alfalfa the soil fertility must be high. In warm climates, humidity should be low because hot and humid conditions promote diseases (FAO, 1996).

Propagation

Alfalfa is propagated by seed. The seed is sown not deeper than 6–12mm into a firmly packed seedbed (Huffaker, 1994). Alfalfa can be sown either in the early spring or in late summer and autumn. The time of alfalfa seeding is influenced by precipitation, temperature, and cropping patterns. Spring seeding allows the first harvest during the seeding year, but weed control is usually required. Competition from weeds and unfavourable summer temperatures and moisture conditions can be avoided if alfalfa is sown in the late summer or autumn. The remaining time must allow adequate seedling development prior to the onset of winter to minimize loss of stand from winter injury. Spring seeding should be made early enough to allow the formation of good root systems before high temperature and low moisture conditions slow the growth rates.

The seed must be in contact with moist soil. The seedlings are unable to emerge from the soil if they are planted too deep. The autotetraploid alfalfa originated from a diploid species, native to an area south of the Black and Caspian Seas (FAO, 1996).

Many cultivars are available, offering specific characteristics for climatic and soil conditions and disease problems. The seedling forms a taproot that is later used for the storage of nutrients (IFFS, 1995). This taproot enables the plant to survive drought conditions, because it allows the plant to get water from deep soil layers.

Selection for disease resistance is important in Alfalfa breeding. For alfalfa intended for use as an energy crop, specially adapted cultivars have to be developed. The proportion of the stems relative to the whole crop should be higher than for the fodder type.

Crop management

A seeding rate of 13–17kg/ha planted in 15cm rows is usually sufficient for a good stand. Alfalfa is a perennial crop, lasting up to 6 years, but is usually used for 3–4 years. With increasing stand age the yield and stand density usually decrease due to competition between the plants, weed incursion, diseases, and poor harvest management (Sheaffer *et al.*, 1988). The growing period depends on the climate and ranges from 100 to 365 days per year. The seed needs to be inoculated with an effective strain of *Rhizobium melioti* that requires a soil pH of at least 6.0. Lime-coated, inoculated seeds have been used with success on more acid soils (FAO, 1996).

Alfalfa lives, like other legumes, in symbiosis with soil bacteria (*Rhizobium melioti*) that are able to fix nitrogen from the air. The bacteria may be present in the soil, but to be sure that the plants are provided with an effective strain, the seed should be inoculated with a fresh (commercial) inoculum immediately prior to seeding. The freshness of the inoculum is more important than the rate. A complete fertilizer programme is essential to a long-lived stand. If the pH value of the soil is below 6.2, lime application is necessary. Nitrogen fertilizer is not required because of the fixation of atmospheric nitrogen by the bacteria in the nodules. Addition of nitrogen fertilizer will reduce the effectiveness of the natural nitrogen fixing mechanism (IFFS, 1995). Potassium (K), phosphorus (P) and boron (B) are most often limiting for alfalfa production (Barnes and Sheaffer, 1995). Potassium is especially important for winter survival and particularly for nitrogen fixation (Duke *et al.*, 1980). In symbiosis with *Rhizobium melioti*, the N_2 fixation ranges from 50 to 463kg/ha with an average of 200kg/ha (Vance *et al.*, 1988).

More than 20 diseases are serious problems for alfalfa (the examples below are from the USA). These include fungal and bacterial wilts, leaf spots, crown and root rots, viruses, and nematodes. Important wilts are bacterial wilt (*Corynebacterium insidiosum* (McCull.) H. L. Jens), fusarium wilt (*Fusarium oxysporum* Schlecht. f. sp. *medicaginis* (Weimer) Snyd. & Hans.), and verticillium wilt (*Verticillium alboatrum* Reinke & Berth). The most serious leaf spots are common leaf spot (*Pseudopeziza medicaginis* (Lib.) Sacc.), lepto leaf spot (*Leptosphaerulina briosianna* (Pollacci) J. H. Graham & Luttrell), stemphylium leaf spot (*Stemphylium botryosum* Wallr.), and summer blackstem (*Cercospora medicaginis* Ellis & Everh). Important crown and root rots include anthracnose (*Colletotrichum trifolii* Bain & Essary), *Aphanomyces* spp. root rot, spring blackstem (*Phoma medicaginis* Malbr. & Roum. var. *Medicaginis*), phytophthora root rot (*Phytophthora megasperma* Drechs.), rhizoctonia diseases (*Rhizoctonia solani* Kuehn.), and sclerotina crown and stem rot (*Sclerotina trifoliorium* sensu Kohn). Alfalfa mosaic (Alfalfa Mosaic Virus complex), is the primary virus disease. Alfalfa stem nematode

(*Ditylenchus dipsaci* (Kuhn) Filpjev), root-knot nematodes (*Meloidogyne* spp.), and root-lesion nematodes (*Pratylenchus* spp.) are the most prevalent nematode species on alfalfa. Resistant cultivars are available for most of the diseases and nematodes.

There are a number of insects pests on alfalfa (again the examples are from the USA). The insect pests that interfere with forage production include the potato leafhopper (*Empoasca fabae* (Harris)), the alfalfa weevil (*Hypera postica* (Gyll.)), the spotted alfalfa aphid, the pea aphid (*A. kondoi Shinji*), the alfalfa plant bug (*Adelphocoris lineolatus* (Goeze)), and the meadow spittlebug (*Philaenus spumarius* L.). The potato leafhopper is the most problematic pest and causes damage throughout most alfalfa producing areas in the eastern and central USA. It causes yellowing of the foliage and stunting of stems. The damage results in significant losses in yield and forage quality, especially loss in carotene (IFFS, 1995).

Biological control of insect pests is possible in some cases. The aphids can be suppressed with significant success by the ladybird. There are experiments to control the alfalfa weevil with parasitic wasps (Bambara, 1994).

Alfalfa is an integral component of many crop rotations because of its ability to fix nitrogen, improve the soil structure and tilth, and to control weeds in subsequent crops (IFFS, 1995). The crop can be used as green manure to provide the soil with nitrogen and organic material. It can provide winter wheat as the following crop with 80–100% of its needed nitrogen. If alfalfa is re-established immediately after a previous alfalfa crop, or if old alfalfa stands are thickened, autotoxicity effects are sometimes observed, in which case the alfalfa shows lower germination, lower production and poorer establishment. These effects are attributed to plant exudates and decomposition products. To avoid the effects, an interval of at least 2 or 3 weeks between ploughing or herbicide killing, respectively, and seeding is necessary (IFFS, 1995).

The cropping plan for the supply of a planned biomass power-plant in Minnesota is a 7 year cycle with 4–2–1 years of alfalfa, corn and soybeans. Immediate environmental benefits as a result of this include increased yield from other rotation crops, reduced external inputs of nitrogen, reduced potential for nitrate leaching, reduced soil erosion, and improved soil tilth (Downing *et al.*, 1996)

Production

Alfalfa stands can be cut several times per year. The dry matter yields may be 10–20t/ha (FAO, 1996). Production decreases with increasing age of the stand.

Processing and utilization

Alfalfa is not only usable fresh; it can be dried to alfalfa hay or processed into silage, pellets or meal. The hay is pressed into bales and can be stored. The time of cutting determines the features of the hay. If cut in the early bud stage, the plants have a high proportion of leaves that are rich in protein. If cut when the plants are mature, the stems constitute a larger proportion of the plant, the protein content decreases, and the fibre content increases; the total yield is also higher. The timing of mowing the stand has a profound influence on the productivity and the life of the alfalfa stand (Figure 7.2). The large taproot is a storage organ for food reserves that are needed to

Figure 7.2. Alfalfa plantation, USA: leaves for livestock feed and stems for power (Moon, 1997; photo: NSP).

regenerate the plant in the spring or after cutting. The maintenance of food reserves is necessary to keep the stand vigorous and productive. The quantity of reserves increases if the interval between the cuttings is prolonged to 35 days, after which it declines. Cutting more frequently than 28 days or continuous grazing will weaken the stand (IFFS, 1995).

Alfalfa is usually grown as a forage crop. It is rich in proteins, minerals and vitamins. The plant can be used fresh or dried to alfalfa hay. Research is under way to use alfalfa not only as a fodder plant, but also for energy purposes – for example, by using the protein rich leaves as fodder and the coarse, fibrous stems as an energy source. In Minnesota, USA, the Northern States Power Company (NSP) plans to build a 75MW power plant that runs on biofuel made from alfalfa stems. This procedure takes advantage of the different characteristics of the plant fractions, because the stems are relatively low in nutrient value whereas products made of the protein rich leaf material have numerous market opportunities. If the stems are used for energy purposes their value is higher than if they are used as fodder. After gasification, the gas will drive a 50MW combustion turbine and the heat is used to produce superheated high pressure steam that will drive a 29MW steam turbine. The net output of the plant is 75MW, and the plant efficiency is 40%. The direct combustion of the stems is also possible and would make the construction of the power plant cheaper, but the efficiency would then be lower (Campbell, 1996).

The plant is dried before the stems and leaves are separated. The separation of leaf and fibrous material not only raises the value of the fodder but also removes the fuel bond nitrogen. This lowers NO_x production in the turbine. Calculated energy balances have provided total system efficiency for the production of leaf meal and electricity from alfalfa of 1:3 – that is, for each unit of energy input, the system produces three units of energy output (Downing *et al.*, 1996).

Selected references

Bambara, S. (1994) via Internet, http://sunsite.unc.edu/london/orgfarm/biocontrol/

Barnes, D.K. and Sheaffer, C.C. (1995) Alfalfa. In: Barnes, R.F. Miller, D.A. and Nelson, C.J. (eds), *Forages*, Vol. 1: *An Introduction to Grassland Agriculture*, Iowa State University Press, Ames, Iowa, pp. 205–16.

Campbell, K. (1996) The Minnesota agri-power project. In: *Bioenergy '96: Proceedings of the Seventh National Bioenergy Conference*, 15–20 September 1996.

Downing, M., Bain, R. and Overend, R. (1996) Economic development through biomass systems integration. In: *Bioenergy '96: Proceedings of the Seventh National Bioenergy Conference*, 15–20 September 1996.

Duke, S.H., Collins, M. and Soberalske, R.M. (1980) Effects of potassium fertilization on nitrogen fixation and nodule enzymes of nitrogen metabolism in alfalfa. *Crop Science* 20: 213–19.

FAO (1996) Ecocrop 1 Database. Rome.

Franke, W. (1985) *Nutzpflanzenkunde*, 3rd edn. Georg Thieme Verlag Stuttgart, New York.

Huffaker, M. (1994) via Internet, http://sunsite.unc.edu/london/organfarm/cover-crops/

IFFS, *International Forage Factsheet Series* (1995) Alfalfa (*Medicago sativa* L.), several authors, via Internet http://www.forages.css.orst.edu/topics/species/legumes/alfalfa/

Moon, S. (1997) Sowing seed, planting trees, producing power. *Solar Today*, pp. 16–19.

Sheaffer, C.C., Tanner, C.B. and Kirkham M.B. (1988) Alfalfa water relations and irrigation. In: Hanson, A.A., Barnes, D.K. and Hill Jr, R.R. (eds), *Alfalfa and Alfalfa Improvement*, American Society of Agronomy, Monograph 29, Madison, Wisconsin, pp. 373–409.

Vance, C.P., Heichel, G.H. and Phillips, D.A. (1988) Nodulation and symbiotic dinitrogen fixation. In: Hanson, A.A., Barnes, D.K. and Hill Jr, R.R. (eds), *Alfalfa and Alfalfa Improvement*, American Society of Agronomy, Monograph 29, Madison, Wisconsin, pp. 229–57.

ANNUAL RYEGRASS (*Lolium multiflorum* Lam.)

Description

Annual ryegrass has its origins in the Mediterranean region. It grows in semi-erect to erect clumps. The stems grow to a height of 30 to 80cm, though the plant may grow as high as 1m or more. The leaf blades are 4–10cm wide, 21–24cm long, grooved and light green with a glossy underside. The inflorescence is a two-rowed lax ear. Annual ryegrass flowering occurs between June and August.

Ecological requirements

Temperate and subtropical climate regions are most appropriate for growing annual ryegrass. The plant prefers a fresh, nutrient rich, medium loamy soil and thrives in humid climates with plenty of rainfall. A sufficient water supply is necessary for optimal yields. Annual ryegrass is known not to tolerate drought, wet soil and salt. Crop damage has been seen in areas with heavy snowfall and during cold winters.

Propagation

Annual ryegrass is propagated using seed.

Crop management

The annual ryegrass seed can be sown in the late summer (beginning of September), following the harvest of the previous crop, or also in the spring. The amount of seed sown is approximately 35kg/ha.

Table 7.3 shows the suggested fertilizer rates from Germany for annual ryegrass crops with more than one harvest per year. Nitrogen fertilizer is applied according to the individual growth stages of the plant and can be in the form of up to 75% liquid manure. The fertilizer rates vary with soil composition.

Among the diseases most damaging to annual ryegrass crops are mould (*Fusarium nivale*) and rusts (*Puccinia* spp.).

Insertion of annual ryegrass into the crop rotation is easy. Because annual ryegrass has not been shown to have negative effects on crops coming after it, and because no crops have been shown to have negative effects on annual ryegrass, there is currently no recommended crop rotation.

Production

Field tests at four locations in Germany where annual ryegrass was harvested once a year at the onset of flowering provided yields averaging between 7.1 and 11.8t/ha dry matter. The plant's dry matter content averaged about 33–35%. The heating value averaged approximately 17MJ/kg.

Additional tests in Germany comparing different annual ryegrass varieties showed an average fresh matter yield of 105.5t/ha and an average dry matter yield of 18.5t/ha. The varieties with the highest fresh matter yield were Adrina and Lolita. Lolita was also the variety with the highest dry matter yield.

Processing and utilization

Once the crop has reached maturity it may be harvested up to six times per year. This corresponds to about once every 4 weeks from the beginning of May through until the end of October.

Table 7.3. Recommended fertilizer levels for annual ryegrass as used in Germany.

Fertilizer	Amount (kg/ha)
Nitrogen (N)	300–350
Phosphorus (P_2O_5)	80–100
Potassium (K_2O)	400–420

Although annual ryegrass is known as a very valuable fodder plant, much of the present research on it has the goal of utilizing the entire plant as a biofuel for the direct production of heat and the indirect production of electricity.

Selected references

Feuerstein, U. (1995) *Erzeugung standortgerechter zur Ganzpflanzenverbrennung geeigneter Gräser für die Nutzung als nachwachsende Rohstoffe*. GFP-Project F 46/91 NR-90 NR 026, Deutsche Saatveredelung Lippstadt – Bremen GmbH.

Kaltofen, H. and Schrader, A. (1991) *Gräser*, 3rd edn. Deutscher Landwirtschaftsverlag, Berlin.

Rehm, S. and Espig, G. (1991) *The Cultivated Plants of the Tropics and Subtropics*. Verlag Joseph Margraf, Priese GmbH, Berlin.

Siebert, K. (1975) *Kriterien der Futterpflanzen einschließlich Rasengräser und ihre Bewertung zur Sortenidentifizierung*. Bundesverband Deutscher Pflanzenzüchter e.V. Bonn.

Wurth, W. and Kusterer, B. (1995) *Ergebnisse der Landessortenversuche mit Futterrüben und Welschem Weidelgrass 1994. Information für die Pflanzenproduktion*. Landesanstalt für Pflanzenbau Forchheim.

ARGAN TREE (IRONWOOD) (*Argania spinosa* (L.) Skeels; syn. *A. sideroxylon* Roem. & Schult)

Description

The ironwood is an evergreen olive-like tree native to Morocco. It reaches 4–10m in height, occasionally attaining 21m with a trunk diameter up to 1m. The ironwood tree flourishes under very dry conditions, forming olive-like berries on its gnarled branches and thorny twigs. The berries contain 1–3 relatively small seeds (Gliese and Eitner, 1995). The kernels contain about 68% oil. The timber is very hard, heavy and durable, and suitable for agricultural implements and building poles in addition to making a good charcoal. The tree can be planted as a shade tree and for soil conservation and windbreaks (FAO, 1996). The argan tree is cultivated in those parts of Morocco where the olive tree cannot be grown (Franke, 1985).

Ecological requirements

Argania spinosa can be found from sea level to an elevation of 1500m. The tree is drought resistant but not adapted to drifting sands. It can shed its foliage and remain in a state of dormancy for several years during prolonged droughts. An annual rainfall of 200–300mm is the optimum; the minimum of precipitation is 100mm and the maximum 400mm. The tree grows best at temperatures from 20–30°C; minimum and maximum temperatures are 10°C and 35°C respectively. The tree is killed by frost (–2°C). It needs a medium deep, well drained soil with a pH value of 6.5–7.5. Bright light is required for growth (FAO, 1996).

Production

The endosperm of the seed contains approximately 68% oil. The oil is rich in oleic acid and edible with a nutty taste. The tree may start to bear fruit after 5–6 years and reaches maximum production after about 60 years. These long living trees may reach an age of 200–400 years. The average fruit yield may be about 8kg per tree.

Processing and utilization

The oil is edible and is also used as fuel (for illumination) and the production of soap (Gliese and Eitner, 1995). The wood can be used as fuelwood or be processed to charcoal.

Selected references

FAO (1996) Ecocrop 1 Database. Rome.
Franke, W. (1985) *Nutzpflanzenkunde*, 3rd edn. Georg Thieme Verlag Stuttgart, New York.
Gliese, J. and Eitner, A. (1995) *Jatropha Oil as Fuel*, DNHE-GTZ, Project Pourghère.

BABASSU PALM (BABAÇÚ) (*Orbignya oleifera* Burret)

Description

The babassu palm is a perennial evergreen palm that grows to a height of 15 to 20m. It has a slender trunk, and the crown is formed of 15–20 huge feathery leaves measuring up to 7m that stand at upright angles. It is native to the dry semi-deciduous forests of central Brazil (FAO, 1996), where it grows wild in large populations. The babassu palm flowers and bears fruit over the whole year, but the yield is higher during the summer (Lennerts, 1984). In Brazil the fruit ripens from July to November. The male and female inflorescences with a length up to 1m hang down from the leave axils. Each inflorescence carries between 200 and 600 oblong nuts the size of goose eggs, each weighing 80 to 250g and containing 4–6 kernels (Gliese and Eitner, 1995). The kernels are surrounded by a pulp and a hard woody shell, nearly 12mm thick, similar to a coconut shell. The fruit bunches weigh from 14–90kg; one to four fruit bunches are produced per year.

Related species that offer a commercial potential as oilseeds:

- *Orbygnia martiana* Barb. is the other main species of babassu found in Brazil. It grows in the wet forest areas of the Amazon basin. The preceding descriptions apply to this species, too.
- *Orbygnia cuatrecasana* A. Dugand grows in the Chocó region on the Pacific coast of Colombia. The seeds are edible and are used for the same purposes as coconuts.
- *Altalia speciosa* Mart. (formerly known as *Orbygnia speciosa* (Mart.) Barb.-Rodr.) grows in tropical South America, especially in Brazil. The fruits are smaller and

contain fewer seeds than the previously mentioned species. The palm provides an excellent oil.

- *Orbygnia cohune* Mart., the cohune palm, grows in the lowlands of Yucatan/ Mexico, Belize, Honduras and Guatemala. Each palm bears from 1000 to 2000 fruits per year, but only a small amount is harvested (National Agency of Sciences, 1975). The fruit flesh is edible and is also used as animal fodder (Rehm and Espig, 1991).

The seeds of palms like *Astrocaryum vulgare* Mart. and the *Scheelea* species and of the American oil palm *Elaeis oleifera* (H. B. K.) Cortes are collected in a similar way.

Ecological requirements

All of the species mentioned require tropical climates with warm temperatures, ample sunshine and moist soils. Babassu palms are most abundant along rivers and valley floors, but they also grow in semi-arid areas (Gliese and Eitner, 1995). An annual rainfall of 1200–1700mm is the optimum, but 700–3000mm is acceptable. The optimum temperature is 25–30°C. The palm accepts a wide range of soil textures, but the soil should be well drained and not waterlogged (FAO, 1996). The climate in the area in which the babassu palm is found is marked by short and irregular periods of heavy rainfall (Lennerts, 1984).

Propagation

The wild palms reproduce by seeds, which remain germinable for a long time. Clearings in babassu forests are quickly overgrown with babassu palms again. Because of the high number of seeds, the stock of babassu palms is rather dense. The germination has a special mechanism. First, a pregermination of the seed occurs: a 30–50cm long carrier is formed, growing vertically into the soil and carrying the embryo at the tip. There the embryo generates roots and the cotyledon (Lennerts, 1984).

Breeding: high yielding varieties should be selected from the wild stands and collected in a germplasm bank. The reason for the great differences in yield in various regions should be investigated (National Agency of Sciences, 1975)

Crop management

Most babassu palms are not cultivated in plantations. Only a few plantations have been established. The fruits of the wild palm are collected. With arranged harvesting and cultivation of the palm in plantations, the yield could be increased (Franke, 1985). In wild stands the plant density is too high for good yields. There are some indications that 100 palms per ha will yield from 50 to 500kg of oil. The yield obtained from this heliotrophic plant is related directly to the available space. With increasing stocking density, the average number and weight of infructescences decrease. By merely considering parameters like the stocking density and age structure of babassu woods, enormous yield reserves could be opened by purposeful stocking alone (Ortmaier and Schmittinger, 1988). However, there is a complete disregard for domestication:

babassu oil has been substituted by other oils, and the labour to harvest the babassu kernels competes with other activities (Lieberei *et al.*, 1996).

Production

The babassu palm is slow to mature. It begins to yield after 8 years and reaches full production after 10–15 years (National Agency of Sciences, 1975). It is said to be productive for 200 years. The kernels contain 64–67% fatty oil; this oil is similar to palm oil and coconut oil. The high quality oil stores well. Oil yields peaking at about 100–200l/ha are possible (Gliese and Eitner, 1995). Various levels of yields are reported for individual trees. One palm may produce 1t of nuts per year, which corresponds to 90kg of kernels. In cultivated plantations 1.5t of nuts per year can be achieved (National Agency of Sciences, 1975).

The yield potential has hardly been tapped: the stock of palms has a yield potential of 883,680t of kernels, but only 281,344t are collected and used for oil production (Ortmaier and Schmittinger, 1988). A lack of infrastructure and transport facilities and the use of wild, uncultivated babassu palms prevent higher yields. The fat-containing kernels make up less than 10% of the fruit, while the share of the shells alone is more than 50%. Table 7.4 shows the composition of fruits and kernels. Although the kernel has the most important properties for commercial use, it accounts for only 10% of the fruit. Therefore, it is not usually economic to transport whole fruits to cracking and separating centres, since 90% of the fruit has only incidental by-product value. Cracking and separating should be done locally.

Table 7.4. The composition of babassu palm fruits and kernels (Lennerts, 1984).

The fruits consist of:	The kernels contain:
10% fibrous husk	67% fat
24% fruit flesh	9% protein
58% hard nut shell	10.5% N-free extracts
7–8% kernels	6.5% fibre
	2% ash

Processing and utilization

There is no well organized harvest technique. In the wild stands, the ripe fruits fall off and are collected and brought to a collecting and storage point. After collection the fruit is dried to facilitate the removal of the shell from the kernel. If the kernels are not well dried, damage to the kernels initiates enzyme action and the oil may become rancid. The cleansed kernels can be stored or exported.

Babassu oil has a melting point between 21 and 31°C. It consists predominantly of glycerides of lauric acid and myristic acid (Franke, 1985). The oil is almost colourless and does not easily become rancid. The press cake contains up to 27% protein and gives a valuable fodder (National Agency of Sciences, 1975). The profitable use of the babassu palm requires a practical utilization of the by-products. If the babassu palm is used as an energy plant, it delivers three forms of raw material: the oil, the fruit flesh and the nut shells. The difficulty is to develop an economic process to divide the kernels from the very hard shell. Amongst the uses for nutrition, the oil is used for technical purposes (lubrication) and as fuel. For the production of 200t oil, 2000–2500t of shells are needed. The shells are used directly as an excellent solid fuel. To a large extent they are used to produce charcoal (Rehm and Espig, 1991). The ripe fruit flesh can be milled and is used as a source of starch. In Brazil, research was conducted to produce ethanol from the fruit flesh for use in the Brazilian ethanol-as-fuel programme: 1t of fruit flesh gives 80l ethanol (Wright, 1996).

Selected references

FAO (1996) Ecocrop 1 Database. Rome.

Franke, W. (1985) *Nutzpflanzenkunde*, 3rd edn. Georg Thieme Verlag Stuttgart, New York.

Gliese, J. and Eitner, A. (1995) *Jatropha Oil as Fuel*. DNHE-GTZ, Project Pourghère.

Lennerts, L. (1984) *Ölschrote, Ölkuchen, pflanzliche Öle und Fette*. Verlag Alfred Strothe Hannover, Germany.

Lieberei, R., Reisdorff, C. and Machado, D.M. (1996) *Interdisciplinary Research on the Conservation and Sustainable Use of the Amazonian Rain Forest and its Information Requirements*, GKSS-Forschungszentrum Geesthacht GmbH, Germany.

National Agency of Sciences (1975) *Underexploited Tropical Plants with Economic Value*, Washington DC.

Ortmaier, E. and Schmittinger, B. (1988) Possibilities of tapping the natural yield potential of the babassu forests in northeastern Brazil by environment-orientated resource management. *Quarterly Journal of International Agriculture* 27(2): 170–85.

Rehm, S. and Espig, G. (1991) *The Cultivated Plants of the Tropics and Subtropics*. Verlag Joseph Margraf, Preise GmbH, Berlin.

Wright, R.M. (1996) *Jamaica's Energy: Old Prospects New Resources*. Twin Guinep (Publishers) Ltd / Stephenson's Litho Press Ltd.

BAMBOO (*Gramineae, subfamily Bambusoideae*)

Description

Bamboos are a group of giant woody grasses that make their natural habitat roughly between the 40° southern and northern latitudes, excluding Europe. Currently, more than 130 genera of woody bamboos and 25 grass bamboos have been named worldwide, though many other estimates place these numbers lower. These bamboos are made up of over 1300 species of the tribe *Bambusoideae* that are distributed among tropical, subtropical and mild temperate zones of the world. Countries such as Bangladesh, China, India and Indonesia along with other Asian countries depend heavily on bamboo.

Bamboo plants, which are usually perennial, consist of an underground root system and rhizome mat from which rise erect aerial stems, referred to as culms. Culms are usually hollow, smooth and round with a brown or yellow-brown to yellow-green colour and consist mainly of cellulose, hemicellulose and lignin. They can have a diameter of up to 20cm or more and reach a full grown height of 10–40m in about 3–4 months. One genus, *Phyllostachys*, studied as a possible energy plant, grows to heights of 2–10m. Branches are often produced from lateral buds near the top of the culm once the culm reaches its full height. The branches are capable of secondary and tertiary branching, with the ultimate branches bearing leaves. The long fibres, which can be from 1.5 to 3.2mm in length, comprise 60–70% of the culm's weight and are the characteristic that makes bamboo desirable for paper production. Bamboo provides a high biomass yield, is strong and has a similar amount of energy to wood.

Bamboos can be put into two categories with respect to their rhizome system: sympodial or monopodial. Sympodial bamboos grow in clumps and the rhizome's neck has no buds or roots. Monopodials, which grow spread out from each other, have roots and buds on the rhizomes, with axillary buds leading to the growth of new culms at intervals of distance. Tropical bamboos are sympodial, while temperate bamboos can be of either category.

Bamboos of the same species, which in the same habitat would look the same, can take on different characteristics when grown in different climates. The majority of the bamboo species seldom flower, and even then some do not fruit. The flowering cycle can be anywhere from 15 to 120 years, with the average falling at about 30 years. Many bamboos or bamboo populations die once they have flowered.

Ecological requirements

Although bamboos are able to adapt quite well to varying environments, most species need warm and humid climates. An annual mean temperature between 20 and 30°C is preferred. Several species have the cold resistance necessary to be able to live in higher latitudes and elevations, where temperatures are as low as –30°C, but it is recommended that the temperature not fall below –15°C and the elevation remain under 800m. Figure 7.3 shows a bamboo stand growing in China.

In its original habitats bamboo usually needs precipitation levels of 300mm per month during the growing season and 1000–2000mm annually. There are drought resistant strains, such as *Dendrocalamus strictus* in India which can survive on a minimum of 750–1000mm annual precipitation.

Bamboos can help to bioremediate saline soils and also help to prevent soil erosion. Bamboo has proved to grow well in all of Europe, but has shown a preference for areas similar to its native habitats. Areas such as Italy, Portugal, Spain, and southern France are where bamboo has thrived. Bamboos prefer light, well drained sandy loam to loamy–clayey soils, with abundant humus and nutrients. The soil should not dry quickly. Soils too high in salt content should be avoided. The optimal soil pH range is between 5 and 6.5. It is also recommended that the bamboo crop be protected from the wind as much as possible.

Propagation

Two categories for methods of propagating bamboo presently exist: conventional propagation and *in vitro* propagation. Conventional propagation has two traditional forms: seed and vegetative. Seed, though less expensive and easier to transport and handle, also introduces negative factors: in general, seeds have low viability, do not store well and are of questionable availability. Seed propagation usually involves the seed and the seedling first being reared in a laboratory or nursery, then being

Figure 7.3. Bamboo plantation, China.

transplanted into the field. It takes the seedling approximately 2–3 months to reach a height of 5–10cm, at which time it is ready for transplantation. Sowing the seed directly into the field is confronted by the yearly necessities of weeding and thinning of the crop.

The main forms of vegetative propagation involve the planting of offsets, culm cuttings or branch cuttings. Propagation by offsets consists of cutting 30–50cm lengths of 1 or 2 year old rhizomes with nodes and buds present, then burying them in the ground on the assumption that within a few weeks new shoots will appear. This method is not suitable for large plantations because of the limited quantity of rhizomes per clump of bamboo culms.

Propagation by culm cuttings involves planting 0.5–1.0m lengths of culm with nodes, 7–15cm deep into a rooting medium. Because of the large diversity of rooting capabilities across the different bamboo species it is difficult to generalize this method for all types. Culm cuttings from thick walled bamboos have displayed a success rate of 45–56%.

Propagation by the planting of branch cuttings involves the direct planting of branch lengths into the ground in the field. This method has not proven very successful; it takes several months for the branch to produce roots and 1 to 3 years for it to form rhizomes. During this time it is very susceptible to negative weather conditions.

Vegetative propagation has the advantage that the characteristics of the new plant are known to be those of the parent plant. But vegetative propagation also has disadvantages in that these methods can only be performed at certain times of the year, the success rates are low, and the amount of work and costs involved make the approach infeasible for large plantations.

The other category, *in vitro* propagation, can be divided into two forms: somatic embryogenesis and micropropagation. Somatic embryogenesis involves plantlet production using somatic embryos, which are non-gametic fusion-formed cells that have the shoot and root pole. Somatic embryo cells can be obtained from the callus of either seeds or young inflorescences. Germination of the cells leads to plant formation. The work is performed in a laboratory on a tissue culture medium until mature plantlets reaching heights of 8–10cm are obtained, at which time the plantlets are transferred to pots containing soil, sand and manure. During this acclimation period the plants will spend 2–3 weeks in an acclimation chamber, then 2 months in a glasshouse or PVC greenhouse. After the 2 months the plants are transferred to bags, where they are kept until they are approximately 8 months old. These are then handed over to the foresters who plant them when they are about 1 year old.

Micropropagation begins with the selection of the starting material. This is usually from seedlings or mature culms. It is currently recommended that the axillary buds from young non-flowering stems be used. The buds have the ability in tissue culture to grow into complete plantlets. Experiments using *D. strictus* have resulted in over 10,000 plantlets from a single seedling in a year. After 6–7 weeks in media culture the nodal explants each produced 8–10 shoots. The shoots were then grown in flasks until rooting occurred. Upon reaching a height of 50–60mm the plantlets were transferred into pots with soil and sand. After acclimation, the plants could eventually be transferred into the field. Micropropagated plants have shown changes such as earlier flowering, earlier culm production and increased growth rates, as compared to those from conventional propagation methods.

In vitro propagation offers advantages in that a large number of plants can be propagated, and in a short period of time. The disadvantages of this form of propagation are its high cost; its labour intensive character; the fact that characteristics of the starting material are often not known; and that at present it is not possible to link qualities of *in vitro* growth with desirable qualities of mature plants, so the production of inferior plants is often the result.

Crop management

The period of new growth is from approximately April/May until the end of August. Planting density tests of *Dendrocalamus strictus* in India concluded that the highest biomass production of 27t/ha, during an 18 month period, was seen at a density of 10,000 plants/ha.

Because little work has been performed with growing bamboo as a crop in Europe, there is very little information covering diseases and pests that attack bamboo. Even in other countries that depend on bamboo, the subject of diseases and pests is not extensively researched. In its native habitat bamboo is subject to disease attack on the roots, rhizomes, culms, foliage, flowers and seeds. The difference between the climate of Europe and that of bamboo's native climate means that time and research will be necessary to determine the effects of diseases and pests on bamboo. In addition, it is likely that bamboo will be affected by new diseases and pests native to Europe.

Bamboo is subject to a number of common diseases in its native habitats: damping-off caused by *Rhizoctonia solani*, rhizome bud rot caused by *Pythium middletonii* and *Fusarium* spp., rhizome decay caused by *Merulis similis*, root infections, basal culm decay caused by *Fusarium moniliforme* and *Fusarium* spp., bamboo blight, culm rot suspected to be caused by *Fusarium equiseti* and *Fusarium* spp., culm necrosis, culm canker caused by *Hypoxylon rubiginosum*, culm stains, inflorescence and seed infections, foliage infections, and rusts. Many of the fungi that attack bamboo are native to tropical areas, where bamboo is made susceptible to them because of excessive humidity and lack of sunlight. Allowing bamboo clumps to become run down makes them more prone to diseases. Excessive humidity, lack of sunlight and allowing clumps to become run down, by tropical climate standards, are three circumstances that are unlikely to be experienced in Europe.

Bamboo is attacked by several common pests in its native habitats. The larvae from *Estigmina chinensis* tunnel throughout the internodes, causing shortened and crooked internodes. The weevil *Cyrtotrachelus longipes* attacks the developing tip of new culms. The plant louse *Asterolecanium bambusae* may attack the buds. Insects such as *Atrachea vulgaris* and *Chlorophorus annularis* bore into bamboo to lay their eggs. Overmature and stored culms are vulnerable to the beetle *Dinoderus minutus*, whose larvae feed on the culm's parenchyma tissue. Aphids, locusts and termites have been shown to attack bamboos, along with other animals that are known to eat bamboo shoots or chew on the rhizomes such as goats, deer, porcupines, rats and squirrels.

Because of the proposed use of bamboo in Europe as biomass for energy production, bamboo plantations in Europe will have an advantage over those of bamboo's native habitats. Many of the diseases and pests that attack bamboo cause changes such as discoloration and crooked growth, which negatively affect the quality of the bamboo in terms of its usage for food, construction materials, furniture, decoration, etc. When bamboo is grown for biomass production the appearance and durability are not important factors.

In India, research was conducted to determine the influence of planting densities and felling intensities, among other factors, on the performance of the bamboo *Dendrocalamus strictus*. The cultivation took place on marginal land with red silt loam soil. The mean annual precipitation was 827mm and temperatures ranged from 13 to 36°C. Planting densities are related to the purpose for which the bamboo is intended. This is demonstrated by the observations that high density planting leads to the production of a greater biomass yield and more plants but with smaller culms, while as the planting density decreases so also does the number of culms, but the size of the individual culms increases.

Fertilizer application tests in India resulted in recommendations for fertilizer amounts for the area in which it was tested. These are shown in Table 7.5. These fertilizer amounts resulted in a threefold increase in biomass production over non-fertilizing. Doubling the amounts showed a fourfold increase in biomass production over non-fertilizing. Fertilization has been shown to lead to an increased number and weight of rhizomes. A nitrogen rich, fast release compound fertilizer should be used in the spring, about a month before sprouting, but avoided in the autumn. Studies in China have concluded that for every 1000kg of bamboo vegetable matter produced, there needs to be in the soil 2.7kg of nitrogen, 3.6kg of potassium and 0.36kg of phosphorus.

Table 7.5. Suggested fertilizer levels for bamboo resulting from tests in India.

Fertilizer	Amount (kg/ha)
Nitrogen (N)	100
Potassium (as K_2O)	50
Phosphorus (as P_2O_5)	50

Once established, the bamboo crop can be maintained for many years. As it is a perennial plant species, a crop rotation is not valid for bamboo.

Production

Until now, bamboo has had a very wide range of uses, but only a little is used as firewood. Bamboo's principle uses for processed products mean that a market and market value do not yet exist for using it as an energy and fuel source.

The yield is greatly affected by the species and the growth conditions (Table 7.6). The value of bamboo as a fuel source is connected to its dry matter content.

Processing and utilization

Bamboo is currently harvested manually with knives, which is the most economical, though labour intensive, way of bamboo harvesting. Mechanized harvesting methods need to be developed, but this would also depend on the extent of annual harvesting and the species of bamboo to be harvested. In addition, the many differences between the geography, climate, labour, technology, etc. of Europe and the countries of bamboo's natural habitat will need to be taken into consideration.

It was previously believed that clear felling was harmful to bamboo stands, but tests in India have shown that clear felling of the stand led to vigorous regrowth of the clumps. Removal of all of the culms can lead to an increase in the number and size of culms, which means an increase in the amount of biomass per year, while providing more room for the growth of new shoots.

Table 7.6. Annual biomass yield of air dried bamboos as found in their native habitats.

Bamboo species	Native country	Yield (t/ha)
Bambusa tulda	India	3
Dendrocalamus strictus	India	3.5
Melocanna baccifera	India	4
Melocanna bambusoides	Bangladesh	10–13
Phyllostachys bambusoides	Japan	10–14
Phyllostachys reticulata	Japan	5–10
Phyllostachys edulis	Japan	5–12
Phyllostachys edulis	Taiwan	8–11
Phyllostachys makinoi	Taiwan	6–8
Phyllostachys pubescens	China	5–10
Thyrsostachys siamensis	Thailand	1.5–2.5

In its native habitats bamboo is transported manually, by animals, carts or trucks, or floated down river. This is usually to a paper mill or a place for storage or further processing. Storage usually amounts to piling the bamboo culms in hills where, while waiting to be used, they are often attacked by disease and pests, which leads to deterioration of the culm's quality. Processing produces a large variety of durable, long lasting products.

Bamboo is considered a possible candidate for bioenergy production because of its high biomass yield and its high heating value of about 4600cal/g (wood has a heating value of 4700–4900cal/g). Bamboo is used to make furniture, building materials, paper, instruments, weapons and decorative ornaments, along with many other products. It is also used as a source of food and fibre. Although, to a small extent, bamboo is burned for heat production, this does not approach the scale of using bamboo as a substantial energy or biofuel producer.

The use of bamboo as an energy and biofuel source is currently being researched. Which species, under what conditions, and to what use have yet to be determined for Europe and the Mediterranean region. In addition, a considerable amount of research is currently under way to genetically create hybrids with desirable energy crop characteristics.

Selected references

Gielis, J. (1995) Bamboo and Biotechnology. *European Bamboo Society Journal* May, pp. 27–39.

Liese, W. (1985) *Bamboos: Biology, Silvics, Properties, Utilization*. Deutsche Gesellschaft für Technische Zusammenarbeit (GTZ), Schriftenreihe Nr.180, 132 pp. TZ Verlagsges., Rossdorf 1.

Liese, W. (1995) Anatomy and Utilization of Bamboos. *European Bamboo Society Journal* May, pp. 5–12.

Mohanan, C. and Liese, W. (1990) Diseases of bamboos. *Int. J. Trop. Plant Diseases* 8: 1–20.

Nadgir, A.L., Phadke, C.H., Gupta, P.K., Parasharami, V.A., Nair, S. and Mascarenhas, A.F. (1984) Rapid multiplication of bamboo by tissue culture. *Silv. Genet.* 33: 219–23.

Patil, V.C., Patil, S.V. and Hanamashetti, S.I. (1991) Bamboo farming: An economic alternative on marginal lands. In: *Proceedings 4th International Bamboo Workshop*, Chiangmai, Thailand. FORSPA Publication 6, pp. 133–5.

Prutpongse, P. and Gavinlertvatana, P. (1992) In Vitro Micropropagation of 54 Species from 15 Genera of Bamboo. *Hort. Science* 27(5): 453–4.

Rao, I.U., Ramanuja, I.V. and Narang, V. (1992) Rapid propagation of bamboos through tissue culture. In: Baker, F.W.G. (ed.), *Rapid Propagation of Fast-Growing Woody Species*. CAB International for CASAFA, pp. 57–70.

BANANA (*Musa* x *paradisiaca* L.)

Description

Bananas are all cultivated in the tropics, and in many of the subtropical regions. The wild forms originated in South-east Asia. They were brought to Africa in the first millennium BC. In America they were first introduced after 1500 AD. The banana is a herbaceous perennial with an underground rhizome, from whose sprouts the new fruit bearing shoots are formed. The sprouts are also used for propagation. The sheaths of the 4–6m long and 1m wide leaves form a pseudostem. The inflorescence develops 7–9 months after the appearance of the sprout. After the fruiting shoot has grown through the hollow pseudostem, it bends and grows more or less vertically downwards. In the axils of the red-violet bracts, groups of flowers are formed, first females, then hermaphrodites, then males. The fruits develop from female flowers without pollination. The hermaphrodite flowers do not form fruits and can be absent (Rehm and Espig, 1991). The fruits are 10–25m long with a green, yellow or brown colour, and with seeds or seedless (FAO, 1996). The cultivated forms are diploid, triploid or tetraploid, based on the wild species *M. acuminata* Colla (genome A) and *M. balbisiana* Colla (genome B). So their genomes (each with many cultivars) are:

2 n: AA or AB, 3 n: AAA, AAB, ABB, BBB, 4 n: AAAA, ABBB

Ecological requirements

Bananas require a uniformly warm climate with an annual average temperature above 20°C, a large amount of sunshine and evenly distributed precipitation of about 2500mm/a. The banana prefers more than 60% humidity. Depending on the cultivar it may be killed by frost and damaged if the temperature falls below 7–12°C for more than several hours. The dwarf varieties cultivated in the subtropics endure temperatures as low as 0°C (Rehm and Espig, 1991). The species *Musa japonica*, originating from the Japanese Ryukyu archipelago, can be grown as a ornamental shrub even in Middle Europe. The stem and the leaves of this species may be killed by the frost, but the plant will regenerate in the following spring if the frost does not reach the roots (van der Palen, 1995). The ground should be loose, well drained and rich in organic material in the top 20 cm. Fertile volcanic or alluvial soils are preferred. The optimum pH value is 5–7, but both higher and lower values are tolerated. For high yields the uptake of nutrients is high, 1t of bananas contains 2kg N, 0.3kg P, 5kg K, 0.4kg Ca and 0.5kg Mg.

Dwarf bananas also thrive in the subtropics and withstand temperatures as low as 0°C, though this brings growth to a standstill. Dwarf bananas are planted close together and develop very quickly, so they can produce yields similar to the tropical cultivars if they are sufficiently irrigated and receive high levels of mineral fertilizer. The banana is very susceptible to wind and hail damage (FAO, 1996).

Propagation

Propagation of the banana is carried out vegetatively with suckers. Preferred are suckers with leaves that are not unfurled. These ‘sword suckers’ as well as ‘maiden suckers’ (young sprouts with unfurled leaves) or older sprouts with leaves are used for propagation, but from the latter the leaves should be removed and the terminal growing point cut out, with the aim that a new sprout develops from an axillary bud. The sets should be treated against nematodes with hot water (56°C for 5 minutes) and with pesticides against pests and diseases. Propagation by way of meristem culture has also been used commercially for several years.

The *in vitro* production of meristem cultures provided a great advance in the rapid vegetative multiplication of *Musa*. Taken from the central growing point and lateral shoots, the tissue is placed on a growth medium and incubated. From 15 to 20 shoots are produced from each meristem in about 30 days. The shoots are transferred into fresh tubes of medium and take about 50 days to develop sufficient roots for a transfer to soil.

The somaclonal variation resulting from tissue culture is a problem with respect to plant storage, but it offers a means to generate mutants for resistance screening to major diseases and pests (Stover and Simmonds, 1987). Resistance to Panama and sigatoka diseases is an aim of breeding. Furthermore, fruit characteristics and tillering capability are of importance.

Crop management

Planting distances are 3m or more for tropical cultivars and 2m for 'Dwarf Cavendish'. Only enough side shoots are allowed to grow and to build a fruit bunch every 4–6 months; excess side shoots are cut off, or killed with chemicals. From the appearance of the tip of the new daughter plant to the ripening of the fruit takes 12–14 months. The cultures last for 15–20 years before they need renewing (Franke, 1985). The most important feeding roots lie close to the surface, so the working of the soil for weed control should be kept to a minimum to avoid damaging the roots. A 20cm thick mulch layer on the ground restricts most weeds. For the preparation of this layer, grass, banana remnants or other organic materials are used. This will also contribute to the supply of potassium. Fertilization with nitrogen compounds is always necessary.

The leaf spot diseases yellow sigatoka (*Mycosphaerella musicola* Leach) and black sigatoka (*M. fijiensis*) are controlled by regular spraying with fungicides. Where the banana wilt (Panama disease) caused by *Fusarium oxysporum* Schlecht var. *cubense* is extensively present, sensitive cultivars cannot be grown. The selection of healthy planting material and disinfection of the sets, together with orchard sanitation and the removal of infected plants, are the most important preventive measures against other fungus, bacteria and insect diseases.

Production

Yields can vary considerably. Peak yields can reach 50t/ha per year. The dwarf forms produce similar yields per hectare because they are planted closer together than the tall cultivars and develop very quickly. The yield of plantains is similar to the yield of bananas: in well kept plantations yields are 38–50 t/ha, though with less care this can fall to 15–20t/ha (Rehm and Espig, 1991).

The production of bananas under cover is well established, and common practice in some desert-like regions (such as Tenerife). This system allows very high productivity and reduced water and nutrient inputs.

Processing and utilization

To harvest the bananas the trunk is broken down and the fruit bunches are cut off. At the packing stage they are washed and treated with disinfectants. For export they are harvested green-ripe and shipped in cold storage chambers.

Bananas are mainly used for nutritional purposes, either for fresh consumption or dried or processed to banana flour. The carbohydrate content in 100g of the edible part of fresh bananas is about 20%. During the ripening period the starch is converted into sugar, though in the cultivars known as plantains this process does not take place. Starch is produced from unripe bananas and from plantains (Franke, 1985).

Because of its sugar and starch content the banana is also a promising energy plant. Unsaleable or surplus bananas could be used to produce energy. Sugar and starch can be used as source of ethanol, with utilization as a biofuel, solvent or raw material for the chemicals industry. The fermentation of sugars is a well known process for producing alcohol. The sugar can be directly fermented to form ethanol. The starch has to be broken down into sugars before fermentation, using acids or hydrolytic enzymes. Another source of energy is certainly required, but the main cost is the raw material (Wright, 1996). The development of continuous working fermentation techniques and the use of better varieties of yeast have improved efficiency and lowered the costs (FAO, 1995).

The trunks are cut down to harvest the bananas, so a great quantity of organic material remains as a by-product of the harvest. It should be possible to use the harvested remains of trunks and leaves, which are now only waste, for energy purposes – for example, after anaerobic fermentation to obtain biogas. The residue can be used as a fertilizer, as are the complete trunks today. In 1993 Costa Rica alone generated 1,750,346t of waste tree stems and leaves, 317,592t wasted substandard fruits, and 143,528t raceme stem waste (Hernandez and Witter, 1996). The methanogenic bacteria of the anaerobic fermentation can convert about 90% of the feedstock energy content into biogas that is a readily usable energy source. The biogas contains approximately 55% methane. The sludge produced in this process is non-toxic and odourless, and has lost relatively little of its nitrogen and other nutrients content (FAO, 1995).

Selected references

FAO (1995) *Future Energy Requirements for Africa's Agriculture*, Food and Agriculture Organization of the United Nations, Rome.

FAO (1996) Ecocrop 1 Database, Rome.

Franke, W. (1985) *Nutzpflanzenkunde*, 3rd edn, Georg Thieme Verlag Stuttgart, New York.

Hernandez, C.E. and Witter, S.G. (1996) Evaluating and managing the environmental impact of banana production in Costa Rica: A systems approach. *Ambio* 25(3): 171–8.

Rehm, S. and Espig, G. (1991) *The Cultivated Plants of the Tropics and Subtropics*. Verlag Joseph Margraf, Preise GmbH, Berlin.

Stover, R.H. and Simmonds, N.W. (1987) *Bananas*, Longman Group / John Wiley.

van der Palen, J. (1995) Musa basjo, *Groei & Bloei*, June 1995.

Wright, R.M. (1996) *Jamaica's Energy: Old Prospects New Resources*, Twin Guinep (Publishers) Ltd / Stephenson's Litho Press Ltd.

BLACK LOCUST (*Robinia pseudoacacia* L.)

Contributed by: K. Jakob

Description

Robinia pseudoacacia L. is endemic to North America between the latitudes 43°N and 35°N. Its natural range is divided into two zones. Its main zone of distribution is the Appalachian Mountains of central Pennsylvania to northern Alabama and Georgia. The second zone covers the Ozark Plateau of southern Missouri and parts of northern and western Arkansas and eastern Oklahoma.

There are many species within the genus *Robinia* used for floriculture, but only *Robinia pseudoacacia* L. plays an important role concerning forestry and wood production. *Robinia pseudoacacia* L. is also known as black locust, white locust, false locust, common locust, false acacia and white acacia.

Black locust is a multi-purpose tree with interesting characteristics. It is a pioneer plant with very rapid juvenile growth. Nonetheless, the wood is very dense, high in decay resistance, has excellent colour and makes a good pulp. This combination of high growth rate and high wood density is not so often seen with other trees (Hanover *et al.*, 1991). In addition, black locust is a leguminous plant that can fix atmospheric nitrogen. This gives it the ability to tolerate low fertility site conditions. Locust uses the C_3 photosynthetic pathway for utilizing solar energy. The leaves are high in nitrogen and can be used as animal feed. A very aromatic honey can be produced from the flowers.

Robinia pseudoacacia L. reaches a height of 15–35m. The stalk is characterized by a very high tendency towards warping. The wood is very heavy, hard, extremely durable and elastic. The fresh wood is green-yellow. Black locust is characterized by a very high net photosynthetic rate, high light saturation, low stomatal diffusion resistance, long leaf retention time and a high transpiration rate (Hanover *et al.*, 1991).

The inflorescence of black locust is raceme-like when developed. Normally, time of flowering is in May to June after leaves have appeared. Pollination is by insects only. Black locust flowers at an early age of about 6 years. Seed maturity is at the end of October. The fruit is a red-brown pod containing 6 to 10 seeds. The seeds are kidney-shaped, brown-grey to black, and fall off beginning in February. The seeds are characterized by a very high skin hardness.

The root develops a turnip shaped taproot during the juvenile growth. Black locust produces strong lateral roots with increasing age. The whole root system is characterized by intensive, wide branched and dense fibrous upper roots. Roots near the soil surface can reach a length of 20m and more (Göhre, 1952). There is a rapid extensive root growth of strong roots in combination with the development of intensive fine roots during the first years. Thus, black locust is able to bind loose soils in a short time. Black locust is a so-called pioneer plant and has a very high root suckering capacity. Its root cuttings show vigorous sprouting.

Black locust is cultivated in many countries. Keresztesi (1983) lists some of the European countries with the largest areas of black locust stands as of 1978: Hungary (275,000ha), Rumania (191,000ha), the then Soviet Union (144,000ha),

France (100,000ha), Bulgaria (73,000ha) and the then Yugoslavia (50,000ha). Other European countries with black locust stands are the Czech Republic and Slovak Republic (28,000ha together) and Spain (3000ha), as well as Austria, Belgium, Greece and Ireland. Outside Europe, South Korea has cultivated about 1,017,000ha of black locust forests on infertile soil for firewood, erosion control and feedstocks. The present area of black locust growth is estimated to be about 3.25 million ha worldwide (Hanover *et al.*, 1991). This is the third largest area in the world within the group of fast growing trees, after eucalyptus and poplar.

Ecological requirements

The climate of origin of black locust has a mean annual precipitation between 1000 and 1500mm. The maximum temperatures in July are 30–38°C, and the minimum temperatures in January lie between −10°C and −25°C.

Black locust is adaptable to a wide range of sites because it does not make specific demands on its environment. It has shown to be resistant to environmental stresses such as drought, high and low temperatures, and air pollutants. It grows on a large number of soils, but not on very dry or heavy soils. Black locust prefers sites with loose structural soils, especially silty and sandy loams. Soil aeration and water regime are the most important soil characteristics for good black locust growth. Also of importance is the soil's rootable depth. Compacted and waterlogged subsoils are not suitable for black locust cultivation. Black locust has the ability to flourish rapidly on disturbed sites, and has therefore been widely planted in erosion control programmes and for soil improvement in the United States (Hanover *et al.*, 1991).

Black locust is a leguminous plant whose roots are infected by nitrogen fixing nodule bacteria. The nitrogen fixing capacity of symbiotic bacteria is difficult to determine because it depends not only on the effective combination of the amount and phyla of nodule bacteria but also on the photosynthesis of infected plants and soil conditions.

Experiments using black locust for reclamation on a spoil mound site in a lignite mining region showed that nitrogen fertilization reduced the growth of black locust trees. Potassium and phosphorus application improved nitrogen uptake, which resulted in increasing nitrogen content of the leaves. In addition, potassium positively affected the nodule growth of nitrogen fixing bacteria (Hoffmann, 1960; Heinsdorf, 1987). Studies from Reinsvold and Pope (1987) have shown that the addition of phosphorus increased nodule number and nodular dry weight. With the addition of nitrogen, however, nodulation was reduced with the lowest and with the highest levels of added nitrogen. Pope and Anderson (1982) found an increased nodular biomass under trees fertilized with phosphorus and potassium in comparison to unfertilized trees.

Zimmermann *et al.* (1982) reported that black locust plants treated with nitrogen at 50kg/ha had significantly higher fresh nodule weight than the unfertilized control plants. According to Reinsvold and Pope (1987), a minimal level of mineral N is needed, at least initially, to support nodulation.

Nitrogen fertilization in two applications during the growing season at a rate of 113kg/ha increased biomass production after one year by more than 30%, but only by 12% by the end of the third year (Bongarten *et al.*, 1992). The author explains these results by stating that the nitrogen fixation consumes more energy than the assimilation of nitrate, but as trees increase in size and nitrogen demand there are not enough developed nitrogen fixing bacteria, and soil nitrogen sources are not adequate for the increased biomass as the fertilization level is held constant.

It is known that the black locust positively affects the nitrogen content of the soil. In experiments with sterilized cultures, nitrogen output from roots or nodules of black locust could not be found. Otherwise, black locust positively influenced the growth of mixed forest trees. In black locust stands where the litters of the trees were removed this positive effect could not be proved. These results are in agreement with results from Keresztesi (1988). He found that old black locust forest stands (16–20 years) enriched the upper 50cm of soil with 590kg/ha of nitrogen, but in young stands (5–10 years) this could not be observed. The conclusion to be drawn from this is that nitrogen fixing bacteria do not directly influence the nitrogen content of the soil. Nonetheless, they increase the nitrogen content of the plant material. It is litter decomposition that causes total nitrogen enrichment of the soil.

Hoffmann (1960), reported that 35kg/ha of nitrogen are fixed in a two year old black locust stand (28,000 plants/ha) from which nearly 15kg/ha in the form of litter returns to the soil. The remainder is fixed in the stem, roots and bark. Nitrogen levels after three years were 116kg/ha fixed with 30kg/ha returning to the soil as litter and after four years 305kg/ha fixed with 200kg/ha returning to the soil as litter.

The soil pH value can influence not only the distribution of *Rhizobium* bacteria but also the availability of the nutritive material. Wide distribution of the soil's calcium carbonate places certain restrictions on soil acidity. Therefore, available phosphorus and pH value are the main factors influencing nitrogen fixation (Deng and Liu, 1991).

Propagation

Propagation of black locust is possible through root cuttings, greenwood cuttings, seedlings or micropropagation. Propagation using root cuttings and greenwood cuttings provides a guaranteed quality, but is more expensive than seed propagation. Optimal time for obtaining root cuttings seems to be spring, though it is possible to obtain root cuttings throughout the year (Göhre, 1952). Root pieces are cut with a length of about 8–10cm and diameter of 2 to 5cm and are planted vertically 1–2cm under the surface with the 'root side' downwards. They should not be put into the soil below a depth of 5cm. Plant spacing in the row should be 5–8cm, with 80–100cm between the rows. The best period for planting is April, and after 3 weeks the root cuttings should begin to sprout. Propagation using greenwood cuttings is also possible. After cutting, the pieces are prepared with a phythohormone and put into a special soil substrate for root development under controlled conditions in a greenhouse (Müller, 1990).

Propagation by seed is not a problem because of black locus's frequent and abundant seed production and the high survival rate of the seedlings. The germination rate is between 40% and 60%, but can be increased by special methods like scouring. The seedlings are produced in a nursery and planted into the field in spring or autumn.

Micropropagation of black locust is still at the research stage, though the first methods of *in vitro* shoot propagation have been established. The success of regeneration processes depends on the genotype and age of the basic tree. Plants produced with the help of *in vitro* propagation can be used as basic plants for propagation by cuttings (Ewald *et al.*, 1992).

Crop management

For the production of fast growing stands, one year old seedlings are preferred. When choosing a location for black locust stands it is important to remember that *Robinia* is shade intolerant. Both in its young and mature forms, black locust cannot survive in shady conditions.

The most popular plant spacing is 2.4m by 0.7 to 1.0m. This spacing requires 4000 seedlings per ha. Other studies show that plant spacings from 1.00m × 1.00m (10,000 plants/ha) and 0.50m × 0.25m are necessary for high biomass yield from short rotation forestry (Baldelli, 1992). It is important to control weeds in the first season after planting to prevent the weeds from overgrowing the young trees. The density and planting design should be chosen to fit weed control and harvest options. One of the aims should be to try to achieve a closed canopy as soon as possible to maximize production and minimize weed competition. A stand that is too dense will be self-thinned. If the intention is to establish a black locust plantation, stems and roots have to be cut to ensure tree development. The results of different experiments concerning plant density have shown that plant spacing does not affect biomass production. It seems that black locust stand density does not depend on the initial plant spacing because the plantation will adjust its density through self-thinning.

Phosphorus and potassium should be added if soil analyses have shown these to be present in extremely low quantities. In most cases, nitrogen fertilization is not recommended. It is probably better, for economic and environmental reasons, to take advantage of the plant's nitrogen fixing ability.

Black locust stands have been observed to be attacked by numerous diseases and pests throughout Europe. The diseases and pests vary greatly with climate and European region.

The cultivation of black locust stands, especially on plantations, is designed to encompass numerous harvests carried out over many years. Because of this, there is no need to develop a crop rotation. In addition, because of the tree's excellent coppicing ability, a stand established for energy production can be successfully regenerated from sprouts several times over without the need for replanting crops.

Production

In comparison to other wood species, black locust produces the highest biomass yield as a result of its early growth and high density. In field experiments in Austria, annual dry matter production between 5 and 10t/ha was reached by three and four year rotations of black locust stands from 10,000 trees/ha (Müller, 1990). The four year rotation reached an increase of yearly dry matter yield that was some 1.4 times higher than that for the three year rotation. The moisture content of the wood ranged between 30% and 38%. All moisture contents were below the maximum moisture content for maintenance of combustion.

Examinations of different black locust stands have shown that a nine year old black locust stand produced the highest annual woody biomass yield of 3.43t/ha compared with stands aged from two to ten years. The annual biomass production ranged from 1.1 to 2.99t/ha within two to six year rotations. After six years the annual woody biomass production increased over3 t/ha with a small loss in the tenth year. On a five year rotation, black locust can produce 57.7GJ/ha per year (Stringer and Carpenter, 1986). After two years 43.5GJ/ha per year can be reached, with a maximum from a nine year old stand at 76.8GJ/ha per year.

One three year rotation produced woody biomass that ranged from about 3 to 8t/ha per year. The maximum of about 24t/ha after three years was reached by the best black locust families with irrigation and nitrogen fertilization (Bongarten *et al.*, 1992). All the yield data are only examples and not comparable because of the very different growing conditions of the stands with regard to planting density, soil properties, fertilization, climate, etc.

Processing and utilization

Results indicate that the current annual increment of black locust peaks at 7–8 years. But by this time the stems may be too thick to harvest using available techniques. The grower should examine the economic options before deciding to harvest, taking into account the ease and possibility of harvesting. On short rotation plantations, the small stem diameters necessary for mechanized harvesting require a shorter rotation period of between 3 and 5 years. Black locust's yearly average of dry matter yield lies between 12 to 15m^3/ha. In fact, after coppicing, the tree sprouts from the main stumps and also from the roots, which are diffused in various directions.

A study of its morphological characteristics has shown differences between black locust and other fast growing trees such as willow and poplar. Black locust has thorns, which make it difficult to handle manually, and it is thus preferable to chip it in the field. Black locust can sprout from the roots, so after the third or fourth cut regrowth will also occur between the rows. In addition, black locust has harder wood than other fast growing trees. Therefore the cutting apparatus must be more durable and powerful than that used normally. There are different machines for direct harvesting (Baldelli, 1992).

Black locust was used primarily as fuel and for agricultural purposes. A common use has been for fence posts. The hardwood contains flavanoids which make it extremely decay resistant. The good durability and strength of black locust mean that the timber industry has uses for this species, especially in mining and hydraulic applications. Other products made from black locust are sawn logs, sawn wood, pit props, parquet, railway sleepers, roundwood framing and interior fittings for agriculture buildings. Black locust may also serve as a source of pulp for the paper industry. The final felling of black locust takes place after 30–40 years and produces high quality sawn wood.

Although black locust has many uses, its highest value can be in biomass for energy feedstock. Many characteristics of black locust show interesting qualities for biomass production in energy plantations. The wood has excellent properties for timber production and for energy use. The heating value of the wood and bark is high and it burns well even when it is green. In Austria, heating values of about 17.82 for wood and 20.08MJ/kg for bark were measured (Müller, 1990). Experiments have shown that the heating value of the bark is higher than that of the wood: Stringer and Carpenter

(1986) reported average heating values of 20.81MJ/kg for bark compared with 19.4MJ/kg for wood. In addition, the moisture content of bark, 34.7%, was lower than that of the wood (55.5%). The leaf tissue had a moisture content of 60.6%. The total moisture content of whole black locust trees ranged from 46.1% for two year old trees to 39.1% for nine year old trees.

The tree's nitrogen fixing ability produces high nitrogen levels in the plant material, an advantage that makes black locust suitable for fodder production. But the nitrogen content of plant material for energy use plays an important role in combustion. Demand for low NO_x emissions requires a low nitrogen content in the plant material, so the nitrogen content of the combustion material must be as low as possible. Analysis of leaf material of black locust shows that nitrogen content, especially of the leaves, is very high compared with other trees. Different authors reported nitrogen contents of leaves between 2.3% and 5.5%.

Selected references

Baldelli, C. (1992) Black locust as a source of energy. In: Hanover, J.W., Miller, K. and Plesko, S. (eds), *Proceedings of the International Conference on Black locust: Biology, Culture & Utilization*, Dept of Forestry, Michigan State University, East Lansing, Michigan, USA, pp. 237–43.

Blümke, S. (1960) Die Robinie in Deutschland. *Forstwirtschaft, Holzwirtschaft* 4: 8–16.

Bongarten, B.C., Huber, D.A. and Apsley D.K. (1992) Environmental and genetic influences on short-rotation biomass production of black locust (*Robinia pseudoacacia* L.) in the Georgia Piedmont. *Forest Ecology and Management (Netherlands)* 55 (1–4): 315–31.

Deng, T.X. and Liu, G.F. (1991) Mathematical model of the relationship between nitrogen fixation by black locust and soil conditions. *Soil Biology and Biochemistry* 23(1): 1–7.

Ewald, D., Naujkos, G., Hertel, H. and Eich, J. (1992) Hat die Robinie in Brandenburg eine Zukunft? *Allg. Forstzeitschrift* 14: 738–40.

Göhre, K. (1952) *Die Robinie und ihr Holz*. Deutscher Bauernverlag, Berlin.

Hanover, J.W., Mebrathu, T. and Bloese, P. (1991) Genetic improvement of black locust: a prime agroforestry species. *Forestry Chronicle* 67: 227–31.

Heinsdorf, D. (1987) Ergebnisse eines Nährstoffmangelversuchs zur Robinie (*Robinia pseudoacacia* L.) auf Kipprohböden. (Results of a nutrient deficiency study on robinia (*Robinia pseudoacacia* L.) on spoil mound soil. *Beitrage f. Forstwirtschaft* 21(1): 13–17.

Hoffmann, G. (1960) Untersuchungen über die symbiontische Stickstoffbindung der Robinie (*Robinia pseudoacacia* L.). Dissertation, Eberswalde.

Keresztesi, B. (1983) Breeding and cultivation of black locust (*Robinia pseudoacacia* L.) in Hungary. *Forest Ecology and Management* 6: 217–44.

Keresztesi, B. (1988) Black locust: the tree of agriculture. *Outlook on Agriculture (UK)* 17(2): 77–85.

Müller, F. (1990) Die Robinie als Biomasseproduzent in Kurzumtriebsplantagen. *Österreichische Forstzeitung* 5: 22–4.

Pope, P.E. and Anderson, C.P. (1982) Biomass yields and nutrient removal in short rotation black locust plantations. In: Müller, R.N. (ed.), *Proceedings of the Fourth Central Hardwood Forest Conference*, University of Kentucky, Lexington, pp. 244–59.

Reinsvold, R.J. and Pope, P.E. (1987) Combined effect of soil nitrogen and phosphorus on nudulation and growth of *Robinia pseudoacacia*. *Can. J. For. Res.* 17(8): 964–9.

Stringer, J.W. and Carpenter, S.B. (1986) Energy yield of black locust biomass fuel. *Forest Sci.* 32(4): 1049–57.

Zimmermann, R.W., Roberts, D.R. and Carpenter, S.B. (1982) Nitrogen fertilization effects on black locust seedlings examined using the acetylene reduction assay. In: Thielges, B.A. (eds), *Proceedings of Seventh North American Forest Biology Workshop*, Lexington, pp. 305–9.

BROOM (GINESTRA) (*Spartium junceum* L.)

Contributed by: C. Baldelli

Description

The genus *Genista* (Leguminosae) includes a variety of characteristic plants, from creeping ground covers and massing shrubs to one small tree. With a single exception, all species bear bright golden yellow flowers in late spring or summer (*Genista monosperma* or *Retama monosperma* has white flowers). Among the many varieties (70–90 species of *Genista*) it is important to list *Spartium junceum* (ginestra), *Genista anglica*, *G. germanica*, *G. pilosa*, *G. sagittalis* and *G. tinctoria*. *Genista aetnensis* (of Mount Etna, Sicily, Italy) is a small tree.

Genista are typically Mediterranean species but they are also present in some North American areas (west coast, south-east, Pacific north-west and mid-Atlantic) and Mexico. In South America *Genista* is present in Brazil. There are also many occurrences in northern Europe and in Asia (the natural range of *G. tinctoria* even reaches into western Siberia).

Spartium junceum, broom, has several advantages, including its suitability for several marginal Mediterranean lands (from the south European Mediterranean to North Africa), its favourable industrial conversion, and its presence in the family of Leguminosae as a nitrogen fixer. Broom produces good quality biomass and precious fibre. The botanical name of the species (*Spartium*) comes from the Greek word 'sparton', which means rope.

The leaves of broom are sparse, small and deciduous, but the plant's green, reed-like twigs suggest the appearance of evergreens in winter. The plant is in fact defined by its thin, bright green, graceful and useful, nodding reed-like branches, which change very little during the season when the plant is not flowering. They contain an interesting natural fibre, which will be described below.

Broom has a very strong root apparatus, which explains the plant's application to soil protection. The influence of broom in erosion control is great. It helps to cut down the runoff and wash down of soil particles, which contain significant amounts of nutrients, so has a direct influence on the productivity of the soil. Moreover, thanks to its strong, twined root system and light above-ground weight, broom encourages an efficient soil consolidation. Since time immemorial, broom has been used for this purpose because of its ability to block landslides on steeply sloping soils. The most significant demonstration of this property is seen in broom's large scale employment along roadside slopes.

Broom is an old fibre crop. The Greeks and Romans used its fibre for textiles and ropes. During the Middle Ages, cloths, carpets and furniture coverings were made from broom. A major development of this use began in Italy in the 1930s, when government policy led Italian scientists and developers to increase study into and the availability of this crop for fibre production. A significant amount of research work was carried out.

The requirements of environmental protection, rural restructuring and new local jobs have led to a re-evaluation of earlier experiences of broom cropping, but utilizing the technological innovations now available. In particular, as will be discussed below,

SRF (short rotation forestry) applied to broom could offer an interesting way to relaunch broom cropping and utilization.

Broom's suitability is mainly demonstrated by its good productivity in terms of biomass; it accepts very closely timed cutting cycles and is capable of fast regrowth after cutting. (The need for short cycles is the result of harvest restraints: mechanical harvest can be carried out only if the trunk diameter is below a certain size). Other favourable characteristics are: good physical-chemical qualities for industrial conversion; large scale availability; no legal harvesting restraints (in Europe it is very difficult to cut and harvest biomass trees); good biomass production in the order of 6 to 10t/year of dry matter; and very low cultivation costs.

Besides these characteristics, this species belongs to the legume family and is a rhizobial nitrogen fixer. Thus nitrogen fertilization can be avoided, preventing the pollution of aquifers, which is a problem today. In addition, it does not require chemical pest or weed control, an environmentally friendly feature that is in marked contrast to other fibre crops such as cotton, hemp, flax, kenaf, *Miscanthus*, etc.

Ecological requirements

Broom is at its best in full sun, but tolerates some shade. Soil quality is also important. Like most other legumes, broom thrives in calcareous soils with low fertility, but performs adequately over a wide range of soil acidity (pH 5 to 8). Just as broom is a good choice for dry, infertile soils, so its continued health depends on the avoidance of excessive water and moisture, soggy soils, and heavy fertilization. Broom requires very free soil drainage, but is otherwise highly adaptable to many pedoclimatic situations, and to poor and even stony soils.

Propagation

Broom is not difficult to propagate. It can be planted using cuttings, seed or seedlings. Softwood cuttings taken in early summer root easily; semi-hardwood cuttings taken later in the summer and hardwood cuttings taken in the autumn have also been successful. These cuttings are then planted in October or February.

Direct seed propagation of broom is similar to that of many other legumes with hard seed coats. A half-hour soak in concentrated sulphuric acid is sufficient to make the seed coats permeable to water. Boiling water treatment may also work well, though some species or seed lots may not require either. A rapid mechanical treatment with a special knife mill is also possible.

Nursery and seedling transplantation can use various spacings. The most positive results have been seen with 1m × 1m spacing (a density of 10,000 plants/ha).

Crop management

Propagation by seeds is the easiest and cheapest way to cultivate broom. The best timing for the sowing is late autumn (November) or late winter (February). If a strong weed infestation is foreseen, late sowing is advisable. Sowing can be performed with the normal mechanical grass sowing machines with suitable regulation of parameters (density, diameter of seeds). The procedure includes good soil surface preparation with removal of stones and production of a fine tilth. Seeds have to be prepared before sowing, as mentioned above.

Seedlings are easy to prepare in the open air as well as in a greenhouse. Cuttings require more care, and it is better to cut pieces in late autumn and preserve them under refrigeration. Hormonal application before planting is preferable. Planting by cuttings or seedlings requires deep ploughing to ensure good contact between roots and deep soil, and surface preparation and stone removal are less important. Before planting, it is necessary to plough the rows deeply at the selected spacing (1m × 1m is a good average). All planting operations can be mechanized using existing planting machines. The best time for planting is from late January through all of March, when there are no problems with weeds.

If the soil is very stony, with large stones present, it is easier and quicker to plant into holes. The seedlings or cuttings are placed in holes 50–100mm deep and covered with soil to close the holes. These operations must be performed by hand. Planting in holes is less mechanized and slows the initial growth of young plants. This technique is favourable if the spring rains are light, because the thin ground cover over the holes retains more water than deep soil, and this benefits the roots. Obviously, seedlings and cuttings start to grow a year sooner than seeds and are more uniformly organized in the row.

Short rotation forestry (SRF) technology is the core of the proposed cultivation technique. The cropping methods follow the main aims of mechanizing the entire production cycle and minimizing agricultural problems, so it will be possible to reduce the cost of the final product (wood chips). To achieve this objective it is necessary to work with a very high plantation density (from 10,000 to 20,000 plants/ha) and with very short cropping cycles (3 years). Spacing in broom plantations is from 1m × 1m to 0.25m × 0.25m (10,000–80,000 plants/ha).

With regard to the environmental impact of broom cropping, a pre-summer harvest will help to overcome fire problems. Harvesting during the hottest and thus highest fire hazard period means that vegetable material will no longer remain in the fields and therefore there is nothing to catch fire. The coppiced plants sprout after 20–25 days, and in the following spring the new root suckers are already about 1m high, favouring the subsequent cycle. This procedure has already been verified.

Broom is a typical early succession species. Many people consider it a weed, and thus a dangerous plant. If the aim is to maximize biomass production, an easily established species is preferable, given the low amount of chemical input necessary.

Few diseases have been reported on broom plants, and none is a very frequent problem. The only true problem is vermene spot associated with plants in old age. The cutting cycle of every 3–5 years avoids spot problems.

Following are the important major characteristics of broom grown using SRF methods:

- no fertilizer is needed (broom is a nitrogen fixer);
- there is no weed control (the species is self-weeding);
- there is no agricultural work after planting, only harvesting every three years; and
- harvesting is possible in any season.

No crop rotation is recommended, because of the extended life of the crop.

Production

To offer some values, it can be considered that the optimum cycle is 2–3 years, in the context of vegetative cycles of over ten years, with productions in the order of 6–10t/year of dry matter (40% maximum moisture).

Yield trials have shown that broom, three years from planting seedlings (year 1990) or three years from coppicing (year 1993), at a spacing of 1m × 1m (10,000 plants/ha), reaches an average height of 3.00m and produces about 9.0t/ha of dry matter per year after the third year (moisture 7%).

The efficiency of broom is very high, considering that three years after planting most fast growing forestry plants produce only an insignificant quantity of dry matter. Other trials performed on an old, wild broom plantation, with a natural density of 8000 plants/ha, gave an average dry matter yield of 8t/ha per year.

The conclusion is that the yield of broom has the advantage of short harvesting cycles once roots are established. In this case, the yield can reach 10t/ha per year of dry matter, which is the target foreseen by forestry technicians for SRF applications.

Processing and utilization

Broom should be harvested in 2–3 year cycles. The optimum coppicing cycle is 3 years from planting, when the trunks have a diameter of about 5cm. In this condition, it is appropriate to call this kind of cultivation 'wood grass'.

As already mentioned, mechanical harvesting can be carried out only if the diameter of the trunks is below a certain value: the small diameter of the trunks of young plants permits mechanized harvesting by a special combine harvesting machine (under construction) that will overcome this main bottleneck of all forestry production.

Broom alone or mixed with other raw materials can produce, through special processes, several marketable final products such as biofuels, paper pulps, textiles, composite materials and agrocompost.

The state of the art of broom exploitation has created innovative paths, from cropping to final treatment, for the cultivation of extensive broom plantations on marginal lands suitable for the production of industrial raw materials, ecological, smokeless solid fuels, and vegetable fibre suitable for biocomposites and/or paper pulps. Another option is the utilization of broom biomass for the production of agrocomposts suitable for upgrading soil quality.

Broom production and its industrial conversion seem to be conducive to sustainable development in such areas as environmental protection, economic yield and new job creation. The plantation's long life, its several coppicing cycles, and the absence of chemical inputs contribute to a good potential from the point of view of conservation and biodiversity. This potential will be maximized if degraded land is used and if some parts of natural woodlands are included among the new plantations.

Table 7.7 summarizes the results of a chemical-physical analysis of dry broom chips. These data support the great interest in this species as a possible biofuel. Further positive attributes can be seen in the results obtained from a pilot pyrolysis plant (Table 7.8). The analyses confirm broom's good qualities as a biofuel, and in particular the virtual absence of sulphur.

Table 7.7. Chemical and physical characteristics of broom.

Characteristic	Value
Ash (%)	4.63
Volatile substances (%)	71.05
Fixed carbon (%)	24.32
High heating value (kcal/kg)	3.906
Fibre length (mm)	8.50
Fibre diameter (μm)	30.20

Table 7.8. Broom pyrolysis performance element analysis.

Characteristic	Charcoal	Bio-oil
C (%wt)	80.96	61.90
H (%wt)	1.70	6.00
N (%wt)	1.45	1.05
S (%wt)	0.01	0.03
Ash (%wt)	7.34	1.50
Moisture (%wt)	0.34	14.60
O by diff.	8.20	14.92
Char content (%wt)		9.20
Viscosity at 50 °C (cP)		200
Viscosity at 70 °C (cP)		55
HHV (kcal/kg)	7060	6290
LHV (kcal/kg)	6956	5980
Specific gravity (15/4 °C)		1.195

There is also interest in the possibility of converting broom twigs into a fibre for making clothes and textiles. New techniques (mechanical fractionating, heating, etc.) are presently being studied.

Composite materials using vegetable rather than fossil derived fibres as reinforcement constitute the most promising application of broom fibre at present, and some research and industrial laboratories are already working on this matter with broom and other fibre crops. In particular, the car and aircraft industries are very interested in developing new materials suitable for external and internal bodywork that utilize vegetable fibres.

Selected references

Baldelli, C. (1987a) Robinia and ginestra from weed plants to an ecological fuel. *Proceedings of the 4th EC Conference on 'Biomass for Energy and Industry'*, Orléans, France, 11–15 May 1987, pp. 355–68.

Baldelli, C. (1987b) *Agricultural Raw Materials for Energy and Industry on Marginal and Poor Lands*, UNITAR/UNDP, Rome, Italy, Newsletter No. 9: 3–5.

Baldelli, C. (1988) Energy crops on marginal lands (biomass perugia). In: *Proceedings of the International Congress, 'Euroforum new energies'*, Saarbrücken, Germany, 24–28 October 1988, pp. 465–7.

Baldelli, C. (1989) Cropping bio-fuels. In: *Proceedings of the International Conference on 'Biomass for Energy and Industry'*, Lisbon, Portugal, 9–13 October 1989, Vol. I, pp. 499–502.

Baldini, S. (1987) Forest biomass for energy in EEC countries from harvesting to storage. In: Ferrero, G.L., Grassi, G. and Williams, H.E. (eds), *Biomass Energy: From Harvesting to Storage*, Elsevier Applied Science, London, pp. 22–40.
Baldini, S. (1988) Potenzialità e possibilità di recupero della biomassa forestale in Italia. In: *Accademia Economico-Agraria dei Georgofili*, Vol. 35, pp. 3–19.
Barnett, P.E. and Curtin, D.T. (1986) *Development of Forest Harvesting Technology: Application in Short Rotation Intensive Culture (SRIC) Woody Biomass*, Muscle Shoals, Alabama, USA, Tennessee Valley Authority, Technical Note B58, 89 pp.
Bartolelli, V. and Mutinati, G. (1991) Economic convenience of cotton, jojoba and broom cultivation. *Proceedings of the International Conference on 'Biomass for Energy, Industry and Environment'*, Athens, Greece, 22–26 April 1991, pp. 317–20.
Chianese, L. (1940) La ginestra e le sue specie nel sistema dell'agricoltura italiana. *Cellulosa* 2: 63–9.
Collins, R.P. and Jones, M.B. (1985) The influence of climatic factors on the distribution of C_4 species in Europe. *Vegetatio* 64: 121–9.
Harrison, F. (1992) Genista. *American Nurseryman*, Dec. pp. 57–61.
Rousseau, S. and Loiseau, P. (1982) Structure et cycle de développement des peulements a *Cytisus scoparius* dans la claine des domes. *Acta Oecologie Applica* V(2): 121–5.
Shen, S.Y. Yvas, A.D. and Jones, P.C. (1984) Economic analysis of short rotation and ultra-short rotation forestry. *Resources and Conservation* 10: 255–70.
Tabard, P. (1985) Une plante energetique a cycle court. La genet: *Cytisus scoparius*. In: *Proceedings of 3rd EC Conference on 'Energy from Biomass'*, Venice, Italy, 25–29 March 1985, pp. 283–7.
Zsuffa, L. and Barkley, B. (1984) The commercial and practical aspects of short rotation forestry in temperate regions: A state-of-the-art review. *Proceedings of Conference on 'Bio Energy'*, Goteborg, 15–21 June 1984, Vol. 1, pp. 39–57.

BROWN BEETLE GRASS (*Leptochloa fusca* Kunth)

Contributed by: A. K. Gupta

Description

Leptochloa fusca Kunth (Syn. *Diplachne fusca* (L.) Beauv) is a C_4 graminaceous plant that probably originated in salt marshes. It is a tropical grass, extensively found in Asia, central Africa (Vesey-Fitzgerald, 1963) and central Australia (Lazarides, 1970). In India and Pakistan, where it is established over many kilometres in saline-alkali habitats, it is known by the vernacular name Kallar ghas, meaning salt grass. The plant grows naturally in low, wet ground and brackish swamps. If the swamp dries up, even for a period of years, the grass mat remains permanent. The plant is an erect perennial growing to 150cm high. The culms are tufted, robust and branched. The branches are covered with narrow leaves and long loose sheaths. The grass forms deep rooted tussocks opening up the impermeable sodic soils. The roots harbour nitrogen fixing bacteria. These qualities make it a suitable crop for introduction into salt affected lands where it would not only be used as an energy crop but would also ameliorate the problematic soil.

Ecological requirements

As a tropical and C_4 plant, brown beetle grass thrives well in hot climates. It occurs over a wide range of regions receiving an average rainfall from 500mm up to 1100mm or even more in coastal areas. It is tolerant of flooding and shows excellent adaptability to saline habitats (Dabadghao and Shankarnarayan, 1973). It can grow on a variety of soils but prefers clay loams or clays (Skerman and Riveros, 1990).

Propagation

No special land preparation is required for propagating this grass. The field is ploughed once and the grass is sown. It can also be propagated without tillage, in which case the field is irrigated and the root clumps are transplanted. Vegetative propagation is preferred since the seed is difficult to handle because of its minute size, and to overcome germination problems arising from poor viability of seed. The best time for propagation is just prior to the summer season, so that the grass can grow profusely in the following summer. Field scale cultivation is done by spreading root stubbles or stem cuttings (having 2–3 nodes) on flooded land in rows with a spacing of 45cm × 45cm; then a person walks over them so that the nodes are buried. About 700kg of cuttings are needed to plant one hectare (Malik *et al.*, 1986).

Crop management

The crop does not respond to applied nitrogen and phosphorus fertilizers. However, application of zinc at 20kg/ha improves the growth in sodic soils. The water requirement of the grass is high in dry periods. Frequency of irrigation may have to be increased to twice per week depending upon weather conditions. The grass can be irrigated with brackish water. No plant protection measure is necessary as the crop is not attacked by any major pest.

Brown beetle grass is a perennial in nature and normally no rotation is recommended, but it is often used as a biological tool to reclaim alkali soils. The growth of grass helps lower the pH of the top layer of soil through the addition of organic matter from decaying roots and leaf litter and also because of the root exudates (Malik *et al.*, 1986). After the grass has been grown for a few years, it can be rotated with some less salt tolerant crops like barley, rapeseed/mustard, etc. The grass is ploughed under thoroughly and allowed to decompose for a few weeks, and later the land is prepared for the rotation crop.

Production

Because brown beetle grass is a photosynthetically efficient plant (C_4), its yield is also high. As many as five cuttings can be taken in a year, giving a total biomass yield of about 50t/ha. The growth of the plant is retarded in winter, affecting the yield of biomass, which is resumed after the rise in temperature.

Processing and utilization

The crop is harvested by mowing. The first harvest takes place when the crop is approximately 50cm high. The harvested material should be dried to make it storable and usable as a biofuel.

Table 7.9. Composition of different parts of Leptochloa fusca *(adapted from Malik* et al., *1986)*

Constituent	Roots (%)	Stem (%)	Leaves (%)
Dry matter	53	38	36
Lignin	19	16	20
Hemicellulose	5	6	5
Cellulose	16	23	19

Table 7.10. Total energy output from Leptochloa fusca.

Particulars	Amount
Green matter production	40t/ha
Dry matter production	16.8t/ha
Total digestible nutrients (TDN)	7.9t/ha
Potential for methane production (0.18m^3/kg dry matter)	3024m^3/ha/year
Sludge (0.72 kg/kg dry matter)	12t
Nitrogen in sludge	240kg
Total energy yield from the grass derived fuel	15x10^6 kcal/ha

Brown beetle grass is most commonly used as a forage crop. Its potential for making pulp, biogas and alcohol has also been recognized. A glance at Table 7.9 reveals that the biomass from this grass is a suitable substrate for microbial degradation and can be converted to alcohol. Malik *et al.* (1986) estimated the total energy obtainable from this grass (Table 7.10). In the light of the plant's great potential (63GJ/ha), technology is needed to utilize the large amount of biomass and biofuel to produce energy in the form of heat and electricity.

Selected references

Dabadghao, P.M. and Shankarnarayan, K.A. (1973) *The Grass Cover of India*. Indian Council of Agricultural Research, New Delhi, 636 pp.

Lazarides, M. (1970) *The Grasses of Central Australia*. Australian National University Press, Canberra, 384 pp.

Malik, K.A., Aslam, Z. and Naqvi, M. (1986) *Kallar grass – a plant for saline land*. Nuclear Institute for Agriculture and Biology, Faisalabad, Pakistan, 93 pp.

Skerman, P.J. and Riveros, F. (1990) *Tropical Grasses*, FAO Plant production and protection series no. 23. Food and Agriculture Organisation of the United Nations, Rome, 832 pp.

Vesey-Fitzgerald, D.F. (1963) Central African grasslands. *J. Ecol.* 57: 243–74.

BUFFALO GOURD (*Cucurbita foetidissima* Kunth ex H.B.K)

Description

The buffalo gourd is a vigorous perennial. It grows wild on wastelands in the deserts of Mexico and south-western USA. It forms large, dahlia-like tubers reaching up to 5m deep to obtain and store water. The plant is covered with a wax coating to avoid water loss. The fruits are spherical, yellow and hard shelled, and contain pulp and flat white seeds. The fruits have a diameter of about 8cm; the seeds are 12mm long and 7mm wide. Wild gourds, belonging to the family Cucurbitae, are a potential source of oil and protein. Several species are highly drought tolerant, particularly the buffalo gourd. On barren land the buffalo gourd may match the performance of traditional protein and oil sources like peanuts and sunflower, which require more water. The buffalo gourd is not yet commercially cultivated anywhere. Great yield differences occur between individual plants. The size of the fruit varies. Some plants are essentially barren, whereas others are prolific. The size of the fruit varies. Some plants have a preponderance of male, others of female flowers (NAS, 1975).

Ecological requirements

Buffalo gourd requires long periods of warm, dry weather. The optimum temperature is 20–30°C. The vines are frost sensitive and may be killed by frost, though the roots may survive winter temperatures as low as –25°C. An annual rainfall of 250mm is the minimum; the optimum is 400–600mm. The soil should be well drained with a pH of 6–7 (FAO, 1996).

Propagation

The plants can be propagated vegetatively from nodal roots. The long running vines are stapled to the soil and watered (NAS, 1975).

Crop management

The roots are reported to live for more than 40 years, but the herbaceous vines are annual (FAO, 1996). The vines are highly resistant to cucumber beetle and squash bug (NAS, 1975).

Production

Yield estimation: each fruit contains about 12g seed. On the basis of 60 fruits per plant, 1ha of buffalo gourd can produce 2.5t of seed. The seed contains 30–65% protein and 34% oil (NAS, 1975). The crude oil can be extracted by pressing or by a solvent process. Linolenic acid is present in the oil at about 50–60%. The remaining seed meal contains about 45% protein and 45% fibre, and may be used as animal fodder. The vines grow along the ground and because of their protein content of 10–13% and digestibility may have forage value (Johnson and Hinman, 1980). The plant forms a big tap root, which can weigh 30kg (70% moisture) after only two growing seasons. The root contains starch. The root, leaves and fruits contain bitter tasting glycosides. However, the starch can be separated from them by soaking in a dilute salt solution (NAS, 1975).

Processing and utilization

The fruit can be mechanically harvested. The flesh dries completely in arid areas, so threshing is possible (NAS, 1975).

Selected references

FAO (1996) Ecocrop 1 Database. Rome.
Johnson, J.D. and Hinman, C.W. (1980) Oil and rubber from arid land plants. *Science*, 208, 2 May 1980.
NAS (National Academy of Sciences) (1975) *Underexploited Tropical Plants with Economic Value*, Washington DC.

CARDOON (*Cynara cardunculus* L.)

Contributed by: J. Fernández González

Description

Cardoon is originally from the Mediterranean area, where it was known by the ancient Egyptians, Greeks and Romans. Nowadays it can be found spontaneously growing in the riverside countries of the Mediterranean, both in the continental zone and in the isles. It is also found in south Portugal, and on the Canary Islands and the Azores. Naturalized cultivars are also found in California, Mexico and in the southern countries of South America (Argentina, Chile and Uruguay), where it is known as 'Cardo de Castilla' (Castilian thistle). This species is also naturalized in Australia.

The leaves of the basal rosette are petiolate, very large (more than 50cm × 35cm), subcoriaceous, and bright green. They are usually deeply divided. Segments are ovate to linear-lanceolate, with rigid, yellow spines 15–35mm at the apex and clustered at the base. The intensity of the spiny character changes among the different varieties. The leaves on the stem are alternate and sessile. The plant can reach a height of more than 2m. The flowers are grouped in large globose capitula (up to 8cm in diameter). Involucral bracts are ovate to elliptical, gradually or abruptly narrowed into an erecto-patent spine (10–50mm × 2–6mm), which can be either glaucescent or purplish. The corolla can be blue, lilac or whitish in colour. The achenes (6–8mm × 3–4mm) are shiny and brown-spotted. The pappus can measure 25–40mm. The chromosome number is 2n=34.

According to Wiklund (1992), the species *Cynara cardunculus* may be divided into two subspecies (ssp. *flavescens* and *cardunculus*) related to the geographical distribution. The ssp. *flavescens* is found in Macronesia, Portugal and the north-west Mediterranean region, whereas ssp. *cardunculus* has mainly a central and north-east Mediterranean distribution. Naturalized cultivars in America and Australia are very similar to ssp. *flavescens*.

Ecological requirements

Cardoon is a species that belongs to the Asteraceae family (Compositae), which also includes artichoke, sunflower, safflower and Jerusalem artichoke. It is a perennial plant that during its natural cycle sprouts in autumn, passes the winter in a rosette form and in spring develops a floral scape that dries in the summer while the remnant roots stay alive. Beginning in the autumn, the buds in the upper part of the roots develop a new rosette in order to continue the cycle for several years. Thanks to its deep root system, it is able to extract water and nutrients from very deep soil zones and as a result, in non-watering conditions, using the rainwater accumulated during autumn, winter and spring, total biomass production can rise to 20–30t/ha/year dry matter, with 2–3t of seeds rich in oil (25%) and protein (20%).

Cardoon is traditionally cultivated in some areas as a horticultural plant but its cultivation cycle is completely artificial in the Mediterranean area, where it is sown at the end of spring and stays in a vegetative state during the summer . Consequently, it needs watering during this period. After a bleaching period, which usually takes about one month, it is harvested at the beginning of the winter. The enlarged petioles of the basal leaves are the commercial product.

Cardoon is a characteristic species of the Mediterranean climate. It is quite sensitive to frost in the seedling state. Winter frost may have a significant effect on the rosette's leaves both in the first and successive years. It can cause tears in the leaves; these leaves will die, but the plant remains alive, recovering from the harm as soon as the period of freezing is over.

For good development of the plants, rainfall during autumn, winter and spring months should be about 400mm or more. With lower levels of precipitation, biomass production decreases substantially.

This species requires light, deep and limy soils, with the capacity of retaining winter and spring water in the subsoil (1–3m).

Crop management

Autumn sowing should be performed as soon as conditions allow (soil is humid after rainfall) in order to let the plant develop the cold-hardier rosette before the first frost (1–2 months, depending on growth speed). Note that the crop can tolerate temperatures below –5°C once the seedlings have four leaves. Production is low during the first year but beginning with the second year production rises, reaching a steady level during the second year, depending mainly on weather conditions. Spring sowing is recommended for those areas where the first autumn frosts are very early. In this case, sowing can take place as soon as there is no more risk of frost.

Soil preparation is analogous to land preparation for cereal sowing. Before sowing cardoon, it is recommended that an adequate basal dressing is applied, depending on soil fertility. During subsequent years, use restoration fertilization to replace the nutrients exported with the harvest. Because cardoon is a great biomass producer, it is a crop that consumes a considerable amount of nutrients. It is estimated that a 20t/ha harvest of the aerial parts extracts 277kg/ha of N, 56kg/ha of P and 352kg/ha of K from the soil. These figures and the soil fertility can be used to calculate the required fertilizer dose.

Rows should be sown approximately 1m apart, though this distance may vary according to the desired density. A pneumatic sowing drill can be used. Optimal final density might be established with approximately 10,000 plants/ha. This can be increased up to 15,000 plants/ha if the ground is fresh/humid and does not lack water, or decreased to 7500 plants/ha if the winter water reserve is too low; 3 to 4kg of seeds are required for 1 hectare.

Weeds can be controlled with herbicides (trifluralin, alachlor, linuron, etc.) or by passing with the cultivator twice until the rosettes have covered the ground. This task is very important during the first year of crop establishment – that is, during sprouting and development of the seedlings – because at first a large portion of the ground is still free of the crop. As the rosettes keep growing, they cover the soil, making it harder for weeds to develop.

In the second year of cultivation, the quick regrowth and the development of a larger rosette of basal leaves in early autumn gives weeds little opportunity to establish themselves, so it can be said that weeds will be no trouble from the second year onwards.

For obvious environmental reasons, mechanical weed control is preferable. Where this is not possible a cheap, effective herbicide is a mixture of alachlor and linuron (4 litres of 48% alachlor and 1.4kg linuron per ha) dissolved in 300–400l/ha of water.

Among the main pests that might attack cardoon are aphids (*Aphis* ssp.), the stem-borer (*Gortyna xantenes* Germ.), the leaf borer (*Apion carduorum* Kirby) and the leaf miner (*Sphaeroderma rubidum* Graells), as well as cutworms (*Agrostis segetum* and *Spodoptera litoralis*) and several flies (*Agromyza* ssp., *Terellia* spp.) and moths (*Pyrameis cardui* L.). They can be treated with either specific or broad spectrum insecticides. Organophosphates work well in most of the cases, but biological control methods are available.

Among the main fungal diseases are downy mildew, powdery mildew and botrytis blight rot. Against the downy mildews, treatments based on copper or Zineb, Maneb or Captan are recommended. Powdery mildew and botrytis blight rot are successfully controlled with sulphur or Benomyl based treatments. Fungal diseases can also be controlled through no irrigation or decreased irrigation of the crop.

Production

Aerial biomass production of cardoon depends mostly on water availability during the active growing period (spring).

For a 450mm average rainfall, distributed in accordance with the Mediterranean climate pattern, the average yield of harvestable biomass is estimated at approximately 20t/ha dry matter. At harvesting time, the moisture content of the biomass is rather low (10% to 15%), because harvesting takes place when the aerial biomass is dry. The average distribution of this biomass among the different plant parts, according to data obtained at the Universidad Politécnica de Madrid over several years, is presented in Tables 7.11 and 7.12.

Table 7.11. Distribution of biomass among cardoon plant parts.

Plant part	Value (%)
Basal leaves	21.0
Stem leaves	12.1
Stems and branches	21.9
Capitula	45.0
Total	100.0

Table 7.12. The main components of the capitula as proportions of the total biomass.

Capitula parts	Value (%)
Receptacle	9.5
Bracts	13.2
Pappi	9.1
Seeds	13.2
Total	45.0

Processing and utilization

Aerial biomass is harvested in summer (from July to September), as soon as it is dry, and always before seed dissemination. Two scenarios are worth considering: harvesting seeds and other biomass separately, or together.

A combine harvester can be used if seeds are to be harvested. The remaining biomass is then swathed and baled. The whole biomass can be harvested with a self-propelled baler, otherwise two operations will be needed: cutting the biomass with a swath mower followed by baling. If the mower is not able to make rows, swathing should be performed before baling.

The dry aerial biomass of cardoon can be used as raw material for fuel in large scale combustion plants, either for electricity production or for heating applications. The heating value of the different components of *Cynara* biomass are presented in Table 7.13. The average values can vary between 3714 and 4000kcal/kg of dry biomass.

Table 7.13. Heating values of the different fractions, and of each one of them relative to cardoon total biomass.

Fraction	%	kcal/kg fraction		kcal/kg total biomass	
		GHP	LHP	GHP	LHP
Basal leaves	21.0	2655	2449	558	514
Stalk leaves	12.1	4096	3809	496	460
Stems and branches	21.9	4204	3914	921	857
Capitula (45%)					
receptacle	9.5	3605	3333	342	316
bracts	13.2	4181	3878	551	512
pappi	9.1	4353	4043	396	368
seeds	13.2	5576	5208	736	687
Total	100.0			4000	3714

These values refer to 1kg dry matter, either for each fraction (kg fraction) or for the total biomass (kg total biomass). GHP = gross heat power, LHP = low heat power.

The seeds, having a 25% oil content, appear to be an interesting source of oil; they represent a high percentage of the total harvested dry biomass (13.2%), which is about 2640kg/ha. Linoleic acid is the main component (59.0%), followed by oleic (26.7%) and palmitic (10.7%) acids.

Silmarin is present, which is an important characteristic from a nutritional point of view because it can act as a regenerator for hepatic cells. Oil is easily extracted by cool pressing (at 20/25°C), which is really useful for dietetic applications since cool pressing does not appreciably alter the oil's components.

Table 7.14 lists the main characteristics of cardoon oil with regard to its possible utilization as a fuel. The most significant characteristics are its high cetane number and the low pour point, which would be advantages for its use as an engine fuel, either unmixed in indirect injection diesels or mixed with gasoil for use in normal diesel engines.

Several laboratories have studied the possibility of using cardoon biomass for paper pulp production, among them Instituto Papelero de España (Spain), Ordinariat für Holztechnologie at Hamburg University (Germany), Departamento Florestal del Instituto Superior de Agronomía de Lisboa (Portugal) and l'Institut National

Table 7.14. Characteristics of cardoon's oil.

Characteristic	Value
Density (g/ml)	0.916
Viscosity (mm^2/s at 20°C)	95
Melting point (°C):	–21
Heat power (MJ/kg):	32.99
Cetane number	51
Flash point (°C):	350
Iodine value	125
Saponification value	194

Table 7.15. Fibre content of the different parts of cardoon biomass (% dry matter basis).

Part of the plant	Cellulose	Hemicellulose	Lignin
Thin stems	46.4	24.1	7.5
Thick stems	49.3	21.5	13.2
Average stems	47.8	22.8	10.3
Branches	41.0	21.3	5.9
Receptacle	23.6	15.9	7.1
Bracts	38.5	23.8	6.6
Pappi	59.7	26.5	2.6

Polytechnique de Tolouse (France). Although investigations are still being carried out to optimize the different processes, the outlook for this use of cardoon biomass is promising.

Table 7.15 shows the cellulose, hemicellulose and lignin content of the different parts of cardoon, excluding the leaves and the seeds.

Selected references

De Los Santos, J.A. (1987) *Aprovechamiento celulósico del cardo*. Investigación Técnica del Papel, Madrid, pp. 568–606.

Fernández, J. and Manzanares, P. (1989) *Cynara cardunculus* L., a new crop for oil, paper pulp and energy. *Proceedings 5th European Conference on Biomass for Energy and Industry*, Lisbon, 9–13 Oct 1989, Elsevier, pp. 1184–9.

Fernández, J. (1992) *Production and Utilization of* Cynara Cardunculus *L. Biomass for Energy, Paper Pulp and Food Industry*. Project JOUB0030-E C.C.E (DG XII), May 1990 to August 1992. Final report. Brussels.

Wiklund, A. (1992) The genus *Cynara* L. (Asteraceae-Carduaceae). *Botanical Journal of the Linnean Society, London* 109: 75–123.

Fernández, J. and Curt M.D. (1995) Estimated cost of thermal power from *Cynara cardunculus* biomass in Spanish conditions. Application to electricity production. *Proceedings of the 8th European Conference on Biomass for Energy, Environment, Agriculture and Industry*, Vienna 1994, Pergamon, Vol 1: 342–50.

Fernández, J. and Marquez, L. (1995) Energetisches Potential von *Cynara cardunculus* L. aus landwirtschaftlichem Anbau in Trockengebieten des Mittelmeerraumes. University of Kassel RFA. *Der Tropenlandwirt* (Journal of Agriculture in the tropics and subtropics) 53: 121–9.

CASSAVA (*Manihot esculenta* Crantz)

Contributed by: H.K. Were and S.G. Agong

Description

Cassava (*Manihot esculenta* Crantz) originated and was domesticated in South America in about 4000–2000 BC, though only recently has it been distributed worldwide. The Portuguese began importing cassava into the Gulf of Guinea in Africa in the sixteenth century. In the eighteenth century, they introduced it to the east coast of Africa and the Indian Ocean islands of Madagascar, Reunion and Zanzibar. Portuguese ships probably carried cassava to India and Sri Lanka after the middle of the eighteenth century. Cassava was little accepted at first, but since the nineteenth century it has extended rapidly across Africa and is now grown in 39 countries.

Cassava is a dicotyledonous plant belonging to the botanical family Euphorbiaceae, and, like most other members of that family, the cassava plant contains latifers and produces latex. The normal cassava plant has a chromosome number 2n=36. Polyploids are not common. Numerous cassava cultivars exist in each locality where the crop is grown. The cultivars have been distinguished on the basis of morphology (for example, by leaf shape and size, plant heights, petiole colour, etc.), tuber shape, earliness of maturity, yield, and the cyanogenic glucoside content of the roots. This last characteristic has been used to place cassava cultivars into two groups: the bitter varieties, in which the cyanogenic glucoside is distributed throughout the tuber and is at a high level, and the sweet varieties, in which the glucoside is confined mainly to the peel and is at a low level. The flesh of the sweet varieties is therefore relatively free of the glucoside, though it still contains some (Purseglove, 1968). The photosynthesis pathway is C_3 II (FAO, 1996).

Ecological requirements

The crop is cultivated throughout the year up to a maximum altitude of 2000m. It does well in a warm moist climate where mean temperatures range from 25–29°C. It performs poorly under cold climates and at temperatures below 10°C, where growth of the plant is arrested. It tolerates drought, but grows best where annual rainfall reaches 1000–2000mm. When moisture availability is low, the plant ceases growth and sheds some of its older leaves, thereby reducing its transpiring surface. When moisture is again amply available the plant quickly resumes growth and produces new leaves, making cassava a valuable crop in places where and at times when the rainfall is low or uncertain or both. It is only during the first few weeks after planting that the cassava plant is unable to tolerate drought to an appreciable extent.

The best soil for cassava cultivation is a light, sandy loam soil of medium fertility and good drainage. On clay or poorly drained soils, root growth is poor, hence the tuber to shoot ratio is considerably reduced and root rot is increased readily. Gravelly or stony soils tend to hinder root penetration so these soils and saline soils are unsuitable. Cassava can grow and yield reasonably well on soils of low fertility where production of most other crops would be uneconomical. Under conditions of very high fertility, cassava tends to produce excessive vegetation at the expense of tuber formation. Continuous light delays tuberization and lowers yields (Mogilaer *et al.*, 1967). For this reason, cassava is most productive when daylight duration is up to 12 hours a day and at latitudes between 30°S and 30°N.

Cassava tends to grow slowly and to yield poorly when grown at altitudes above 1000m; most of its present day cultivation is at lower altitudes. It does reasonably well in windy regions.

Propagation

In agricultural production, cassava is propagated almost exclusively from stem cuttings. Ripe wood with 4–6 eyes is used and the planting depth is 5–15cm. Deeper planting may be useful in dry regions, but the deeper development of the tubers makes the harvest more difficult. In nature and in the process of plant breeding, propagation by seed is quite common. The main aims of breeding are the development of types that are resistant to the mosaic virus and bacterial blight, and to further the root form and quality (such as starch content, low fibre and linamarin content). The procedure is based mainly on selection and combination of suitable local strains, but crossing with other *Manihot* species is also possible. Experiments for better storage quality are being carried out in several countries (Rehm and Espig, 1991).

Spontaneous sexual and asexual polyploids occur in cassava. Some polyploidal breeding strategies use these processes. Some of the polyploids are very vigorous. Several tetraploids have performed as well as improved varieties, and some triploids out-yielded, by over 200%, the best improved varieties, indicating that triploids are more promising than tetraploids (Hahn *et al.*, 1994). Since the introduction of improved varieties of cassava, developed from crosses between South East Asian and Latin American germplasm, yields have increased by 20–40%. The roots of the new varieties also have higher starch contents (CIAT, 1997).

In each inflorescence, the female flower opens first; the male flowers do not open until about a week later. Cross pollination is the rule. Insects are the main pollinating agents. After pollination the cassava ovary develops into a young fruit that requires 3–5 months to mature.

Crop management

In traditional agriculture, seedbed preparation for cassava planting may take various forms, but the most common are planting on unploughed land and planting on mounds. When planting is done on unploughed land, no tillage is done beyond that required to insert the stem cuttings into the soil. The soil is opened up with machete, hoe or dibble and the cutting is inserted vertically, horizontally or at an angle. Horizontal planting seems to yield best results on unploughed land (Takyi, 1974). In Kenya and Uganda, cassava is not planted on mounds but on unploughed land or well prepared land, because it is always accompanied by intercrops such as maize, beans, millets and sesame, among others.

In improved agriculture, land is first ploughed and then harrowed; thereafter cassava may be planted on the flat, on ridges or in furrows. For plantings on the flat, the cuttings are inserted directly in the earth after it has been harrowed. On ridges or furrows, the earth is ridged or furrowed after it has been harrowed. Flat plantings of cassava produce greater yields of tuber than ridge or furrow plantings (CIAT, 1976). However, flat planting is unsuitable on heavy soils because the tubers tend to rot.

The age of the stem cutting has a profound effect on the tuber yield. Generally, cuttings taken from the older, more mature, parts of the stem give a better yield than those taken from the younger portions (IITA, 1974). Therefore, stem cuttings for planting should be as mature as possible. The number of nodes on a cutting is very important because they serve as origins of shoots, and of roots if they are submerged. A cutting with one or two nodes runs the risk of failing to establish the stand if the buds or sprouts from those nodes are destroyed. For this reason, it is generally recommended that cuttings used for planting should have at least three nodes (Krochmal, 1969).

Cassava is normally planted in rows 80–100cm apart with 80–100cm between the plants, depending on cultivars and the growing conditions. Cultivars with a spreading habit should be spaced further apart than those with an upright habit. In areas of high soil fertility and high rainfall the plants tend to be luxuriant in foliage growth, so plants should be spaced further apart.

Cassava is drought tolerant except in the first few weeks after planting. It is therefore important that it should be planted at a time when there is ample moisture, and when the likelihood of further moisture supply is good. In cooler subtropical or high altitude regions, planting should be done early in the warm season to allow for ample growth before the dormant cold season. In traditional agriculture where intercropping is practised, planting is often delayed until the later part of the rainy season when the other intercrops are nearly ready for harvest.

If cultivated on slopes, intercropping with tree crops (such *Leucaena* or *Eucalyptus*) can reduce runoff and soil loss. Erosion can be 70–80% less than with mono crops of cassava, because of the better canopy coverage of the soil surface. The disadvantage of intercropping trees with cassava is that the yields fall after the first year, because the trees are very efficient in removing nutrients from the soil (Ghosh *et al.*, 1996)

Weed competition is most detrimental to cassava during the early stages until 2–3 months, when it has formed a closed canopy.

About 30 diseases of the cassava are described. In many regions the cassava is not normally affected by diseases or pests. However, in others it may be attacked by virus diseases – mosaic, brown streak and leaf curl of tobacco, resident in many parts of Africa – and bacterial diseases – *Phytomonas manihotis* (Brazil), *Bacterium cassava* (Africa) and *Bacterium solanacearum* (Indonesia).

When the soil is fertile and growth is normal, cassava removes large amounts of nutrients from the soil. Cassava has a high requirement for potassium (CIAT, 1976), and a lack of its adequate supply limits cassava yields on many soils. When the potassium level in the soil is low, the response of the crop to nitrogen or phosphorus is poor.

Excessive nitrogen may be disadvantageous, because it results in annual luxuriant shoot growth at the expense of the tuber. Also a high level of nitrogen fertilization produces a high level of cyanogenic glucosides in the tuber (de Brujin, 1971). Tuber glucoside level is greater if nitrogen is applied to the soil than if it is applied as a foliar spray.

Because of its high uptake of nutrients cassava is used in a shifting cultivation system, preferably as the last crop before bush fall. In permanent systems it is good if a green manure plant can be integrated in the crop rotation system.

Production

Cassava is the third largest source of carbohydrates for human food in the world, at an estimated annual yield of 136 million tonnes. Africa is the largest centre of production with 57 million tonnes grown on 7.5 million ha in 1985. On a world basis, average yields are estimated to be about 5–10t/ha. On research experimental stations, yields of 80t/ha have been noted.

Processing and utilization

Young tubers are known to contain much less starch than older tubers, so harvesting must be delayed until an appreciable amount of starch has accumulated in the tubers. However, as the tuber becomes older, it tends to become more lignified and fibrous and the starch content as a percentage of the dry weight of the tuber tends to decrease or remain constant.

Cassava plots are rarely harvested in one pass, either in traditional or improved agriculture, because cassava deteriorates rapidly and it can only be kept in good condition for one or two days after harvesting. The exact time of harvesting depends on the cultivar, but ranges from 10 to 30 months. Cassava may be harvested manually (using hand tools) or by machines. The stem is cut a few centimetres from the ground. Then the ground is loosened using a hoe or a machete, and a pull of the remaining stub of stem is enough to lift out the tubers. Sometimes cassava is replanted in the field as it is harvested: a piece of stem from the plant being harvested is planted in the same hole or very close by. Mechanical harvesting of cassava has met with some success in Brazil and Mexico and is being vigorously developed in other parts of world.

Cassava plays a major role in alleviating food crises in Africa because of its efficient production of food energy, year round availability, tolerance to extreme stress conditions, and suitability for present farming and food systems in Africa (Hahn and Keyser, 1985; Hahn *et al.*, 1987). Traditionally, cassava roots are processed by various methods into numerous products and utilized in various ways according to local customs and preferences.

Constraints on cassava processing range from environmental to agronomic factors. During the rainy season, sunshine and ambient temperatures are relatively low for processing cassava, particularly in the humid lowland areas where cassava is mainly grown and utilized. In other localities, particularly in savanna zones, the water that is essential for processing is not easily available. In the dry season, the soil is so hard that harvesting and peeling cassava tubers for processing are difficult and result in more losses. Furthermore, cassava root shape varies among cultivars. Roots with irregular shapes are difficult to harvest and peel by hand, resulting in great losses of usable root material. Varietal differences in dry matter content, and in starch content and quality, influence the output and quality of the processed products.

Thailand exploits the industrial prospects of cassava on a large scale. About 52% of the country's cassava now goes to starch production. About one-third of this is further processed into various modified starches. Quantities of dried cassava chips and pellets are exported as fodder (CIAT, 1997). When dried chips are produced, 98% of the linamarin is destroyed during the drying process (Rehm and Espig, 1991).

Sun drying of peeled cassava roots is practised in many parts of Africa. Peeled roots are split and cut into smaller pieces to standardize or shorten the drying process. Shortening is necessary to reduce the chance of microbial spoilage in case of humid or rainy weather or food shortage, or to comply with financial and market demands. This method reduces the cyanogenic glucoside levels from 400eq/kg dry weight to 56eq/kg (van der Grift *et al.*, 1995).

Fermentation consists of two distinct methods: aerobic and anaerobic. For aerobic fermentation, the peeled and sliced cassava roots are first surface dried for 1–2 hours and then heaped together, covered with straw or leaves and left to ferment in air for 3–4 days until the pieces become mouldy. They are then sun dried after the mould has been scraped off. The processed and dried pieces (called 'mokopa' in Uganda) are then milled into flour, which is prepared into a 'fufu' called 'kowan' in Uganda. The growth of mould on the root pieces increases the protein content of the final products by three to eight times (Amey, 1987; Sauti *et al.*, 1987).

In anaerobic fermentation, grated cassava for processing into 'gari' is placed in sacks and pressed with stones or a jack between wooden platforms. Whole roots or pieces of peeled roots for processing into fufu are placed into water for 3–5 days. During the first stage of gari production, the bacterium *Corybenacteria manihot* attacks the starch of the roots, leading to the production of various organic acids.

In the second stage, the acidic condition stimulates the growth of the mould *Geotrichum sandida*, which proliferates rapidly causing further acidification and the production of a series of esters and aldehydes that are responsible for the taste and aroma of gari (Odunfa, 1985). The optimum temperature for gari processing is 35 °C increasing to 45 °C. The roots are peeled and boiled before they are eaten, but this does not effectively remove HCN. With processed leaves, the cyanide in pounded leaves ('pond' or 'sakasaka') remains high at 8.6mg/100g, though 95.8% of the total cyanide in leaves is removed through further processing into soup.

The dried root pieces are fermented, dried and then milled into flour by pounding in a mortar or using hammer mills. The dried cassava roots (both fermented and unfermented) are often mixed in a ration of 2–3 parts cassava with one part of sorghum, millet and/or maize and milled into a composite flour. This increases food protein, and enhances palatability by improving consistency.

Tissues may be disintegrated in the presence of excess moisture during grating or fermenting in water permitting the rapid hydrolysis of glucosides, effectively reducing both free and residual cyanide in the products.

During or after fermentation of roots for gari production, the grated pulp is put into jute or polypropylene sacks, and then pressed by stones or jacked between wood platforms. In this way much of the cyanide is lost together with the liquid, through a dewatering process.

Gari is very popular in Nigeria but less so in Cameroon, Benin, Togo, Ghana, Liberia and Sierra Leone. In Brazil, this method is used for the production of 'farinha de madioca'. Fresh roots are peeled and grated, mainly by women and children using a simple traditional grater though the work is done by men if a power driven grater is used.

Besides the described uses, cassava is an energy crop with a high potential, because cassava is one of the richest fermentable substances for the production of alcohol. The fresh roots contain 30% starch and 5% sugars; the dried roots contain about 80% fermentable substances.

For processing in factories, only the outermost cork layer is removed. The starch in the skin layer is also used. The roots are crushed into a thin pulp. Saccharification is carried out by addition of sulphuric acid to the pulp in pressure cookers. Later, the pH value is adjusted with bases and then yeast fermentation is allowed. The alcohol is separated by distillation (Grace, 1977).

In Brazil, the cultivation of cassava has been intensified in recent years to produce alcohol for use in the bioethanol fuel project (Diercke, 1981). One tonne of cassava fruit can be processed to make 180 litres of ethanol (Wright, 1996). The potential yield of alcohol is 8000 l/ha, applying adequate agrotechnics. In 1978 the first Brazilian cassava alcohol distillery began production with a capacity of 60,000l per day. The alcohol production process used in this plant is very simple and is suitable even for farmers, though it is not yet sufficiently developed (Kišgeci *et al.*, 1989). Especially where water availability is limited (not enough for the cultivation of sugarcane), cassava is a preferred feedstock for ethanol production.

The agricultural wastes produced in large amounts during harvest should not be thrown away but utilized instead. It is not only specially grown crops that are suitable for energy production; agricultural wastes should be considered as an energy source too. Energy from the cut off stems and leaves can be obtained by using anaerobic fermentation to produce a methane rich biogas. Methanogenic bacteria convert about 90% of the feedstock energy into biogas. The odourless, non-toxic residue can be used as fertilizer (FAO, 1995). Anaerobic digestion of starchy by-products has several advantages, including the use of the biogas during the transformation cycle (drying), pollution control, and the agronomic use of residues. Cassava by-products mechanization has been evaluated at pilot and industrial scales. The stalk can also be used for fermentation (Kišgeci *et al.*, 1989).

Tables 7.16, 7.17, and 7.18 show, respectively, the composition of the cassava tubers, edible portion, and leaves (Onwueme, 1982).

Table 7.16. Composition of the cassava tuber.

Tissue	Composition (%)
Peel	10–20
Cork layer	0.5–2.0
Edible portion	80–90

Table 7.17. Composition of the edible portion.

Component	Composition (%)
Water	62
Carbohydrate	35
Protein	1–2
Fat	0.3
Fibre	1–2
Ash	1

Table 7.18. Composition of the leaves 100g (edible part).

Component	Amount
Water	80g
Carbohydrate	7g
Protein	6g
Fat	1g
Calcium	0.2g
Iron	0.3g
Vitamins:	
B1 (thiamin)	0.2mg
B2 (riboflavin)	0.3mg
C	200mg
A	1000 IU
Niacin	1.5mg
Calories	50

Selected references

Amey, M.A. (1987) Some traditional methods of cassava conservation and processing in Uganda. Paper presented at the Third East and Southern Africa Crops Workshop, 7–11 December 1987, Muzuzu, Malawi, 7 pp.

CIAT (1976) *Annual Report, 1975*, CIAT (Centro International de Agriculture Tropical, Cali, Colombia).

CIAT (1997) Internet: http://www.worldbank.org/html/cgiar/newsletter/june97/9ciat.html

de Brujin, G.H. (1971) *Etude du caractere cyanogenetique du manioc.* (Manihot esculenta Crantz). Mededelingen Landbowhogeschool, Wageningen, The Netherlands, pp. 17–18.

Diercke (1981) *Weltwirtschaftsatlas* 1. Deutscher Taschenbuchverlag/Westermann, Germany.

FAO (1995) *Future Energy Requirements for Africa's Agriculture*, Food and Agriculture Organization of the United Nations, Rome.

FAO (1996) Ecocrop 1 Database, Rome.

Ghosh, S.P., Mohan Kumar, B., Kabeerathumma, S. and Nair, G.M. (1996) Productivity, soil fertility and soil erosion under cassava based agroforestry systems. *Agroforesty Systems* 8: 67–82.

Grace, M.R. (1977) *Cassava Processing*, Plant production series No. 3, FAO, Rome.

Hahn, S.K. and Keyser, J. (1985) Cassava: A basic food of Africa. *Outlook on Agriculture* 4: 95–100.

Hahn, S.K., Bai, K.V., Chukwuma, Asiedu, R., Dixion, A. and Ng, S.Y. (1994) Polyploid breeding of cassava. *Acta Horticulturae* 380: 102–9.

Hahn, S.K., Mahungu, N.M., Ottoo, J.A., Msabaha, M.A.M., Lutaladio, N.B. and Dahniya M.T. (1987) Cassava and African food crisis. In: *Proceedings of the 3rd International Symposium for Tropical Root Crops*, August, 1986, pp. 24–9.

IITA (1974) *Annual Report*, IITA (International Institute of Tropical Agriculture), Ibadan, Nigeria.

Kišgeci, J. (ed.) *et al.* (1989) *Agricultural Biomass for Energy*, CNRE study No. 4, University of Novi Sad, Food and Agriculture Organization of the United Nations, vol. 21, nos 58–9, Novi Sad.

Krochmal, A. (1969) Propagation of cassava. *World Crops* 21: 193–5.

Mogilaer, I., Orioli, G.A. and Bletter, C.M. (1967) Trial to study topophysis and photoperiodism in cassava. *Bouplandia* 2: 265–72.

Odunfa, S.A. (1985) African fermented foods. In: B.J.B Wood (ed.), *Microbiology of Fermented Foods*, Vol. 2, Elsevier Applied Science, Amsterdam, pp. 155–61.

Onwueme, I.C. (ed.) (1982) *The Tropical Tuber Crops Yams, Cassava, Sweet Potato and Cocoyrus*, English Language Book Society / John Wiley, Chichester.

Osuntokum, B.O. (1981) Cassava diet, chronic cyanide intoxication and neuropathy in the Nigerian Africans. *World Rev. Nutro Diet* 26: 141–73.

Purseglove, J.W. (1968) *Tropical Crops. Dicotyledones 1*, Longmans, Green, London, 332 pp.

Rehm, S. and Espig, G. (1991) *The Cultivated Plants of the Tropics and Subtropics*. Verlag Joseph Margraf, Preise GmbH, Berlin.

Sauti, R.F.N., Saka, J.O.K. and Kumsiya, E.G. (1987) Preliminary communication research note on composition and nutritional value of cassava maize composite flour. Paper presented at the 3rd East and Southern Africa Root Crops Workshop, 7–11 December 1987, Mzuzu, Malawi, p. 6.

Takyi, S.K. (1974) Effects of nitrogen, planting method and seed bed type on yield of cassava (*Manihot esculenta* Crantz). *Ghana J. Agric. Sci.* 7: 69–73.

van der Grift, R.M., Eboug, C. and Essers, A.S.A. (1995) Solid substrate fermentation and sundrying. Two popular cassava processing techniques in Uganda. In: *Roots Newsletter* of the Southern African Root Crops Research Network (SARRNET) and the East Africa Root Crops Research Network (EARRNET), vol. 2, no. 1, September 1995, pp. 17–20.

Wright, R.M. (1996) *Jamaica's Energy: Old Prospects, New Resources*. Twin Guinep (Publishers) / Stephenson's Litho Press.

CASTOR OIL PLANT (*Ricinus communis* L.)

Description

The castor oil plant is native to East Africa, though its present day dispersal is in all warm regions of the world (Gliese and Eitner, 1995). Castor grows wild or in cultivation. Originally it was a perennial plant, a tree growing over 10m high. In the temperate zones castor grows only 1–2m high and dies off at the first frost. The agricultural cultivars are usually grown annually, particularly the dwarf cultivars (60–120cm high), which are best suited for dry areas and for mechanized cultivation. In perennial cultivation the plants are kept 2–3m high and are harvested for 2–4 years (Franke, 1985).

The shoot formation is sympodial (an inflorescence at the end of the shoot and branching under each inflorescence). The leaves are very large, have a long stalk and are formed like a hand. With most types the female flowers are in the upper part of the inflorescence, and the male in the lower. The capsule is composed of three carpels and splits off in wild forms, but it remains closed in recently bred cultivars so the whole crop can be harvested. Short day length causes the increased formation of male flowers. The seeds are egg shaped or oval, and of different sizes and colours. The seed coats cover the white, oil containing kernels. The ratio between seed coat and kernel can vary. The seeds contain 42–56% oil. The seed coats form 20–40% of the weight. Other components of the seeds, mainly concentrated in the seed coats are ricin, a protein (3% of the seed), and ricinin, an alkaloid (0.03–0.15%). The ricinin can be easily removed because of its solubility in water. The ricin produces the toxic effect of the press cakes (Lennerts, 1984). The castor oil plant uses the photosynthesis pathway C_3 II (FAO, 1996).

Ecological requirements

The best conditions for growth are in tropical summer rainfall areas. Castor grows well from the wet tropics to the subtropical dry regions. The optimum temperature is 20–25°C. The warmer the climate of the location, the higher the content of oil in the seeds, but at temperatures over 38°C seed setting is poor. Frost will kill the plant. Dwarf cultivars can be grown in summer rainfall areas with only 500mm annual precipitation, but 750–1000mm is optimal. It can be grown between the latitudes 40°S and 52°N and up to 1500–3000m elevation, but it is best grown between 300 and 1800m. Because of its deep reaching root system the castor oil plant is quite drought tolerant. Short day conditions delay the flowering, but otherwise castor has little sensitivity to day length. The soil should be a deep sandy loam. The castor oil plant has a high demand for moisture, warmth, fertility and calcium content of the soil. The range of the pH value is 5–8, with an optimal pH of 6.

The nutrient removal for 1t of seed (with capsules) is about: 30kg N, 12kg K, 4kg Ca and 3kg Mg. Unbalanced nitrogen fertilization encourages the growth of foliage at the expense of seed formation. Castor does not tolerate salt. The best performance is achieved with a relative humidity of 30–60%; ideally, humidity should be higher during vegetative growth and lower during the period of maturity and harvest. Excessive humidity can lead to increased insect problems.

Propagation

The castor oil plant is propagated by seeds. The aims of breeding are cultivars with non-opening capsules, which allow mechanized harvesting and minimize the loss of seed. Their height and capsule shattering habit mean that the wild forms are not ideal for commercial production. Improved cultivars vary in height between 1 and 4m, depending upon environment, and behave as annuals.

Crop management

The seed is sown 5cm deep in rows with 90cm between the rows and 50 cm within rows. The plants develop quite slowly; weed control is important in the first weeks. Good preparation of the soil and eradication of persistent weeds before sowing are a substantial help. In the main areas of cultivation that have low air humidity diseases do not play a major role. Harvesting can take place 4–6 months after sowing the seed. The yield can be increased by nitrogen fertilization at up to 80kg/ha, though tillage alone leads to an increase of the yields, even in the absence of nitrogen fertilization. Where water is the limiting factor for productivity, improving the productivity of the soil should involve the reduction of runoff, using ridges and furrows, and increasing the water intake rate through *in situ* moisture conversation (Uma Devi *et al.*, 1990). For castor, the bed furrow system has proved to be more beneficial than the ridge furrow system: it was observed that initial plant vigour was greater in the bed furrow system than for ridges. This could be the result of better root penetration and establishment in the 20cm high loose beds (Venkateswarlu *et al.*, 1986).

Insect pests can be controlled by sanitary measures and by crop rotation. Because of its deep root system and resistance to nematodes (*Meloidogyne* spp. and *Striga* spp.), castor is an excellent crop for growing in rotation with tobacco, cotton, maize, millet, etc. (Rehm and Espig, 1991).

Production

Oil yields peak at 1000l/ha of high quality industrial oil (Gliese and Eitner, 1995). The global average yield of seed is 0.6t/ha/year. Under more favourable conditions 1–1.5t/ha/year are possible. In the USA, seed yields of 3t/ha/year have been achieved under irrigation (Rehm and Espig, 1991).

Processing and utilization

Indehiscent cultivars are harvested by picking the capsules by hand or mechanically. Later, the capsules are threshed to separate the seeds from the fruit walls. With cultivars that burst open it is necessary to harvest the infructescence before full ripeness and then letting them dry, to avoid the loss of the seeds which are ejected if the capsules burst open (Lennerts, 1984).

The aim of cultivating castor is to produce oil for several purposes. Nowadays, the oil is almost always used industrially. The seeds are pressed or extracted with solvent to obtain castor oil. Before pressing or extraction, the seeds are usually shelled. Tables 7.19 and 7.20 show, respectively, the glyceride contents of castor oil, and some additional properties of castor oil.

Table 7.19. The glycerides making up castor oil (Diercke, 1981).

Glyceride	Amount (%)
Ricinolenic acid	80–85
Oleic acid	7
Linoleic acid	3
Palmitic acid	2
Stearic acid	1

Table 7.20. Some properties of castor oil (Römpp, 1974).

Property	Value
Density (g/ml)	0.961–0.963
(highest density of all vegetable oils)	
Melting point (°C)	−18 to −10
Iodine number	82–90
Saponification number	176–190

Most castor oil is used for technical purposes, with much smaller quantities going to medicinal and energy uses. Castor oil has some properties that make it an attractive material for chemical and technical purposes. The oil retains its viscosity at high temperatures and it is non-drying, and its adhesion is also strong. Therefore it is used as a lubricant. It does not corrode rubber, so it is used as an additive in the production of rubber and in hydraulic systems (such as brake fluids). In the plastics industry it is used as a plasticizer. It is used in cosmetic products, inks and in the textile industry as a solvent for certain colours. Polymerized castor oil gives permanent adhesion and plasticity to paint material. Because of a double bond on the ninth carbon atom and a hydroxyl group at the twelfth, ricinoleic acid is more reactive than other fatty acids. Dehydration gives a fatty acid with two conjugated double bonds, which can be polymerized. At 275°C in an alkaline medium the oil is split, forming sebacic acid and octanol. Vacuum distillation results in undecenylic acid and heptaldehyde.

The press cake is poisonous because of its ricin content and is therefore only used as fertilizer. The shells and stems are used as fuel or for mulching (Rehm and Espig, 1991). If castor is grown annually, a large amount of agricultural wastes is generated. Advanced energy production techniques such as gasification should be considered for the shells and stems.

Selected references

Diercke (1981) *Weltwirtschaftsatlas* 1. Deutscher Taschenbuchverlag/Westermann, Germany.

FAO (1996) Ecocrop 1 Database, Rome.

Franke, W. (1985) *Nutzpflanzenkunde*, 3rd edn, Georg Thieme Verlag Stuttgart, New York.

Gliese, J. and Eitner, A. (1995) *Jatropha Oil as Fuel*, DNHE-GTZ, Project Pourghère.

Lennerts, L. (1984) *Ölschrote, Ölkuchen, pflanzliche Öle und Fette*, Verlag Alfred Strothe, Hannover, Germany.

Rehm, S. and Espig, G. (1991) *The Cultivated Plants of the Tropics and Subtropics*, Verlag Joseph Margraf, Preise GmbH, Berlin.

Römpp, H. (1974) *Römpps Chemie Lexikon*, 7th edn, Franckh'sche Verlagshandlung, Stuttgart.

Uma Devi, M., Santaiah, V., Rama Rao, S., Prasada Rao, A. and Singa Rao, M. (1990) Interaction of conversion and nitrogen fertilization on growth and yield of rainfed castor. *Journal of Oilseeds Research* 7: 98–105.

Venkateswarlu, J., Vittal, K.P.R. and Das, S.K. (1986) Evaluation of two land configurations in production of pearl millet and castor beans on shallow red soils of semi-arid tropics. *International Agrophysics* 2(2): 145–52.

COCONUT PALM (*Cocos nucifera* L.)

Description

The coconut palm grows up to 25m high. The trunk is slender, unbranched and leafless, and from a height of 1.5m to the top has a diameter of 25–35cm. The crown is built of 2.5–5.5m long, 1–1.7m wide leaves. About 10–15 leaves are formed new every year, and the same number are shed. The coconut palm originated in the south Pacific region, and is distributed from the latitudes 15°N to 15°S (Rehm and Espig, 1991) at elevations below 750m (Gliese and Eitner, 1995), even at the equator. The fruits can float for weeks in salt water without losing fertility (demonstrably, as far as 4500km), so the coconut palm is found at the coasts everywhere in the tropics (Franke, 1985). It is an evergreen tree.

The inflorescences can develop in every leaf axil. Male and female flowers are formed in the same inflorescence, the female on the lower part of the panicle branches. The male flowers open 10–20 days before the female, so cross pollination is the rule. Pollination is possible by insects and by wind. It takes 12–14 months from the time of pollination to the ripeness of the fruits. Dwarf forms open their male and female flowers at the same time; for them, self-pollination is the rule (Franke *et al.*, 1988). Most have smaller fruits that ripen 3 months earlier than larger types.

The fruits are ovoid and 15–25cm long. The outer covering is a thick fibrous husk. A bony endocarp is lined with albumen and has three eyes at the apical end. Flowering begins at an age of 5–7 years (dwarf varieties 3.5–4.5 years). The palm reaches full bearing after 10–12 years, maturity at about 15 years. It will live up to 60–100 years in the wild state and 50–70 years under cultivation. Flowers and fruits are formed during the whole year. There is a 360–365 day yield cycle. The inflorescence is initiated 16 months before the spathe opens, and the nut takes about 1 year to mature from time of pollination. After 250 days, the fruit has reached the final size and the nutshell is formed; now the fruit flesh begins to develop. The photosynthesis pathway is C_3 (FAO, 1996).

Ecological requirements

The coconut palm requires a mean annual temperature in the vicinity of 26°C and low diurnal variations of temperature; this is the cause of the absence of the coconut palm in regions with higher variations, even in the equatorial zones. The lowest temperatures that are tolerated for longer periods are for the palm 10°C, for the leaves 15°C, and for the flowers 25°C. It will withstand a small amount of frost. The palm needs a large amount of sunshine (Rehm and Espig, 1991).

If sufficient groundwater is available, the coconut palm grows even in dry regions. Where it depends on rainfall, a precipitation of 1250–2000mm is regarded as necessary for optimal growth (Gliese and Eitner, 1995). The palm needs well drained and aerated soil, but otherwise it is not very demanding on the soil. Groundwater in dry areas should be available at a depth of 1.0–2.5m. The tree tolerates up to 1% salt in the groundwater. An air humidity of 84–97% is required for good production; 63% is the absolute minimum. The monthly mean should not fall below 60%.

Propagation

Whole fruits are put horizontally into a seedbed, the narrowest end downwards and covered with earth so that the top edge is above the surface. First roots are formed inside the fibrous hull. After 4–5 months, more robust roots are developed that penetrate into the soil. The major aims of breeding are earlier yielding, and resistance against 'lethal yellowing'. Very few cultivars are truly tolerant to the lethal yellowing, a disease that is linked to phytoplasma-like organisms. Several hybrids look promising, but the most tolerant material will have to be tested on a larger scale (Mariau *et al.*, 1996). To attain these goals, tall palms are crossed with dwarf types. The F1 hybrids between selected dwarf and tall growing parents have better productivity than the original strains. They are already cultivated in substantial numbers.

Propagation by callus culture is possible but has not yet reached the stage of practical application. After 4–6 months the seedlings can be transferred to their final position. The development of the dwarf types is quicker.

Methods are being developed to collect and transport samples of coconut cultivars and wild forms as embryo culture (Ashburner *et al.*, 1996). Investigations are under way to achieve vegetative propagation by *in vitro* culture techniques (Verdeil *et al.*, 1993).

Crop management

Tall and short varieties require a spacing of 9m and 6–7m, respectively (Gliese and Eitner, 1995). One hundred fruits contain: 10kg N, 2kg P, 10kg K; the use of a potassium fertilizer is unnecessary if the fibrous husks are returned to the soil. The coconut palm needs an adequate supply of Cl; chlorosis is prevented in areas far from sea by giving each plant 2kg of cooking salt per year. If there is enough rainfall, cropping between plants is possible. High yields result if legumes are planted as ground covering plants (Rehm and Espig, 1991). Especially where the soil is impoverished after decades of coconut palm monoculture, the planting of legumes between the palms helps to regenerate the soil, though soil fertility and the organic and mineral content have to be increased to support the crops. Experiments have shown that suitable plant species are *Acacia mangium* and *A. auriculiformis* (Dupuy and Kanga, 1991).

The greatest dangers to fully grown palms are lethal yellowing in Jamaica, Cuba and Haiti; Kerala disease in India; and cadang-cadang in the Philippines. The causal factors of these diseases are not yet sufficiently clear. Lethal yellowing is a mycoplasm; Malayan dwarfs are resistant against this disease. Of the fungus diseases, bud rot (*Phytophthora palmivora*) is particularly dangerous in regions with high rainfall.

Insect pests are the rhinoceros beetle (*Oryctes rhinoceros* L.), the palm weevil (*Rhynochophorus* spp.), and other leaf eating beetles and caterpillars. Sanitation (e.g. pesticides, fumigation) of the plantation is the most important action to be taken to control pests. Insecticides are only necessary for severe attacks. Rats and shrews have to be controlled, as do squirrels and bats that eat the growing fruit (Rehm and Espig, 1991).

Production

Each palm produces 30–50 nuts per year; that means 8000 nuts per hectare (Gliese and Eitner, 1995). Hybrids of dwarf and tall forms yield 200–600 smaller nuts per year.

High yields are achieved up to the 40th year, after which production markedly decreases.

Processing and utilization

Harvesting for oil production takes place when the fruit is fully ripe. The fruits are cut off with a knife attached to a bamboo pole, or someone climbs up to harvest the fruits. In Indonesia, trained monkeys are used for harvesting. They manage to pick up to 1000 nuts per day. It is also possible to wait for the nuts to fall, and then to collect them. The nuts can be stored for several weeks before processing. At the processing stage, the fibrous husks are removed and the nuts are broken. The opened nuts are dried for a short period, and then the meat is removed and dried to get copra. Drying takes place in the sun or in drying ovens. Copra is an intermediate product that can be used for oil production or for export. Copra contains 65–70% oil. Peak yield is 9t/ha/year of copra, which gives 6t/ha/year of oil (Gliese and Eitner, 1995). The copra has a dry matter content of 94% and contains 66% fat.

The copra is pressed to obtain the oil. The remaining press cake contains about 20% protein and is used as fodder. The pressing of the undried fruit flesh has not yet replaced the production and pressing of copra. The oil contains very small amounts of unsaturated fatty acids and does not turn rancid. Tables 7.21 and 7.22 show the glyceride content of coconut oil, as well as other properties. The main part of the coconut oil is used for nutritional purposes. Other parts are utilized in the soap and stearin industry.

In the tropics coconut oil is used as fuel. The oil is a source of liquid fuel that can be used for the running of special engines. Transesterified oil is a substitute for diesel. The shells are used as solid fuel – for example to run the ovens in which the fruit flesh is dried into copra. Other parts are processed to charcoal (Diercke, 1981).

Table 7.21. Composition of the coconut oil (Diercke, 1981).

Glyceride	Amount
Caprylic-, lauric- and myristic acids	50–60%
2 x lauric- and myristic acids	15–20%
(Mixed) glycerides of other fatty acids, like oleic acid and palmitic acid	Small amounts

Table 7.22. Some properties of the coconut oil (Römpp, 1974).

Property	Value
Density (g/ml)	0.88–0.90
Melting point (°C)	20–23
Iodine number	7.5–9.5
Saponification number	255–260
Acid number	2.5–10

Selected references

Ashburner, G.R., Faure, M.G., Tomlinson, D.R. and Thompson, W.K. (1996) Collection of coconut (Cocos nucifera) embryos from remote locations. *Seed Science and Technology* 24(1): 159–69.

Diercke (1981) *Weltwirtschaftsatlas* 1. Deutscher Taschenbuchverlag/Westermann, Germany.

Dupuy, B. and Kanga, A. N'G. (1991) Utilisation des acacias pour régénérer les anciennes cocoteraies. *Revue Bois et Forets des Tropiques* 230, 4th quarter.

FAO (1996) Ecocrop 1 Database, Rome.

Franke, G., Hammer, K., Hanelt, P., Ketz, H.-A., Natho, G. and Reinbothe, H. (1988) *Früchte der Erde*, 3rd edn, Urania Verlag Leipzig, Jena, Berlin.

Franke, W. (1985) *Nutzpflanzenkunde*, 3rd edn, Georg Thieme Verlag Stuttgart, New York.

Gliese, J. and Eitner, A. (1985) *Jatropha Oil as Fuel*, DNHE-GTZ, Project Pourghère.
Mariau, D., Dery, S.K., Sangaré, A., N'Cho Yavo, P. and Phillipe, R. (1996) Coconut lethal yellowing disease in Ghana and planting material tolerance. *Plantations, Re-cherche, Développement* 3(2): 105–12.
Rehm, S. and Espig, G. (1991) *The Cultivated Plants of the Tropics and Subtropics*, Verlag Joseph Margraf, Preise GmbH, Berlin.
Römpp, H. (1974) *Römpps Chemie Lexikon*, 7th edn, Franckh'sche Verlagshandlung, Stuttgart.
Verdeil J.L., Buffard-Morel, J., Magnaval, C., Huet, C. and Grosdemagne, F. (1993) Vegetative propagation of coconut (*Cocos nucifera* L.): Towards coordinated European research in conjunction with producer countries. In: *La recherche européenne au service du cocotier*, Proceedings of the Seminar held 8–10 September, Montpellier.

COMMON REED (*Phragmites australis* (Cav.) Trin. ex Steud.; syn. *P. communis* Trin.)

Description

Common reed is a perennial wetland grass that produces rhizomes. It is most often found in temperate zones, subtropics and tropical highlands growing naturally in swamps, drains and on moist headlands. The stem is erect and partially branched and has a diameter of up to 20mm. the plant is known to grow to heights of 2–4m. The smooth, flat lanceolate leaf blades are light green to grey-green in colour, and vary in length from 15–45cm and width from 1–5cm. The inflorescence is a 20–50cm long open panicle. The plant flowers between July and September. Common reed effectively prevents water erosion along drains and river banks (Kaltofen and Schrader, 1991; Tothill and Hacker, 1973).

Ecological requirements

Firm mineral clays are the optimum soils for common reed. The plant prefers damp soils rather than wet soils. The appropriate rainfall range for common reed crops is between 750 and 3000mm, with seasonally wet locations being suitable. The plant's most favoured temperature range for growth is 30–35°C with full sunlight (Skerman and Riveros, 1990). Common reed is also salt tolerant.

Weeds should not be a significant problem to the crop, because common reed is often considered a dominant weed itself.

Propagation

Common reed can be propagated using seed or rhizomes.

Crop management

In the spring, the seed can be sown or the rhizomes planted. In its most favoured habitat, sunny wetlands, there is no need for maintenance of the crop. Because common reed is an invasive plant it usually crowds out other weeds.

The plant is perennial and is often considered undesirable, so there is no established crop rotation recommendation.

Production

Although common reed has proved to have a fairly large variance in dry matter yield, ranging from 5t/ha up to 43t/ha, average yield is about 15t/ha (Hotz *et al.*, 1989). Common reed crops in Sweden have yielded 7.5–10t/ha dry matter (Björndahl, 1983).

Processing and utilization

Once the crop has established itself, harvesting can take place in the autumn. At this time the food reserves in the leaves and stems are relocated into the rhizomes and the plant has a higher dry matter content.

Traditionally, common reed has been used for grazing (but only in its very young stages), thatching, matting, primitive construction and as a solid fuel for heat production, with a heating value of 14–15MJ/kg (Hotz *et al.*, 1989). More modern applications include the use of extracts from reed wastes and rhizomes to produce alcohol and pharmaceuticals. The high cellulose content of about 40–50% and the long fibres make reed components useful for the paper and chemical industries (Skerman and Riveros, 1990; Cook, 1974).

Additionally, common reed has been planted for water purification and the protection of lake and river banks. Common reed, along with a few other plants, is currently being used as plant cover for wastewater treatment in the Czech Republic. This is part of a new constructed wetlands technology (Vymazal, 1995).

Selected references

Björndahl, G. (1983) Structure and biomass of *Phragmites* stands. Dissertation, Göteborg.
Cook, C.D.K. (1974) *Water Plants of the World*, Dr. W. Junk Publishers, The Hague.
Hotz, A., Kolb, W. and Schwarz, T. (1989) Nachwachsende Rohstoffe- eine Chance für die Landwirtschaft? *Bayerisches Landwirtschaftliches Jahrbuch* 66(1).
Kaltofen, H. and Schrader, A. (1991) *Gräser*, 3rd. edn, Deutscher Landwirtschaftsverlag, Berlin.
Skerman, P.J. and Riveros, F. (1990) *Tropical Grasses*. Rome, FAO.
Tothill, J.C. and Hacker, J.B. (1973) *The Grasses of Southeast Queensland*, University of Queensland Press, St Lucia.
Vymazal, J. (1995) Constructed wetlands for wastewater treatment in the Czech Republic – State of the art. *Water Science and Technology* 32(3): 357–64.

CORDGRASS (*Spartina* spp.)

Contributed by: D. G. Christian

Description

There are 16 species of *Spartina*. All are temperate halophytes, rhizomatous perennials and major components of salt marsh ecosystems (Figure 7.4). Distribution is mainly along the European and American seaboards of the Atlantic ocean. *S. pectinata* is also found in fresh water marshland and dryland habitats in the eastern USA from the east coast to the great plains (Hitchcock, 1935). A classification and distribution

*Figure 7.4. Cordgrass (*Spartina pectinata*), UK.*

of *Spartina* was made by Mobberley (1956), and a good summary of the distribution of species can be found in Long and Woolhouse (1979). Many species form monotypic communities dominating a habitat; roots and rhizomes help to collect and stabilize the mud. This characteristic has made it important as a marshland reclamation plant. *S. anglica* was introduced into China for this purpose (Chung, 1993).

Cordgrasses have the C_4 photosynthetic pathway. The benefits of better light conversion efficiency compared to C_3 plants make C_4 plants more productive (Monteith, 1978). Recent studies of *S. cynosuroides* and *Miscanthus* x *giganteus* have shown that the superior light conversion efficiency of C_4 photosynthesis also functions under cool (temperate) conditions (Beale and Long, 1995). As a forage plant *Spartina* has little economic value. Shoots emerge late in spring and palatability is lost early in the growing season.

Production

Three productive species of cordgrass have been investigated as potential biomass crops: *S. anglica*, *S. pectinata* and *S. cynosuroides*. *Spartina anglica Hubbard*, a fertile amphidiploid, is thought to have evolved about 1870 in Southampton Water, southern England, by a natural crossing between *S. alterniflora Lois* and *S. maritima Fern.* The hybrid *S. x townsendii* H. & J. Groves evolved from *S. anglica Hubbard* (Gray *et al.*, 1990).

The productivity of primary stands is variable and depends on location and season (Long *et al.*, 1990). The time of harvesting can affect long term production. In one study, yield fell from 16t/ha/year to 8t/ha/year over a three year period following summer harvests which resulted in the loss of some plants. The crop did not respond to fertilizer applications (Scott *et al.*, 1990). The decline in yield may be prevented by including a rest year after two years (Callaghan, 1984).

Dryland production of *S. anglica* has not been successful (Ranwell, 1972). Two *Spartina* species more suited to dryland, *S. pectinata* and *S. cynosuroides*, were successfully established in southern England and Ireland in the 1980s (Jones *et al.*, 1987; Jones and Gravett, 1988). Selected sites were on productive soils in England and on marginal soils in Ireland. The experiments were established with plants produced by ramets from a single clone. These species produce fertile seed but seed set can be variable with both species, and special methods of seed storage are required (Jones *et al.*, 1987). *S. pectinata* did not grow as well as the other species on the marginal land sites and all crops had low yields in the first season.

The yield of both species has been studied for seven years at two sites in eastern England (Potter *et al.*, 1995). Yield was low in the first year and both species produced their heaviest yields in the following year, 20t/ha and 15t/ha on a clay soil and on a lighter, less fertile soil, respectively. Although yields were less in subsequent years they showed no evidence of decline with time. Averaged over all years, *S. pectinata* yielded 12t/ha and 14t/ha for non-fertilized and fertilized treatments respectively. *S. cynosuroides* yielded less. The yield of both species was significantly greater on the heavier soil than on the lighter soil. Fertilization improved yield by a small but non-significant amount.

In another study, observation plots of *S. pectinata* were grown on a fertile peaty loam over fen clay in eastern England and on a low grade silty clay over chalk at another site approximately 100km further south. Both crops received 60kg/ha of nitrogen in the spring. Yields at the peaty loam site were 4.9t/ha and 14.0t/ha in successive years and 2.8t/ha and 7.8t/ha at the silty clay site. Yield improvement between years was approximately the same at each site, suggesting that the yield reflected differences in soil quality (Bullard *et al.*, 1995). In northern Germany, *S. pectinata* achieved a yield of 12.8t/ha/year (El Bassam and Dambroth, 1991). Although some of the studies show that fertilizer is of little importance in the long term, nutrients may need to be applied to compensate for harvest losses.

Processing and utilization

Harvesting of cordgrass occurring naturally in coastal salt marshes is not considered worthwhile, though it is technically feasible using reed harvesters (Callaghan *et al.*, 1984). With dryland planting harvesting takes place after stems have died in the winter. Dieback may not occur in mild climates. Dead stems could be cut and collected using standard grass harvesting or forage harvesting machinery, though no references to using these methods have been found. The moisture content of collected material will depend upon season and location.

Cordgrass can be considered a potential crop for producing a solid feedstock which could be converted into a wide variety of biofuels.

Selected references

Beale, C.V. and Long, S.P. (1995) Can perennial C_4 grasses attain high efficiencies of radiant energy conversion in cool climates? *Plant, Cell and Environment* 18: 641–50.

Bullard, M.J., Christian, D.G. and Wilkins, C. (1995) *Quantifying Biomass Production in Crops Grown for Energy*, Interim Report, ETSU Contract B/CR/00387/00/00, Energy Technology Support Unit, Harwell, UK.

Callaghan, T.V., Scott, R., Lawson, G.J. and Mainwaring, A.M. (1984) *An Experimental Assessment of Native and Naturalized Species of Plants as Renewable Sources of Energy in Great Britain. II. Cordgrass –* Spartina anglica. Report No. ETSU-B-1086/P-2, NERC, Cambridge Institute of Terrestrial Ecology.

Chung, C.-H. (1993) Thirty years of ecological engineering with *Spartina* plantations in China. *Ecological Engineering* 2: 261–89.
El Bassam, N. and Dambroth, M. (1991) *A Concept of Energy Plant Farms*. Institute of Crop Science, FAL Bundesallee 50, D-38116 Braunschweig, Germany.
Gray, A.J., Benham, P.E.M. and Raybould, A.F. (1990) *Spartina anglica* – The evolutionary and ecological background. In Gray, A.J. and Benham, P.E.M. (eds), Spartina anglica*: A Research Review*. ITE publication No. 2, NERC, Polaris House, North Star Avenue, Swindon SN2 1ET, UK.
Hitchcock, A.S. (1935) *Manual of the Grasses of the United States*. US Department of Agriculture miscellaneous publication No. 200, Washington DC, 1040 pp.
Jones, M.B. and Gravett, A. (1988) The productivity of C_4 perennials grown for biomass in North West Europe. *Proceedings of the Euroforum, New Energies Congress*, Saarbrücken, Germany, vol. 3, 494–6.
Jones, M.B., Long, S.P. and McNally, S.F. (1987) The potential productivity of C_4 Cordgrasses and Galingale for low input biomass production in Europe – Plant establishment. In: Grassi, G., Delman, B., Molle, J.F. & Zibetta, H., *Biomass for Energy and Industry. Proceedings of the 4th EC Conference*, Orléans, France, Elsevier Applied Science, pp. 106–10.
Long, S.P. and Woolhouse, H.W. (1979) Primary production in *Spartina* marshes. In: Jefferies, R.L. and Davy, A.J. (eds), *Ecological Processes in Coastal Environments*, Blackwell Scientific Publications, Oxford, pp. 333–52.
Long, S.P., Dunn, R., Jackson, D, Othman, S.B. and Yaakub, M.H. (1990) The primary productivity of *Spartina anglica* on an East Anglian estuary. In Gray, A.J. and Benham, P.E.M. (eds), Spartina anglica*: A Research Review*, ITE publication No. 2., NERC, Polaris House, North Star Avenue, Swindon SN2 1ET, UK.
Mobberley, D.G. (1956) Taxonomy and distribution of the genus *Spartina*. *Iowa State College Journal of Science*, 30: 471–574.
Monteith, J.L. (1978) Reassessment of maximum growth rates for C_3 and C_4 crops. *Experimental Agriculture* 14: 1–5.
Potter, L., Bingham, M.J., Baker, M.G. and Long, S.P. (1995) The potential of two perennial C_4 grasses and a perennial C_4 sedge as lignocellulosic fuel crops in NW Europe. Crop Establishment and yields in E. England. *Annals of Botany*. In press.
Ranwell, D.S. (1972) *Ecology of Salt Marshes and Sand Dunes*. Chapman and Hall, London.
Scott, R., Callaghan, T.U. and Jackson, G.J. (1990) *Spartina* as a biofuel. In Gray, A.J. and Benham, P.E.M. (eds), Spartina anglica*: A Research Review*, ITE publication No. 2, NERC, Polaris House, North Star Avenue, Swindon, SN2 1ET UK.

COTTON (*Gossypium* spp.)

Description

The origins of cotton were in Africa (as forms of *Gossypium herbaceum*). Four cotton species of the genus *Gossypium* are now under cultivation: *G. herbaceum* L., *G. arboreum* L., *G. hirsutum* L., and *G. barbadense* L.

Cotton species are potentially perennial, even though they are normally grown for only one year in modern agriculture. The plants form a strong taproot, even at the seedling stage. The roots can reach a depth of 3m. The shoot system is dimorphic and the fruiting branches are sympodial. Each flower is surrounded by three deeply divided bracts and the fibres themselves are single celled hairs. The most useful ones are long hairs (lint), which are more than 20mm long in modern cultivars. The fruit (bolls) grow very quickly after pollination. When the seeds ripen, the hairs die and their wall collapses. The wall of hairs is composed of many layers of cellulose fibres.

Ecological requirements

The optimal temperature for germination is about 25°C. For further development, 27°C is optimum. At temperatures over 40°C, and with strong insolation, the bolls will be damaged and fall off. Cotton is extremely sensitive to frost, and its cultivation is only possible where 200 frost free days can be relied on. The highest yields are achieved in dry areas under irrigation. Ripening should occur in a rainless period, because rainfall after the opening of the bolls damages the quality of the fibres. The plants are drought tolerant to some extent because of their long root system.

Cotton needs deep soil with sufficient drainage. The pH should lie between 6 and 8. The plant is relatively salt tolerant at low salinity levels.

Crop management

The nutrient uptake ability is strong and the nutrient requirements are moderate. Too much nitrogen fertilization encourages vegetative growth and extends the vegetation period. Sufficient potassium is important for attaining food fibre quality and for disease resistance. The calcium requirement is decidedly high. Cotton withstands relatively high boron concentrations in the soil, but adequate boron fertilization (by spraying) is required when its deficiency appears.

Good preparation of the land is especially important before sowing so that the seedlings (which germinate epigeally) can penetrate easily. Before sowing, the fuzz must be removed from the seed mechanically or chemically, because the seeds otherwise cling together. The row spacing lies between 50 and 120cm, the spacing within the row between 20 and 60cm, depending upon local factors. For mechanical harvesting, the spacings are reduced to 15–20cm between the rows and 8–10cm within the rows (*G. hirsutum* is recommended for this practice). The seeds should be sown no deeper than 5cm. Cotton can be sown on level soil, in furrows, or on ridges (in poorly drained soils).

The most economical method of weed control has been the application of a strip of soil herbicides over the row of seeds, and later the flaming of weeds between the rows.

Production

The yield of cotton (cotton seed) can reach 4 t/ha/year under optimal conditions, but in practice it is seldom over 2.5t/ha/year. With primitive cultivars, the yield of fibres (ginning out-turn) is 20–25%; good upland cultivars nowadays yield at least 35% and the best cultivars more than 40%.

Processing and utilization

The seeds that are left over after ginning are also a valuable product because they provide linters (5%), cotton seed oil (24%), oil cake (33%) and hulls (34%). The cake and meal contain 23–44% protein and are valuable animal fodder. The linters have high cellulose fibres, coarse threads used as a material for cushions and in papermaking. Cotton seed hulls can be used as solid fuel, roughage for livestock, as bedding and fertilizer. This waste can be successfully converted to clean hot gas. The dried cotton stems can be used as fuel.

Cotton seed oil is a valuable by-product. The cotton seeds contain 16–24% oil. The oil is light yellow and half drying. It has a high linoleic acid content and congeals between +4°C and –6°C. After refining (which also destroys toxic gossypol) it is used for nutrition, energy or industrial purposes.

If a large amount of cotton by-products are produced they can be used as energy sources. For example, stems, seed hulls and linters are cellulose rich materials, which after pyrolysis and gasification could be prime material for the industrial and energy sectors. Likewise, cotton seed oil could be transformed into biodiesel by the process of transesterification.

The production of charcoal, pellets and briquettes from cotton hulls and stems is now being developed in several countries in Africa and Asia.

Selected references

FAO (1996) Ecocrop 1 Database, Rome.

Franke, W. (1985) *Nutzpflanzenkunde*, 3rd edn, Georg Thieme Verlag, Stuttgart, New York.

Rehm, S. and Espig, G. (1991) *The Cultivated Plants of the Tropics and Subtropics*, Verlag Joseph Margraf, Priese GmbH, Berlin.

CUPHEA (*Cuphea* spp.)

Description

The *Cuphea* species are perennial or annual herbs. Cuphea is a genus with 250 undomesticated species, native to Mexico, Central and South America, and the Caribbean area. One species, *Cuphea viscosissima*, is native to the USA. Several species are adapted to temperate agriculture and have seed oils rich in capric and lauric acids (Knapp, 1990). First attempts to domesticate some species as a new oil seed crop began only a few years ago.

Cuphea is a plant of open, often disturbed, mesophytic areas, roadsides, pastures and rocky road cuttings. It is 50cm to 2m tall, erect or sprawling. The leaves are opposite or whorled and finely scabrous, especially on the underside. Stem and leaves are covered with sticky hairs. In North America it flowers primarily from July to December. The seed capsules are thin walled, the capsule and the persistent floral tube each rupturing along a dorsal line. The placenta and attached seeds emerge through the ruptures. The seeds are bilaterally compressed, and ovoid to orbicular.

Ecological requirements

The vegetative growth of *Cuphea* is aided by warm to hot weather with sufficient moisture, but seed yield decreases under hot and dry conditions. The *C. viscosissima* phenotypes would probably grow well in the mid-west and north-west USA, Southern Africa, south-eastern South America, eastern and northern Australia, and in many parts of Europe (Mediterranean area). *Cuphea* could be cultivated as a summer annual plant, and has the advantage of being a widely adapted warm season species.

Propagation

Cuphea shows characteristic features of a wild plant; domestication began only a few years ago. Wild populations of *Cuphea* are characterized by seed shattering and seed dormancy, and these problems are not yet overcome. In recently collected wild populations of *Cuphea* no differences in fruit morphology and seed retention have been observed. All were seed shattering (Knapp, 1990).

The aim of breeding is the generation of non-shattering, autofertile cultivars that do not show seed dormancy. These varieties may be the base for further improvement, like adaptation to colder climates, increased yields and improved shape. Inducing mutations is complicated because some species are polyploid. The chromosome numbers are: x=8, n=6–54. Often, mutations are not visible because in the polyploid species the effect of the mutation is masked by the non-mutated genes. The discovery of non-shattering phenotypes within the *C. viscosissima* x *C. lanceolata* f. silenoides population VL-119 were reported. This is the only known non-shattering phenotype in the genus. A fully non-dormant inbred line of *C. lanceolata* has been developed and is the first to have been reported.

This brightens the outlook for the further domestication of *Cuphea*; these are the first steps of the development of autofertile, non-dormant, non-shattering *C. viscosissima* x *C. lanceolata* f. silenoides (Knapp, 1993). The *C. viscosissima* x *C. lanceolata* hybrids are fertile. The development of autofertile, non-dormant and non-shattering germ plasm has created a basis for the development of profitable *Cuphea* cultivars.

Insect pollination is required in the allogamus species such as *C. laminuligera*, *C. lanceolata* and *C. leptopoda*. A suitable pollinator for commercial plantings has not been found. Experimental plantings are mainly pollinated by bumblebees. The long floral tubes of the allogamus species prevent honeybees from gaining access to the nectar (Knapp, 1990). Insect pollinated allogamus species of *Cuphea* probably cannot be produced commercially. Even if honeybees or some other domesticated pollinators effectively pollinate these species, their commercial use is impractical and prohibitively expensive. Until the problem of pollination is solved, it is more promising to work with autogamus species like *C. lutea*, *C. viscosissima*, *C. tolucana*, *C. wrightii* and *C. carthagenesis* (Knapp, 1990).

Production

Economical seed yields cannot be obtained without seed retention. The crop could then be harvested after a killing frost and seed yield could accumulate to the end of the growing season. The problem is the capturing of the seed, not the seed yield (Knapp, 1990). Most *Cuphea* species are characterized by sticky or glandular hairs covering their leaves, stems and flowers. This is unpleasant, but it is no barrier to the commercialization of *Cuphea*. Sticky non-shattering *Cuphea* can be harvested like other summer annuals. The hairs may be a defence against insect pests, and it has been no prior aim of breeding to remove them. The wild type crop architecture poses no problem for the harvest: the plant grows upright, and very strongly upright if planted densely. It might be useful to develop determinate flowering species, but indeterminate flowering does not cause a problem for the harvest or seed production of *Cuphea* (Knapp, 1993).

Experimental seed yields of *C. lutea* from swathing on tarpaulins and subsequent threshing ranged from 400 to 1200kg/ha at different harvest dates in 1987 in Medford, Oregon. The harvesting methods are under research and are not standardized or optimized. The yield data are useful for comparing species, but not for predicting yields under commercial cultivation. The single harvest seed yield represents about 10–40% of the yield potential of equivalent populations with seed retention (Knapp, 1990). The seed and oil yields seem to be sufficient for *Cuphea* to compete as an oilseed crop (Knapp, 1993).

Processing and utilization

Some species are rich in specific single fatty acids – for example, *C. painteri* with 73% caprylic acid, *C. carthagenensis* with 81% lauric acid, and *C. koehneana* with >95% capric acid. Some examples of the fatty acid composition are provided in Table 7.23, where they are compared with coconut, one main source of medium chain fatty acids (MCFA). No other temperate oil seed crop supplies these lipids.

As a source of lauric acid, *Cuphea* species have more to offer than coconut oil, because the concentration of lauric acid in the oil is potentially much greater. The isolation of single fatty acids should be easily accomplished, and tailor-made fatty acid compositions should be possible. Oil and protein values were determined on only a few occasions because only a few seeds have been collected from many of the species. In the tested species the oil content of the seeds varied from 16 to 42%. The protein content (N x 6.25) ranged from 15–24%.

The triglyceride analysis of *C. lanceolata* showed that the combination of fatty acids in the ester is not the result of random distribution but the specific combination of certain fatty acids (Kleiman, 1990).

Cuphea has an enormous potential as a source of renewable, safe and economical energy. The range of fatty acids in this genus is unique. Manipulated biosynthesis pathways can produce medium chain acids similar to those from coconut. It can produce unique short chain acids with low viscosity that can be used directly as a diesel

Table 7.23. Fatty acid composition of some Cuphea *seed oils (Kleiman, 1990).*

Species	Distribution (% of total fatty acids)				Others
	Caprylic acid (8:0)	Capric acid (10:0)	Lauric acid (12:0)	Myristic acid (14:0)	
C. painteri	73.0	20.4	0.2	0.3	6.1
C. hookeriana	65.1	23.7	0.1	0.2	10.9
C. koehneana	0.2	95.3	1.0	0.3	3.2
C. lanceolata		87.5	2.1	1.4	9.0
C. viscosissima	9.1	75.5	3.0	1.3	11.1
C. carthagenensis		5.3	81.4	4.7	8.6
C. laminuligera		17.1	62.6	9.5	10.8
C. wrightii		29.4	53.9	5.1	11.6
C. lutea	0.4	29.4	37.7	11.1	21.4
C. epilobiifolia		0.3	19.6	67.9	12.2
C. stigulosa	0.9	18.3	13.8	45.2	21.8
Coconut	8	7	48	18	19

fuel without chemical processing. Uneconomical and environmental unfriendly transesterification is not necessary.

The *Cuphea* species are a source of MCFA, which are useful raw materials. Caprylic acid, capric acid and myristic acid are potentially useful for industrial and nutritional purposes. Lauric acid is an important raw material for detergent products (Table 7.23).

Selected references

Kleiman, R. (1990) Chemistry of new industrial oilseed crops. In: Janick, J. and Simon, J.E. (eds), *Advances in New Crops*, Timber Press, Portland, OR, pp. 196–203.

Knapp, S.J. (1990) New temperate oilseed crops. In: Janick, J. and Simon, J.E. (eds), *Advances in New Crops*, Timber Press, Portland, OR, pp. 203–10.

Knapp, S.J. (1993) Breakthroughs towards the domestication of Cuphea. In: Janick, J. and Simon, J.E. (eds), *New Crops*, Wiley, New York, pp. 372–9.

EUCALYPTUS (*Eucalyptus* spp.)

Contributed by: C. D. Dalianis

Description

Eucalyptus species are native to Australia. During the last two centuries, eucalyptus were spread from Australia into many tropical and subtropical regions of the world. The genus *Eucalyptus*, with more than 550 species, belongs to the Myrtaceae. The name of the genus was derived from the Greek words *eu* and *kalyptos*, meaning well covered in relation to the operculum of the flower bud (Brooker and Kleinig, 1993).

A few eucalyptus species have proved to be excellent commercial crops: *Eucalyptus grandis* and *Eucalyptus tereticornis* in tropical regions and *Eucalyptus globulus* and *Eucalyptus camaldulensis* in temperate climates. The last is also extensively grown in tropical regions. *E. grandis* is a dominant species in Brazil, *E. tereticornis* in India, *E. globulus* in Spain and Portugal, and *E. camaldulensis* in Morocco and Spain. Some other important species are *E. saligna*, *E. urophylla*, *E. citriodora*, *E. viminalis* and *E. deglupta*. *E. citriodora* is being planted on a substantial scale in southern China (Eldridge *et al.*, 1994). Although several eucalyptus species are adapted to certain Mediterranean countries, the two most important species for the Mediterranean climate are *E. camaldulensis* and *E. globulus*.

Ecological requirements

Many eucalyptus species are highly adaptive and grow rapidly in a wide range of climatic conditions. Sensitivity to low temperatures is the most important environmental factor limiting the latitudinal and altitudinal range over which eucalyptus can be planted. Many eucalyptus species can be grown successfully not only in the regions where they occur naturally, but in most parts of the tropics, subtropics and warm temperate zones. About 100 eucalyptus species have been planted outside Australia with various degrees of success.

The amount and distribution of rainfall affects eucalyptus species adaptation. Some species thrive in areas with summer rainfall, others in winter rainfall areas. As a general rule, the transfer of species from a native habitat of winter rainfall areas to summer rainfall areas is unsuccessful (Hillis, 1990).

The genus as a whole grows on old and well leached soils, with a fairly low pH. Although many eucalyptus species can grow on soils of low nutrient status, especially on those deficient in nitrogen and phosphorus, eucalyptus respond well when nutrients are not limiting (Turnbull and Pryor, 1984; Hillis, 1990; Lapeyrie, 1990). Soil drainage and acidity may affect the adaptation of certain eucalyptus species. Some species have been used to assist control of waterlogging, water and wind erosion; others have shown smog tolerance, and others have proved invaluable for planting in arid regions (Hillis, 1990).

Eucalyptus camaldulensis is the most widespread in Australia as well as in several Mediterranean countries (Figure 7.5). Because of its wide distribution, several provenances have been developed that differ in performance and adaptation. As a general rule, *E. camaldulensis* is able to produce acceptable yields on relatively poor soils with a prolonged dry season, exhibits some frost resistance, tolerates periodic waterlogging and some soil salinity, and becomes chlorotic on highly calcareous soils. It is a drought resistant species and grows in areas receiving 200mm rainfall per annum, though growth is better where the annual rainfall exceeds 400mm (Turnbull and Pryor, 1984).

E. camaldulensis is a vigorous coppicer and has several uses. In many countries it is used for fuelwood, charcoal, poles, posts, shelterbelts and hardboard production. In Argentina it is used for charcoal for the iron industry. In Morocco and to a lesser extent in Spain and Portugal it is used for pulp production, though pulp quality is inferior

Figure 7.5. Short rotation and dense populations of E. camadulensis four months after harvesting (in front) and E. globulus sixteen months old (in the background).

to that of *E. globulus*. It is considered as a drought tolerant species suitable for afforestation in arid or semi-arid regions.

Eucalyptus globulus Labill. ssp. *globulus* is widely spread in certain Mediterranean countries. Adult plants attain a height of up to 55m. The juvenile leaves are opposite, sessile, stem clasping, and glaucus to bluish.

This subspecies is widely cultivated in the Iberian Peninsula and other parts of the world. It is well adapted to mild, temperate climates and high elevations in cool tropical regions. Ideal conditions in Europe are along the north-western coasts of Spain and Portugal, where the mean annual precipitation is above 900mm, the dry season is not severe and the minimum temperature above –7° C. It is considered very sensitive to moisture stress. In south-western Spain with 465mm annual precipitation and about four months dry season, it grows on deep soils with available soil moisture. On drier and shallower soils *E. camaldulensis* is superior.

Eucalyptus globulus is well adapted to a wide range of soil types. Although the best development is on deep, sandy clay soils, good growth is also attained on clay-loams and clay soils, provided they are well drained. Insufficient depth, poor drainage and salinity are limiting factors. It is considered as one of the best eucalyptus species for papermaking. It is also highly regarded for high and heavy construction, poles, piles, railway sleepers and as fuel.

Crop management

Most of the eucalyptus plantations established outside Australia are managed as coppice crops. Coppice shoots develop from dormant buds situated in the live bark or from lignotubers, buds found near the junction of root and stem in many eucalyptus species (Jacobs, 1979). Since eucalyptus regenerates itself after each harvest operation, biomass accumulation can be generated over several growth cycles with only one initial cost of site preparation and planting. A number of eucalyptus coppice readily, sometimes up to six or seven times (Hillis, 1990). The number of coppice stems per stool is influenced by the initial planting density, genotype, stump height and climatic conditions immediately before and after harvest. Total volume growth of the coppice may be greater than initial seedling growth (Dippon *et al.*, 1985).

Throughout the EU, several plants are considered as short rotation coppice plants exclusively to provide biomass for energy. In the southern EU, candidates for such a purpose are black locust (*Robinia pseudoacacia* L.) and eucalyptus. Both plants are still in the experimental stage. In Greece, both *E. camaldulensis* and *E. globulus* are being tested under various environmental and soil conditions, very short rotation cycles and dense populations. On fertile fields with a plant density of 20,000 plants/ha (1m × 0.5m) and appropriate irrigation and fertilization management, dry matter yields of two year rotating *E. camaldulensis* of up to 32t/ha/year have been obtained. After each cutting a large number of sprouts (15–25) develop from each stump. However, very quickly a few of them, usually one to three per stump, dominate; the rest either die or remain thin and stunted. In contrast, on marginal and abandoned fields after each cutting the many sprouts are competing with each other and the plantation has a bushy appearance. It is reported (Borough *et al.*, 1984) from Australia that the mean annual increment for above ground woody biomass of densely planted *E. globulus* (0.3m × 0.3m and 0.6m × 0.6m, and with 140g per tree of 3:3:2 NPK fertilization) varied from 11.0 to 14.3t/ha/year on a three year rotation cycle.

Eucalyptus have often been accused of causing nutrient depletion. However, many scientific papers (Bara, 1970; Bara *et al.*, 1985) have proved their positive effects on the soil and their relatively low levels of nutrient removal. An extensive literature review demonstrated that afforestation with eucalyptus improved soil fertility in the long term in several areas of the world (Philliphis, 1956; Karschon, 1961; Ricardo and Maderia, 1985). Miguel (1988) states that as a general rule nutrient removal by eucalyptus plantations in northern Spain is relatively light and can usually be compensated by nutrient inputs from rainfall and rock weathering. In southern Spain some nitrogen and/or phosphorus fertilization is needed.

Taking into account the large volume of biomass produced, eucalyptus are less nutrient depleting than many other agricultural crops. This is partly because the largest proportion of the biomass produced consists of lignocelluloses. The nitrogenous compounds are either absent in wood or limited in quantity, and any nitrate movement downwards, which pollutes underground waters, has an increased chance of being trapped by the rich root system of thick eucalyptus plantations. However, long term sustainability and site fertility still need to be key concerns in any eucalyptus plantation scheme.

Eucalyptus plantations have been accused of absorbing more water from the soil than any other tree species. The uptake of soil water depends mainly on the architecture of the root system and the depth of root penetration (Lima, 1993). Most eucalyptus plantations develop superficial root systems (Reis *et al.*, 1995), though some eucalyptus roots can grow to 30m in depth and extract water from 6 to 15m depth (Peck and Williamson, 1987). However, eucalyptus seem to behave as any other tree plantation or natural forest cover with respect to water dynamics and water balance of the watersheds (Couto and Betters, 1995).

Effects of eucalyptus plantations on soil water balance should also be considered from two other points of view. First, many eucalyptus plantations in certain Mediterranean regions are grown on hilly sites with sandy to sandy loam soils that are vulnerable to erosion and unsuitable for many other crops. Eucalyptus coverage of such hilly sites, beyond providing an income to the farmers and employment to local people, offer valuable protection against water and wind erosion. Second, since eucalyptus are heavy biomass producers compared to slow growing plants (for example pine trees) it is reasonable to absorb larger amounts of soil water in absolute numbers, though water use efficiency, as previously mentioned, is more or less the same. This increased amount of soil water absorption may have negative effects only if it is affecting water tables, springs or wells in lowlands; otherwise there is no reason to worry because eucalyptus are depleting the soil moisture to the benefit of producing more biomass.

Production

Productivity in Brazil is among the world's highest. Dry matter yields of up to 50t/ha/year of *Eucalyptus grandis* have been reported (Rossillo-Calle, 1987). The Brazilian average productivity for eucalyptus is estimated to be the vicinity of 25–30m^3 /ha/year. In the coastal area, average increases have been obtained in the order of 50 to 60m^3/ha. For a long time, *E. grandis* and *E. saligna* were planted almost exclusively. The area per plant is estimated at around 6m^2. The volumetric conversion rate is 1.8–2.0 steres of wood per 1m^3 of charcoal (wood's humidity between 25 to 30%) (Mangales and Rezende, 1989).

Processing and utilization

Heating values of 18.94MJ/kg for wood and 16.46MJ/kg for bark were reported. However, much higher heating values have been reported (Rockwood *et al.*, 1985) for various eucalyptus species, ranging from a low value of 19.7 up to 21.0MJ/kg dry fuel for some heavier species such as *E. paniculata* (Jacobs, 1979).

Biomass from short rotation eucalyptus plantations, beyond stems and branches, includes leaves. Therefore, the raw material for combustion is quite different from the raw material obtained from willows. A full understanding of the physical and chemical properties of eucalyptus biomass is necessary for a proper design of systems utilizing short rotation eucalyptus biomass for energy.

In a two year rotation cycle, high nitrogen concentrations were found in both *E. camaldulensis* and *E. globulus*, probably the result of the previous overfertilization of the field. Among the biomass components, the highest nitrogen concentrations were found in leaves. In contrast, the sulphur content was very small in all plant components. Consequently, harmful SO_x emissions from the combustion of eucalyptus fuel will be practically insignificant, while the nitrogen levels indicate the need for NO_x removal equipment in medium to large scale eucalyptus fuelled energy systems.

Biomass components of both *E. camaldulensis* and *E. globulus* have high heating values (Table 7.24). Ash content is also critical for the type of energy system. In both species, ash content is higher in leaves, lower in stems and intermediate in branches. In the case of *E. globulus*, the ash content of branches was almost double that of stems, and the ash content of leaves was almost double that of branches. The high ash content of leaves requires automatic ash removal equipment in combustion systems.

The fast growth rates of eucalyptus under appropriate conditions, and the good tree form characteristics (tall and slender trees with little branching) combined with adequate wood properties, make some eucalyptus species suitable for pulp production (Pereira, 1992). The use of eucalyptus wood as a source of pulp for paper manufacture has increased markedly in recent decades and this trend is likely to continue. There has been a rapid increase in the world production of chemical eucalyptus pulp, from about 40,000t/year in the early 1960s to nearly 6 million t/year in 1985 (Eldridge *et al.*, 1994).

Presently, over 3 million tonnes of eucalyptus pulp, mostly from Brazil, Spain, Portugal, South Africa and Morocco, are sold annually on the market (Borralho and Coterril, 1992).

Other uses include the less well known utilization of eucalyptus for the production of minor forest products such as floral nectar for honey, bark for tannin, and rutin and leaf oils for pharmaceutical and industrial purposes (Boland *et al.*, 1991).

Among the many uses of eucalyptus wood, probably more is used for fuel than for any other purpose.

A few eucalyptus species are increasingly being planted in some global regions to provide wood for fuel and/or charcoal production for industrial and home use (Hillis, 1990). The high density of eucalyptus wood makes it a good fuel. Eucalyptus charcoal is of major economic importance for the iron and steel industry of Brazil and for

Table 7.24. Fuel analysis of short rotation eucalyptus species.

Elementary analysis (% of dry matter)	*E. camaldulensis* Stems	Branches	Leaves	*E. globulus* Stems	Branches	Leaves
C	44.54	44.00	47.76	44.40	43.18	45.80
H	5.47	5.62	5.92	5.40	5.53	5.86
O	45.58	44.41	37.84	46.59	45.60	37.81
N	1.31	1.22	2.65	1.36	1.19	2.05
S	0.05	0.10	0.10	0.10	0.10	0.11
Residues	3.05	4.65	5.73	2.15	4.40	8.37
Proximate analysis (% of dry matter)						
Volatiles	75.17	77.15	77.29	78.33	75.51	75.50
Fixed carbon	21.85	18.24	17.15	19.67	20.08	16.26
Ash	2.98	4.61	5.59	2.00	4.41	8.24
Heating value (MJ/kg dry matter)						
Gross heating value	18.97	18.45	19.62	19.16	18.61	19.23
Net heating value	17.86	17.31	18.41	18.07	17.48	18.03

domestic use in many other countries (Eldridge *et al.*, 1994). Significant amounts of eucalyptus residues and wastes, such as bark and black liquors produced during the pulp process, are used by the pulp mills to meet their own energy needs.

Selected references

Bara, S. (1970) *Estudio sobre* Eucalyptus globulus. *I. Composición mineral de las hojas en relacion con su posicion en el arbol, la composición del suelo y la edad. Evolución del suelo por el cultivo del Eucalyptus.* IFIE, Comunicación no. 39, Madrid.

Bara, S., Rigueiro, A., Gil, M.C., Mansilla, P. and Alonso, M. (1985) *Efectos ecologicos del* Eucalyptus globulus *en Galicia: Estudio comparativo con* Pinus pinaster *y* Quercus robur. Monografias INIA no. 50, Madrid.

Boland, D.J., Brophy, J.J. and House, A.P.N. (1991) *Eucalyptus Leaf Oils. Use, Chemistry, Distillation and Marketing*, Inkata Press, Sydney.

Borralho, N.M.G. and Coterril, P.P. (1992) Genetic improvement of *Eucalyptus globulus* for pulp production. In: Pereira, J.S. and Pereira, H. (eds), *Eucalyptus for Biomass Production*, Commission of the European Communities.

Borough, C.J., Incoll, W.D., May, J.R. and Bird, T. (1984) Yield statistics. In: Hillis, W.E. and Brown, M.T. (eds), *Eucalypts for Wood Production*, Academic Press, New York.

Brooker, M.I.H. and Kleinig, D.A. (1993) *Field Guide to Eucalypts. South-eastern Australia.* Inkarta Press, Sydney.

Couto, L. and Betters, D.R. (1995) *Short Rotation Eucalypt. Plantations in Brazil: Social and Enviromental Issues*, Oak Ridge National Laboratory, Tennessee.

Dalianis, C., Christou, M., Sooter, Ch., Kyritsis, S., Zafiris, C. and Samiotakis, G. (1994) Productivity and energy potential of densely planted eucalypts in a two year short rotation. In: Hall, D.O. *et al.* (eds), *Biomass for Energy and Industry; Proceedings of the 7th Biomass Conference*, Ponte Press, Bochum, Germany.

Dalianis, C. (1995) *Improvement of Eucalypts Management.* Report submitted to EU in the framework of AIR-CT93-1678 project.

Dippon, D.R., Rockwood, D.L. and Comer, C.W. (1985) Cost sensitivity analysis of *E. grandis* woody biomass systems. In: Smith, W.H. (ed.), *Biomass Energy Development*, Plenum Press, London.

Eldridge, K.G., Davinson, J., Harwood and Wyck, G.V. (1994) *Eucalypt Domestication and Breeding*, Clarendon Press, Oxford.
Hillis, W.E. (1990) Fast growing eucalypts and some of their characteristics. In: Werner, D. and Müller, P. (eds), *Fast Growing Trees and Nitrogen Fixing Trees*, Gustav Fischer Verlag, Stuttgart and New York.
Jacobs, M.R. (1979) *Eucalypts for Planting*. FAO Forestry series 11, Rome.
Karschon, R. (1961) Soil evolution as affected by eucalyptus. Vol. 2, pp. 897–904. In: *Relatorios e documentos, sequnda conferencia mundial do Eucalypto*. Food and Agriculture Organization, São Paulo.
Lapeyrie, F. (1990) Controlled mycorrhizal inoculation of eucalyptus, results and perspectives of applied and basic research. In: Werner, D. and Müller, P. (eds), *Fast Growing Trees and Nitrogen Fixing Trees*, Gustav Fischer Verlag, Stuttgart and New York.
Lima, W.P. (1993) *Impacto Ambiental do Eucalypto*, 2nd edn, Editora da Universidade de São Paulo.
Mangales, R.J.G. and Rezende, M. (1989) Charcoal and by-products from planted forests. A Brazilian experience. Paper presented in the first workshop of CNRE on charcoal production and pyrolysis technologies held in Poros, Norway, 23–25 October 1989.
Miguel, A. (1988) Short rotation forest biomass plantations in Spain. In Hummel, F.C. *et al.* (eds), *Biomass Forestry in Europe: A Strategy for the Future*, Elsevier Applied Science, London.
Peck, A.J. and Williamson, D.R. (1987) Effects of forest cleaning on groundwater. *J. Hydrol.* 94: 47–65.
Pereira, H. (1992) The raw material quality of *Eucalyptus globulus*. In: Pereira, J.S. and Pereira, H. (eds), *Eucalyptus for Biomass Production*, Commission of the European Communities.
Philliphis, A. (1956) Proteción des los cultivos y defensa del suelo. Primeira Conferencia Mundial do Eucalypto, Food and Agriculture Organization, Rome.
Reis, M.G., Kimmins, J.P., Rezende, G.C. and Barros, N.E. (1995) Acumulo de biomassa em una sequencia de idade de *E. grandis* plantado no cerrado em duas areas com deiferents produtividades. *Rivista Arvore* 9: 149–62.
Ricardo, R.P. and Madeira, M.A.V. (1985) *Relacoes Solo-Eucalipto*. Universidade Tecnica de Lisboa.
Rockwood., D.L., Comer. C.D., Dippon, D.R. and Huffman J.B. (1985) Wood Biomass Production Options for Florida. Florida. *Agric. Exper. Sta. Bull* 856.
Rossillo-Calle, F. (1987) Brazil. A biomass society. In: Hall, D.O. and Overend, R.P. (Eds), *Biomass Regenerable Energy*, John Wiley and Sons, New York.
Turnbull, J.W. and Pryor L.D. (1984) Choice of species and seed sources. In: Hillis, W.E. and Brown, M.T. (eds), *Eucalypts for Wood Production*, Academic Press, New York.

GIANT KNOTWEED (*Polygonum sachalinensis* F. Schmidt)

Description

Knotweed is a C_3 perennial shrub that occurs in many varieties. Although knotweed has its origins in east Asia, it is often found growing naturally in meadows around rivers or on the banks of streams and creeks. It was originally brought to Europe as a fodder crop from east Asia around the middle of the nineteenth century. The plant grows to a height of 2 to 5m, depending on the variety (Kowalewski and Herger, 1992; Janiak, 1994). It has a strong hollow stalk with a diameter of up to 5cm (Janiak, 1994). The large ovate leaves are approximately 17cm long and 12cm wide. The leaves have been found to contain a compound that can be extracted and used as a prophylactic treatment on plants to induce resistance against powdery mildew (Herger *et al.*, 1988).

Flowering of the plant takes place from September to October. The underground biomass consists of a rhizome network, with finer roots developing from the rhizomes.

Ecological requirements

Results from three different test sites in Germany have shown that the site with the most soil moisture and highest nitrogen content produced the highest plant growth (Kowalewski and Herger, 1992). For large yields it is important that the crop receives sufficient irrigation and fertilization, especially if it is to be grown in an area with low levels of precipitation and low soil nutrient content. The giant knotweed has an advantage in that the strong root system often enables the plant to obtain sufficient nutrients and water from the soil even during dry years.

The plant is able to grow well through a wide pH range (4–8), but a pH value of about 7 is recommended.

Propagation

Knotweed propagation is most commonly performed using seedlings or rhizomes (Herger *et al.*, 1990). Seed taken from the plant is stored for about 14 days in the refrigerator at 1°C, then sown into pots containing well fertilized, nitrogen rich soil. The pots are stored at 20°C during the day (16 hours) and at 10°C during the night (8 hours) until the seedlings reach a height of about 5cm. At this time the seedlings are then transferred to larger pots and stored in a greenhouse under long day conditions. By the beginning of May they should have reached a height of about 60cm and are ready for planting into the field.

Plant propagation using rhizomes has not been as successful as with seedlings. Test plots using rhizomes have not shown uniform growth of the plants throughout the plots. This is believed to be a potentially greater problem if giant knotweed is ever planted on a large (farm) scale.

Crop management

Seedlings are planted at the beginning of May at densities between 1 plant per 75cm^2 and 1 plant per 1m^2 (Herger *et al.*, 1990). Greater densities result in the plants shading each other, thus causing the plant's lower leaves to die. Along with planting, a layer of mulch should also be placed on the field to help prevent weeds. Weed removal will be necessary primarily in the first year; once the crop reaches maturity it usually crowds out weeds. Any removal of weeds in the first year should be done by hand because heavy machinery may damage the developing rhizome bed.

Tests in New Zealand using varying levels of inorganic fertilizers to grow seedlings showed that increasing fertilizer levels, especially nitrogen, led to increased seedling height and a greater dry matter production from the seedlings. The highest inorganic fertilizer levels of 250kg/ha P, 50kg/ha K and 200kg/ha N led to a plant height of about 75cm and a dry matter yield of about 3.4t/ha. When only 100kg/ha N was used the seedling height was about 65cm and the yield about 2.8t/ha (Herger *et al.*, 1990).

Field observations have shown that knotweed is not significantly affected by disease, but knotweed seedlings and younger plants have shown to be attacked by rabbits and the black bean aphid (*Aphis fabae*) (Herger *et al.*, 1990).

Because knotweed is a perennial plant there is no need to establish a crop rotation. Its rhizomes mean that care should be taken to remove all residues of the knotweed crop before a different crop is planted, otherwise knotweed may appear in the new crop.

Production

Knotweed is a very high yielding crop. In Germany, harvests of the entire plant in August and September have provided dry matter yields of 20 to 30t/ha, while harvests in the winter, after the leaves have fallen off, have produced dry matter yields of 8 to 10t/ha. Above ground yields on the island of Sakhalin in Russia have been as high as 27.2t/ha, with 10.5t/ha of below ground matter (Morozov, 1979). In northern Japan, above ground yields of 11.9t/ha have been seen (Iwaki *et al.*, 1964).

Processing and utilization

As mentioned above, harvesting in the autumn produces the largest dry matter yield because leaves are also harvested, whereas harvesting in the winter produces a smaller yield but with a lower plant moisture content. If heavy machinery is used during the harvest, care should be taken to avoid damaging the underground rhizome system of the crop. This is most important during the first year, at which time it is best to harvest the crop by hand to prevent excessive rhizome damage. Starting with the second year, machinery such as a corn harvester can be used for the harvest. In addition, there is also the possibility of harvesting the crop more than once per year, in which case the successive harvests should be at least four weeks apart (Herger *et al.*, 1990).

For storage or further processing, the moisture content should be in the 8–12% range. If the crop has a low enough moisture content at the time of harvest then no additional drying is necessary, but, if not, the harvested material must be quickly dried and not left in a damp field for any long period of time, otherwise the plant will begin to ferment. The crop can be dried in a facility similar to those used for drying tobacco or grains using circulating fresh air. Once dried the crop can be pressed into bales and transported (Herger *et al.*, 1990).

Fresh or dried leaves can be processed to obtain aqueous or ethanolic extracts, which have been tested to be an effective prophylactic powdery mildew inhibitor on plants (Kowalewski and Herger, 1992). If the leaves are to be harvested for the extraction of this compound, it is advisable to wait until at least the middle of June. Research has shown that the highest amount of the active agent in the leaves is present from about this time (Kowalewski and Herger, 1992).

The unbleached plants as harvested can be processed into paper, insulation or a solid biofuel. A lighter coloured paper can be obtained through bleaching (Janiak, 1994). To be used as a solid biofuel the plant needs to have a moisture content of approximately 12% or less. The compacted plant material can then be burned to produce heat or for the production of electricity. It is possible to combine the plant material to burn with other biofuels, but more research needs to be done to determine the most beneficial combinations.

Table 7.25. Fuel attributes of giant knotweed dry matter (Vetter and Wurl, 1994).

Characteristic (%)	Value
Lignin content	18.9
Ash content	6.3
Volatiles content	75.9
Carbon	47.7
Hydrogen	6.6
Silicon dioxide	9.0
Chlorine	0.22
Nitrogen	0.54
Potassium	0.75
Sulphur	0.17
Heating value (MJ/kg)	17.2

Table 7.25 lists the fuel attributes of giant knotweed dry matter obtained during tests conducted by Vetter and Wurl (1994).

Research has shown that knotweed is able to absorb and store high amounts of heavy metals such as lead and cadmium. Selective breeding and cultivation of plants is under way to enhance this trait. It is believed that these plants could save and restore soil that is not usable as a result of toxic metal contamination. On a contaminated field, the plants would be harvested two or three times a year; then the rhizome and root system would be removed after 2 years (Haase, 1988).

Selected references

Haase, E. (1988) Pflanzen reinigen Schwermetallböden. *Umwelt* 7–8: 342–4.

Herger, G., Kowalewski, A. and Güttler, J. (1990) Untersuchungen über die Möglichkeiten eines feldmäßigen Anbaus von Knöterich-Arten mit fungiziden und insektiziden Eigenschaften und Entwicklung von Anbau-, Pflege-, Ernte- und Aufbereitungs-verfahren. Report, Institut für biologische Schädlingsbekämpfung der Biologischen Bundesanstalt für Land- und Forstwirtschaft, Darmstadt.

Herger, G., Klingauf, F., Mangold, D., Pommer, E.H. and Scherer, M. (1988) Die Wirkung von Auszügen aus dem Sachalin-Staudenknöterich, *Reynoutria sachalinensis* (F. Schmidt) Nakai, gegen Pilzkrankheiten, insbesondere Echte Mehltau-Pilze. *Nachrichtenbl. Deut. Pflanzenschutzd.* 40(4): 56–60.

Iwaki, H., Midorikawa, B. and Hogetsu, H. (1964) Studies on the productivity and nutrient element circulation in Kirigamine grassland, Central Japan. II. Seasonal change in standing crop. *Bot. Mag.* (Tokyo) 77: 918.

Janiak, B. (1994) Personal communication.

Kowalewski, A. and Herger, G. (1992) Investigations about the occurrence and chemical nature of the resistance inducing factor in the extract of *Reynoutria sachalinensis. Med. Fac. Landbouww, University of Gent*, 57/2b, p. 449–56.

Morozov, V.L. (1979) Productivity of tall herbaceous vegetation in the Far East. *Izvest. Sibir. Otd. Akad. Nauk SSSR*, Ser. Biol. Nauk, 2: 32–39; cited in: Walter, H. (1981) Über höchstwerte der produktion von natürlichen Pflanzenbeständen in N.O. Asien. *Vegetatio* 44: 37–41.

Vetter, A. and Wurl, G. (1994) Möglichkeiten der Nutzung von Topinambur and weiteren Pflanzenarten als Festbrennstoff. In: *Workshop on Environmental Aspects of Production and Conversion of Biomass for Energy*, REUR Technical Series 38: 43–53.

GIANT REED (*Arundo donax* L.)

Contributed by: C. D. Dalianis

Description

Giant reed is a wild growing plant in southern European regions (Greece, Italy, Spain, southern France, Portugal) and other Mediterranean countries. It also grows wild in other parts of the world (China, southern USA etc.). Although giant reed is a warm climate plant, certain genotypes are adapted to cooler climates and can be grown successfully as far north as the United Kingdom and Germany.

Giant reed has several attractive characteristics that make it the champion of biomass crops. Certain natural, unimproved populations give dry matter biomass yields of up to 40t/ha. This means that giant reed presents a good starting point in terms of yields, being one of the most productive among the biomass crops currently cultivated in Europe, and that it has a good chance, through selection and genetic improvement, of becoming the leading biomass crop in certain European regions.

Giant reed is also an environment friendly plant:

- Its robust root system and ground cover, and its living stems during the winter, offer valuable protection against soil erosion on slopes and erosion-vulnerable soils in southern European countries.
- It is a very aggressive plant, suppressing any other vegetation under its canopy.
- During the summer it is green and succulent, and has the ability to remain undamaged if an accidental fire (very frequent in southern EU conditions) sweeps across a giant reed plantation.
- It is an extremely pest (disease, insect, weed) resistant crop, not requiring any of the chemical inputs (pesticides) that under certain conditions pollute the environment.

Giant reed is also considered to be one of the most cost-effective energy crops, because it is perennial and its annual inputs, after establishment, are very low. Only harvesting costs will occur and, depending on site and climate, irrigation and/or fertilization costs. Giant reed is also a lodging resistant plant. All these attributes make giant reed a very attractive and promising candidate species for biomass production in European agriculture.

Giant reed, also known as Provence reed or Indian grass, is a grass that belongs to the *Arundo* genus of the Gramineae family. The *Arundo* genus consists of two reed like, perennial species, *Arundo donax* L. and *Arundo plinii* Turra. They have coarse, knotted roots, cauline, flat leaves and large, loose, plumose panicles. Their spikelets are laterally compressed with few, usually bisexual, florets. The glumes are nearly equal, as long as the florets, with 3–7 nerves. The lemmas have 3–5 nerves, with long, soft hairs on the proximal of the back. The rachilla is glabrous.

The two species are quite similar in appearance. However, there are certain morphological and growth differences. In *Arundo donax* L. the spikelets are at least 12mm long with 3–4 florets. The hairs of the lower lemma are almost the same length as the glumes and the lower lemma is two-pointed. In *Arundo plinii* Turra the spikelets

*Figure 7.6. Giant reed (*Arundo donax*) plantation in Greece.*

are not more than 8mm long with 1–2 florets. The hairs of the lower lemma are shorter than the glumes and the lower lemma is entire at the apex. *Arundo plinii* L. differs also from *Arundo donax* Turra in that the plants are shorter, usually less than 2m, with stems that are always slender and with leaves that are rigid and that stick out stiffly from the stem at a right angle or less with tips sharply pointed. In *Arundo donax* L. the chromosome number is 2n=110, 112, while in *Arundo plinii* Turra the number is 2n=72.

Giant reed is probably the largest grass species in the cool temperate regions (Figure 7.6), only exceeded in size by some of the bamboos. It is a vigorously rhizomatous perennial species with a stout, knotty rootstock. Rhizomes are long, woody, swollen in places, covered in coriaceus, scale like sheaths. The stems are stout, up to 3.5cm in diameter and up to 10m tall. The leaves are regularly alternate on the stems and the leaf blades are up to 5cm wide and up to 3.3m long. They are almost smooth, green, and scabrid at the margin. The leaf sheaths are smooth, glabrous, covering the nodes. The largest leaves and most vigorous stems are produced on plants that are cut to ground level at the end of each season.

The inflorescence is a highly branched panicle up to 60cm long, erect or somewhat drooping. Its colour is initially reddish, later turning white. In cool regions the stems will not achieve flowering size.

In the warm Mediterranean regions, the above ground giant reed parts remain viable during the winter months. If plants are not cut, in the following spring new shoots emerge at the upper part of the stem from buds located at stem nodes. After cutting a giant reed plantation, usually in autumn/winter, new growth starts early next spring. New shoots emerge from buds located on the rhizomes and they develop very rapidly. Later in the season, in June–July, peak growth rates up to 7cm per day have been observed.

In fertile fields, new shoots continue to emerge until early August under a huge, well developed canopy. These late shoots develop at a faster rate and attain the same height as the early ones, though the leaves are smaller and the stem diameter is much larger – as much as twice as large.

There are three off-type giant reeds that are used for ornamental purposes:

- *Arundo donax* 'microphylla', in which the leaves are even more glaucous and broader than the basic type, up to 9cm wide;
- *Arundo donax* 'variegata', known in the USA as *Arundo donax* 'versicolor', in which all plant parts are usually smaller and the plants very much more frost sensitive than the basic type, and the leaves are white striped, usually with broad white bands at the margins;
- *Arundo donax* 'variegata superba', a name used to distinguish a superior variegated form in which the leaves are much broader than in *Arundo donax* 'variegata' – the leaf blades are normally as much as 6.5cm wide, about 30cm long and borne on stems that attain heights of up to 1m, and the internodes are shorter and the leaves grow much closer on the stems.

Because giant reed is wild growing and entirely unknown as a cultivar or crop, the 'state-of-the-art' production knowledge is missing. This means that neither selection of wild grown genotypes nor genetic improvement has been attempted so far, and the most appropriate cultural techniques for maximizing biomass yields are unknown.

Ecological requirements

In its wild state, giant reed is usually found along river banks and creeks and on generally moist soils, where it exhibits its best growth. However, it is also found in relatively dry and infertile soils, at field borders, on field ridges or on roadsides, where it grows successfully. Giant reed can be grown on almost any soil type from very light soils to very moist and compact soils. When there is an underground water table it has the ability to absorb water from the table.

Propagation

In nature, giant reed populations spread outwards through their rhizomes' growth. Where farmers have planted giant reed on their field borders to serve as windbreaks, the plant creates problems by spreading into the fields, reducing the available crop land. In such cases, the unwanted rhizomes need to be eradicated every few years so that giant reed growth remains limited to the borders.

Giant reed is a seedless plant. For agricultural purposes it could be propagated either by rhizomes or stem cuttings. Rhizome propagation is implemented early in the spring before the new shoots start emerging in the mother plantation. Propagation by stem cuttings is implemented later in the season when the soil warms up and promotes mobilization of the node buds to develop new shoots.

Giant reed rhizomes are irregular in shape and variable in size and bud bearing. Rhizomes range from 1cm up to 10cm in diameter. Their abundant reserves promote vigorous new growth. Several buds are mobilized and up to 10 stems per rhizome may emerge by the end of the first growing period. However, propagation by rhizomes is labour intensive and very expensive. After collection, rhizomes have to be cut into pieces and sorted according to their bud bearing capacity.

It is much cheaper to use stem cuttings or whole stems. Stem cuttings consist of one node with sections of adjacent internodes. Stem cuttings could be either planted directly in the field or planted in plastic bags for transplanting into the field after they sprout. In the field, stem cuttings are covered to a depth of 4 to 8cm, depending upon the soil temperature and soil moisture.

Whole stems could be used instead of stem cuttings. Stems are laid down into soil furrows at a depth of 6 to 8cm and covered by soil.

Crop management

Giant reed has no special soil preparation requirements. A simple ploughing and/or disk harrowing is considered sufficient.

Natural populations are usually very dense: more than 50 stems per m^2 is quite common. When establishing giant reed plantations with rhizomes, care should be taken that each piece has at least one bud in order to avoid gaps in the field. Distances of 70cm between rows and 50cm within rows result in a relatively thick plantation with an average stem number of up to 10 per m^2 at the end of the transplanting growing period. In the subsequent two years stem density increases and results in a thick stand.

Giant reed plantations established with stem cuttings are much thinner at the end of the growing period. Emerging plants have to develop rhizomes in order to overwinter in case the above ground part is destroyed by low winter temperatures. Plant survival of between 70 and 82% has been reported (Jodice *et al.*, 1995a).

The biomass yields at the end of the establishment year are much lower – less than one-third of the biomass obtained by rhizome planting. because stem cuttings are much cheaper it is advisable to plant them closer within the rows.

Although giant reed can be grown without irrigation under semi-arid southern European conditions, its response to irrigation is significant. However, the effect of irrigation rates on fresh and dry matter biomass yields are insignificant. It was reported (Dalianis *et al.*, 1995) that fresh and dry matter biomass yields of giant reed, averaged over three years for autumn harvests, were respectively 59.8t/ha and 32.6t/ha for the high irrigation rate (700mm/year) and 55.4t/ha and 29.6t/ha for the low irrigation rate (300 mm/year).

Giant reed is a perennial crop that lasts for several decades and is also a high biomass yielding crop, so before establishing a new plantation it is necessary to incorporate sufficient phosphorus into the soil by ploughing – more than 200kg/ha – especially in phosphorus deficient fields. Most fields in semi-arid Mediterranean regions are rich in potassium, so potassium fertilization is not required.

Annual applications of nitrogen at up to 100kg/ha, especially in nitrogen poor soils, are recommended. Applications should be implemented before the new sprouts start emerging early in spring. However, Dalianis *et al.* (1995) reported that high nitrogen rates (240kg/ha) have no significant effect on biomass yields compared to low rates (60kg/ha). This leads to the conclusion that the application of reduced nitrogen rates is justified, at least during the initial growing periods.

Giant reed is one of the most pest resistant plants. So far, no diseases have been reported or observed. Occasionally during the early growth stages of the new sprouts, while they are still in a succulent condition, they may be attacked by *Sesamia* spp. and die. However, very soon new sprouts appear from the rhizome buds and replace the damaged ones.

Giant reed develops a huge canopy that suppresses any weed growth. Even during the establishment year there is no need for herbicide applications if rhizomes are used as planting material. However, if establishment is implemented by stem cuttings, preplanting herbicide applications help the establishment and early growth of the giant reed plantation.

Production

There are only a few references to giant reed biomass yields in the world literature. In an old study, in southern France, giant reed produced 20–25t/ha dry matter (Toblez, 1940), while in northern Italy a survey made of an almost fully grown stand of giant reed determined 35t/ha, taking into account only air dried stems (Matzke, 1988). In recent studies (Dalianis *et al.*, 1995; Jodice *et al.*, 1995a, 1995b; Morgana and Sardo, 1994), giant reed's high biomass yield potential has been confirmed.

It should be stressed that these high yields are obtained from unimproved wild populations with almost no crop management. This indicates the great biomass potential of this biomass plant for the future.

Processing and utilization

Giant reed can be harvested each year or every second year, depending on its use. For example, for constructing sun protection shelters it is usually harvested every second year because the durability of the shelters is increased. For energy or pulp production giant reed is harvested each year and new growth starts in the spring.

In southern European regions the reed could be harvested either in the autumn or in the late winter. However, it should be noted that a significant reduction in biomass yield is observed between the autumn and late winter harvests. The dry matter yield reductions are the result of losses of the leaves and many of the tops, especially if hard winters are accompanied by strong winds. Dry matter losses of up to 30% were reported by Dalianis *et al.* (1995).

In semi-arid Mediterranean climates, the moisture content of the autumn harvested plants ranges between 36% and 49%, and weather conditions are suitable for natural drying in the field after cutting. These results indicate that not only is delaying harvesting time until after November useless, but there is a danger, depending on the prevailing weather, of significant biomass losses (Dalianis *et al.*, 1995). However, the autumn harvest, especially in fertile fields and warm regions, may result in an early sprouting next spring. If a late winter frost occurs these early sprouts are usually destroyed, but are quickly replaced with new ones that emerge from buds at the rhizomes.

A significant advantage of giant reed is its good storability compared to many other biomass crops. It can be stored outdoors without any shelter protection with minor losses. Storage losses occur mainly in the leaf fraction (blades and sheaths), which represents a small percentage, about 10 to 15%, of the total biomass production. Stems can be stored with almost no losses.

Giant reed has not so far been exploited on a commercial basis throughout the world. In certain windy areas it is used as windbreak to protect other crops, while its stems are used for the construction of sun protection shelters or to make baskets. Its rot resistant stems are used as supporting poles for climbing vegetables and ornamental plants. Giant reed stems, being tough and hollow, are also used as a source of reeds for musical instruments.

Recently, giant reed has been considered as a new source of raw material for energy, pulp and/or the construction of woody building materials. Heating values of 3600kcal/kg were determined for giant reed (Dalianis *et al.*, 1994). Based on these values and the dry matter yields obtained so far, the estimated energy potential is up to 11.8toe/ha/year.

Only a few references are available concerning the possibility of exploiting giant reed for pulp production (Arnoux, 1974; Faix *et al.*, 1989). The cellulose and hemicellulose content of its stems are about 45% on a dry matter basis, while its lignin content is about 25%. Because giant reed is a pithless plant (in contrast to *Miscanthus*, for example) it is considered to be a very suitable non-wood plant for pulp production since no depithing is required.

Selected references

Arnoux, M. (1974) Recherches sur la canne de Provence (*Arundo donax* L.) en vue de sa production et de sa transformation en pate a papier. *Ann. Amelior. Plantes Paris*, 24(4): 349–76.

Dalianis, C., Sooter, C. and Christou, M. (1995) Growth, biomass productivity of *Arundo donax* and *Miscanthus sinensis* '*giganteus*'. In: Chartier *et al.* (eds), *Biomass for Energy, Environment, Agriculture and Industry. Proceedings of the 8th EU Biomass Conference*, Pergamon Press, UK, vol. 1, pp. 575–82.

Faix, O., Meier, D. and Beinhof, O. (1989) Analysis of lignocelluloses and lignines from *Arundo donax* L. and *Miscanthus sinensis* Andress., and hydroliquefaction of Miscanthus. *Biomass* 18: 109–26.

Jodice, R., Vecchiet, M., Parrini, F. and Schenone, G. (1995a) Giant reed multiplication and cultivation experiences. In: Chartier *et al.* (eds), *Biomass for Energy, Environment, Agriculture and Industry. Proceedings of the 8th EU Biomass Conference*, Pergamon Press, UK, vol. 1, pp. 686–91.

Jodice, R., Vecchiet, M., Parrini, F. and Schenone, G. (1995b) Giant reed as energy crop: Initial economic evaluation of production costs in Italy. In: Chartier *et al.* (eds), *Biomass for Energy, Environment, Agriculture and Industry. Proceedings of the 8th EU Biomass Conference*, Pergamon Press, UK, vol. 1, pp. 695–9.

Matzke, W. (1988) The suitability of arundo reed (*Arundo donax* L.) as raw material for the paper industry. *Escher Wyss Ltd. Manufacturing Programme*, pp. 40–5.

Morgana, B. and Sardo, V. (1995) Giant reeds and C_4 grasses as a source of biomass. In: Chartier *et al.* (eds), *Biomass for Energy, Environment, Agriculture and Industry. Proceedings of the 8th EU Biomass Conference*, Pergamon Press, UK, vol. 1, pp. 700–7.

Toblez, F. (1940) *Arundo donax* (Pfahlrohr) als Zellstoffquelle. *Faserforschung* 15(1): 41–2.

GROUNDNUT (*Arachis hypogaea* (L.) Merr.)

Contributed by: S. G. Agong

Description

The groundnut (*Arachis hypogaea* (L.) Meir.), also known as the peanut, is a crop of global economic significance, not only in the geographically widespread areas of its production but also in the even wider areas of its processing and consumption (Smart, 1994). The crop belongs to the genus *Arachis*, which belongs to the family Papilionacea, the pea and bean family, and consists of some 22 described and 40 or more undescribed species (Gregory and Gregory, 1976). All the species are native to tropical and subtropical South America.

There is phytogeographical evidence that cultivated *Arachis* progenitors evolved in Eurasia (Leppick, 1971). After much collection and study, Krapovickas (1968) concluded that the cultivated peanut originated in Bolivia, along the eastern slope of the Andes. Krapovickas based this conclusion on linguistic, cytological, genetic, morphological, chemical and geographical information. It is thus difficult to ascertain precisely the origin of the various species of the cultivated *Arachis*, simply because of the antiquity and generality of their worldwide distribution (Candolle, 1976).

Production

Groundnuts constitute more than half of all tropical legume seed production (Ashley, 1984). Asia and Africa have the largest areas under groundnut cultivation and the greatest production out of a world total of more than 25 million tonnes in shell (shelling percentage of about 70%) as reported by FAO (1994). It is estimated that 67% of the world's groundnut production is grown in the semi-arid tropics (SAT), almost entirely by small scale farmers. The average yield of 780kg/ha of dried pods compares unfavourably with the 2900kg/ha grown in countries with developed agricultural practices, and yields of over 3000kg/ha are not uncommon on research farms in the SAT (McDonald, 1984; Smart, 1994). This indicates good potential for improving farmers' yields.

A comparison of peanut and other crops clearly demonstrates that this crop is very rich in energy (Table 7.26).

The quantity of oil production on a crop basis shows that peanut is a key crop that would double as a cash crop as well as food crop. Groundnut is one of the food crops that can be utilized as a source for biofuel, fuels originated from vegetable oils. Biofuels are generally clean and renewable, giving the farmer a competitive alternative considering the need to sustain the agricultural systems in an environmentally sound manner.

Studies performed by Anders *et al.* (1992) can be utilized in confirming the proportions of the groundnut biomass. This biomass may be subdivided into above ground and below ground matter. Both fresh and dry yields are compared in Table 7.27.

From these data it can be estimated with respect to above ground matter that the fresh fodder (2.62t/ha) makes up approximately 52% of the total fresh biomass and the dry fodder (1.88t/ha) makes up approximately 46% of the total dry biomass.

Table 7.26. Oil production for typical oleaginous plant crops.

Crop	Oil yield (kg/ha/year)[a]
Soybean	374
Safflower	653
Sunflower	801
Peanut	887
Rape seed	999
Castor bean	1188
Babassu	1541
Coconut	2260
Oil palm	7061

[a] Based on total annual volume production worldwide.

Table 7.27. Comparison of fresh and dry matter yields of groundnut (t/ha).

Tissue	Above ground	Below ground	
	Stems + leaves	Pods	Roots[a]
Fresh	2.62	2.41	–
Dry	1.08	1.28	–

[a] Harvested but not determined.

A study of groundnut sowing dates conducted by Mayeux (1992) further divides up the above ground biomass into stems and leaves in the distribution of biomass per plant (Table 7.28).

Table 7.28. Distribution of dry mass in stems, leaves and pods with respect to sowing date (Mayeux, 1992).

Sowing date	Total dry mass (g)	Distribution of dry mass (%)		
		Stems	Leaves	Pods
Early	98.1	35.7	29.9	30.3
Mid	67.7	29.6	8.1	57.7
Late	23.2	29.7	18.1	47.0
Very late	14.7	38.8	40.8	13.6

Processing and utilization

Groundnut is an important oilseed crop and grain legume with high oil and protein contents. It is estimated that groundnuts contain 40–50% oil, which consists mainly of the glycerol esters of oleic acid (50%) and linoleic acid (25%). As the nuts contain 30% protein, they are a useful addition to a meatless diet (Vickery, 1976). Norden (1980) also reported a similar high percentage of oil (42 to 52%) and protein (25 to 32%) in groundnut seed. Furthermore, the development of vegetable oil crops in most oil deficient developing countries would help reduce import requirements. Groundnut is also used directly for food and may be processed into oil and animal cake, or used for confectionery products (Musangi and Soneji, 1967). As an edible legume, peanut is popular for its horticultural importance and used largely as peanut butter, salted table nuts or roasted in the shell. As an oil crop the main by-products are protein rich press cakes.

As a renewable energy source, groundnut was the first vegetable oil used in internal combustion engines, dating as far back as Rudolf Diesel (1858–1913) and his experiments with peanut oil, and Fujio Nagao, who achieved operation with pie oil in 1948 (Nagao *et al.*, 1948). When vegetable oils are used as fuel for diesel engines, there is usually a decrease in the smoke and particulate matter released into the atmosphere. The thermal efficiency is much better than when diesel oil is used. Renewable fuels will remain as a way to help where farm policies have failed, or as a source of energy for local consumption in tropical countries suffering from a lack of foreign fuels, or to recycle discarded oils that are difficult to dispose of.

The world community is rapidly growing and the demand for food will continue to escalate. Groundnut's suitability for the industrial production of high quality oil products cannot be underrated. Improving groundnut production in the semi-arid tropics would enhance proportionally the energy yield with respect to shells, vegetative components and nuts. The plant tissues may readily be used for human and livestock nutrition, as in the case of nuts. The shoots may readily be used for feeding livestock and as source for cheap fuel. The shell is also considered a source of cheap fuel, whereas the groundnut residues would always go into improving the soil nutrients and structure. Groundnut oil can be considered as a potential biofuel source in semi-arid regions.

Selected references

Anders, M.M., Potdar, M.V., Pathak, P. and Laryea, K.B., (1992) Ongoing production agronomy studies with groundnut at ICRISAT Center. In: *Proceedings of the Fifth Regional Groundnut Workshop for South Africa*, pp. 111–20.

Ashley, J.M. (1984) Groundnut. In: P.R. Goldsworthy and N.M. Fisher (eds), *The Physiology of Tropical Field Crops*, John Wiley & Sons, Chichester, pp. 453–94.

Candolle, A. de. (1967) *Origin of Cultivated Plants*, Hafner, New York and London, pp. 411–15.

FAO (1994) *Production Yearbook for 1993*, FAO, Rome, pp. 108–9.

Gregory, W.C. and Gregory, M.P. (1976) Groundnut. In: N.W. Simmonds (ed.), *Evolution of Crop Plants*, Longman, London, pp. 151–4.

Krapovickas, A. (1968) The origin, variability and spread of the groundnut (*Arachis hypogaea*). In: Ucko and G.W. Dimbiby (eds), *The Domestication and Exploitation of Plants and Animals*, Duckworth, London, pp. 427–41.

Leppick, E.E. (1971) Assumed gene centre of peanuts and soybean. *Econ. Bot.* 25: 199–294.

Mayeux, A. (1992) Effect of sowing time on groundnut yield in Botswana. In: *Proceedings of the Fifth Regional Groundnut Workshop for South Africa*, pp. 72–9.

McDonald, D. (1984) The ICRISAT groundnut programme. *Proceedings of the Regional Groundnut Workshop, Southern Africa*, pp. 17–30.

Musangi, R.S. and Soneji, S.C. (1967) Feeding groundnut (*Arachis hypogaea* L.) hulms to dairy cows in Uganda. *E. Afri. Forest Journal* 33: 170–4.

Nagao Fs. Ohigashi, M. Kuwal and H. Hisam (1948). *Japan Soc. Mech. Eng. Transactions* 51(354): 92.

Norden, A.J. (1980) Peanut. In: W.R. Fehr and H. Hadley (eds), *Hybridization of Crop Plants*, American Society of Agronomy and Crop Science.

Smart, J. (1994) *The Groundnut Crop: Scientific Basis for Improvement*, Chapman and Hall, London.

Vickery, M.L. (1976) *Plant Products of Tropical Africa*, Macmillan.

HEMP (*Cannabis sativa* L.)

Description

Hemp has its origins in Central Asia. It is an annual, short day, C_3 plant that is wind pollinated. It has a high cellulose and lignin content in its stems, and a high fat and protein content in its seeds. Although some varieties can reach heights of up to 5m, the average height is about 2.5m. The leaves are finger like with serrated edges. Both the stalk and the leaves are covered with glandular hairs. Its root system is dominated by a tap root. The plants vegetative period is about 100 days, with the main growth period in June and July, followed by flowering in August.

Hemp belongs to the oldest group of plants used by humans. Approximately 260,000ha of hemp are cultivated throughout the world, including about 55,000ha in Europe. The seeds contain an oil that is approximately 35% fat and 46–70% linoleic acid. The stalk contains a very strong and durable fibre. The amount of this bast fibre ranges from about 8 to 32% depending on the hemp variety; 15% fibre content is considered the cut off between the high fibre and low fibre varieties (van Soest *et al.*, 1993). The cellulose content of the stems averages about 57–60%, while the lignin content averages about 8–10%.

Because of the presence of delta-9-tetrahydrocannabinol (THC) the cultivation of hemp is completely prohibited in Germany. Hemp usually contains 8–12% THC (Long, 1995), which is produced by the glandular hairs mainly in the leafy parts of the inflorescences of female and hermaphrodite plants. For THC to be used as a drug it needs to be present at over 2%. Because of this, EU regulation permits the cultivation of hemp varieties with THC contents of 0.3% or less. The following French varieties meet this criterion: 'Fedora 19', 'Felina 34', 'Fedrina 74', 'Ferimon', 'Fibrimon 24', 'Fibrimon 56' and 'Futura' (Höppner and Menge-Hartmann, 1994).

Ecological requirements

Hemp exhibits optimal growth in a moderate climate at 13–22°C, and is sensitive to frost during germination. Although the crop thrives on most soils, it prefers deep, humus rich, calcareous soils with a good water and nitrogen supply. Sand, heavy clay and excessively wet soils are less suitable, as are compacted soils. A pH value of 7 is recommended, though slightly basic soils are also appropriate. The soil should be well prepared to a fine tilth and well supplied with water, but not excessively wet (Höppner and Menge-Hartmann, 1994). A precipitation level of 700mm is necessary for a substantial yield.

Propagation

Hemp is propagated using seed. Because of the many varieties, care should be taken when choosing the variety that it is appropriate for the climate and soil where it will be planted, and for its intended use. The varieties are classified according to their fibre, oil and THC contents, along with the time of ripening. One of the main breeding objectives currently under way is to increase the bast fibre content of the plant while still maintaining a low THC content (van Soest *et al.*, 1993).

Crop management

Sowing should be performed between the middle of April and the end of May. Approximately 35–50kg/ha of seed should be planted, and at a depth of 2–4cm. The distance between rows should be 10–20cm. These conditions will lead to a crop density of about 200–350 plants/m^2. A higher crop density, about 300 plants/m^2, is used if the crop is grown for the production of long fibres (Höppner and Menge-Hartmann, 1994). The higher density leads to a reduced production of undesired leaves. The seed has a germination period of 4 to 6 days.

Because of the plant's rapid growth, weeds have not been found to be much of a problem. Herbicides may not even be necessary. Farms in England report that after drilling the seed there is no need for spraying or top dressing: the crop is left alone until it is time to check its progress (Long, 1995).

Several diseases and pests have been known to attack hemp crops. The detrimental presence of *botrytis*, *fusarium* and *sclerotinia* has been observed, along with damage to germinating plants caused by snails and wireworms. The plant fibres may suffer damage from *Spherella cannabis* and *Phoma herbarum*.

Table 7.29 shows the fertilizer levels for hemp as recommended by the Federal Agricultural Research Centre in Germany.

Table 7.29. Recommended fertilizer amounts for hemp cultivation.

Fertilizer	Amount (kg/ha)
Nitrogen (N)	60–100[a]
Phosphorus (P_2O_5)	70–100
Potassium (K_2O)	150–180

[a] Divided into two or three applications.

Because of its deep and extensive root system, hemp makes a good preceding crop in a crop rotation. In England a crop rotation of wheat and hemp has proved to be appropriate (Long, 1995). Hemp can also follow itself in a crop rotation.

Production

Seeding rate research on hemp in The Netherlands led to stem dry matter yields in 1988 and 1989 of 11.9t/ha and 13.6t/ha respectively. The tests concluded that final harvest was not affected by seeding rate (Meijer *et al.*, 1995). Under optimal circumstances a fibre yield of between 2 and 5t/ha could be obtained. Research in Germany has produced similar results, showing whole plant dry matter yields of 8–16t/ha and fibre yields of 2–4t/ha (Höppner and Menge-Hartmann, 1994).

A typical hemp harvest consists of 52% usable refuse, 31% fibre (40% long and 60% short), 9% non-usable refuse and 8% seeds (Karus and Leson, 1995).

The subsidy for hemp cultivation in the European Union in 1995 was 774.74 Ecu/ha. This subsidy is recalculated yearly.

Processing and utilization

Harvesting takes place in September, depending on the variety planted and its intended use. Research is under way to develop an appropriate hemp harvester. Although success has been achieved using a row independent maize harvester, what has been deemed necessary is a harvester that not only cuts the stalks but also removes the fibres. Research to develop an appropriate harvester is currently in progress.

In England, success was achieved with hemp grown for fibre production by cutting the crop in August and turning it twice in a six week period before it was baled (Long, 1995). While the crop is in windrows, the presence of rain helps to loosen the fibres.

The many uses of hemp are related to the four main parts of the plant: bast fibres, leaves, seeds, and processable remains. The bast fibres can be used by the textile industry for making products ranging from clothing to carpet. The fibre can be used by the paper industry to produce a wide range of paper products of varying degrees of quality. The fibres can also be processed to make building materials. The leaves can be used for animal bedding, mulches and composts. The seeds can be used as bird feed and as a cereal ingredient or, after the oil has been extracted, they can be used for animal food and high protein meal. The oil can be processed into products such as fuel, paints, machine oils, soap, shampoo and cosmetics, and is also processed by the food industry – because of the high fat and linoleic acid contents – into edible oil, margarine and food additives. The processable remains can be used to make a range of paper products and construction materials, or, like the leaves, can go to make animal bedding, mulches and composts.

The entire plant can also be used as a solid fuel when compacted (Karus and Leson, 1995). It can be converted to other biofuels by a variety of conversion technologies. The oil from hemp can be used as biodiesel. Hemp's high total biomass means that productivity can be as high as 18t/ha. Hemp can be considered as a potentially important energy crop species. Special oil hemp varieties can produce up to 600l/ha of oil.

Selected references

Bramm, A. and Bätz, W. (eds) (1988) Hanf. In: *Industriepflanzenbau: Produktions- und Verwendungsalternativen*, Bundesminister für Ernährung, Landwirtschaft und Forsten, Bonn, pp. 20–1.

Höppner, F. and Menge-Harmann, U. (1994) Anbauversuche zur Stickstoffdüngung und Bestandesdichte von Faserhanf. *Landbauforschung Völkenrode* 44(4): 314–24.

Karus, M. and Leson, G. (1995) Nova-Institute, Cologne Germany. Cited in: *Textilforum* 2/95, p. 26.

Long, E. (1995) Unravelling new fine fibre markets for hemp. *Arable Farming*, November 14, pp. 29 and 32.

Meijer, W.J.M., Vanderwerf, H.M.G., Mathijssen, E.W.J.M. and Vandenbrink, P.W.M. (1995) Constraints to dry matter production in fibre hemp (*Cannabis sativa* L.). *European Journal of Agronomy* 4(1): 109–17.

van Soest, L.J.M., Mastebroek, H.D. and de Meijer, E.P.M. (1993) Genetic resources and breeding: A necessity for the success of industrial crops. *Industrial Crops and Products* 1: 283–8.

JATROPHA (PHYSIC NUT) (*Jatropha curcas* L.)

Contributed by: A. Riedacker and S. Roy

Description

Jatropha is a shrub or tree that grows to a height of 1.5–8m. It is presumed to be native to South America, but nowadays it is common worldwide in the tropics and parts of the subtropics (Münch and Kiefer, 1989). The 6–15cm long leaves are lobed three or five times and are sometimes heart shaped; young leaves often are reddish.

Male and female flowers are produced on the same plant but at separate sites on the plant. Flowering probably does not depend on the day length, but on the climatic conditions and the variety of the plant. Short day length causes an increased formation of male flowers. In regions with dry seasons, flowering is associated with the rainy season (flowering at the beginning and the end of the rainy season). In the continuous wet tropics, jatropha flowers during the whole year.

The plant is well adapted to drought (stem and root succulence, shedding of leaves) (Heller, 1992). Pollination is by insects (Münch and Kiefer, 1989). The fruit is a capsule containing three seeds. Ripe fruits change colour from green to yellow and open a little. The castor like seeds are black, 1.5–2cm long and 1cm wide, and contain 33% oil. The embryo is embedded between two oil containing halves. The ploidity is 2n=22 (Heller, 1992).

The genus *Jatropha* contains 160 species, most of which can be crossed with *Jatropha curcas*. This opens up opportunities for the breeding of hybrids – for example, some hybrids showed earlier flowering or more strength than *Jatropha curcas* and the other ancestor. Many varieties of jatropha are known, with yellow, pink or green flowers, red or green leaves, poisonous or non-poisonous seed, different ploidity, and different composition of the oil. This is an interesting starting point for breeding (Münch and Kiefer, 1989).

Ecological requirements

Jatropha grows well in arid regions (Lutz, 1992). The soil requirements of jatropha are modest; it can grow well on oligotrophic soil. Well drained soils are preferred. It will not grow on saturated soils. Jatropha enters symbiosis with root fungi (mycorrhizae) as a defence against phosphate deficiency (Münch and Kiefer, 1989). It requires very little water and can survive long periods of drought by shedding most of its leaves in order to minimize the loss of water. Jatropha grows in regions with wide temperature and rainfall ranges. Jatropha has been reported growing between 480 and 2380mm annual rainfall, but for high yields the plant needs 625 to 750mm precipitation (Jones and Miller, 1991).

Drought for several years causes damage. At first, fewer fruits are developed; later, fruit development stops and in the end the plant dies. Jatropha prefers warm temperatures, but manages with lower temperatures, and even slight frost if the plant is tough. Seeds developing under dry climatic conditions have an elevated oil content.

Propagation

There are generative and vegetative methods of propagation, depending on the local situation. The seeds should be selected with attention to size and weight. They are drilled 2–3cm deep. Sowing takes place prior to or at the beginning of the rainy season, or in the perennial humid tropics at any time of the year.

Cuttings should be set well before the beginning of the rainy season – that is, before the plants begin to sprout. The cuttings, with a diameter of 3–4cm and a length of 0.4–0.5m or 1–1.5m, are taken from the base of the branch and planted at a depth of 20cm. The planting distances depend on the soil quality, moisture and the intended use, and can vary widely from 20cm to about 2.5m. Seeds can be planted *in situ* or in nursery beds with later replanting, or wild seedlings can be transplanted. The cuttings too can be planted directly into place or in nursery beds. Precultivated plants and direct planted cuttings show higher rates of survival than directly seeded plants. *In vitro* propagation of jatropha is possible. Regenerated shoots could be rooted and, after hardening, transferred to soil (Sujatha and Mukta, 1991).

Crop management

The plants are cultivated as hedges to prevent erosion while enhancing soil productivity (Rehm and Espig, 1991). One goal is to stimulate the production of jatropha nuts for oil extraction. Only the seedling needs careful nurturing. Young plants are fragile in the rainy season, and in the dry season they have to be protected from termites; 30% of them have to be replanted (Lutz, 1992). Cultivation methods are well known and, in part, long established in many countries. The useful life of jatropha is 50–60 years. Jatropha is very tolerant to pests, the most common of which include scale and wax lice (*Coccina*). The jatropha shows a vigorous root development and good infiltration, and hence improves the soil's water absorbing capacity. It is suitable for soil stabilization and to hinder erosion on slopes and on steep inclines. Jatropha is also a pioneer plant that can be used to cultivate wasteland. Hedges are planted to protect the fields from cattle and other animals and to stop the deterioration of the ecosystem, to stop the loss of agricultural land. Goats do not eat the leaves and twigs because of their content of bitter and poisonous constituents.

Jatropha belongs to the Euphorbiaceae family, like other important cultivated plants such as cassava and castor. Some pests are common to these, which may be important for plant protection if jatropha is planted on a larger scale. In Brazil the pathogen *Spaceloma manihoticola*, a cassava disease, infests jatropha too (Münch and Kiefer, 1989). Fungal diseases of the fully grown trees occur, but are not significant. There are no problems with weeds, because jatropha is competitive. All in all jatropha shows high resistance or tolerance to diseases and pests.

Self-incompatibility does not seem to occur. Protandry and protogyny could not be observed in Cape Verde (Heller, 1992).

Production

Once big enough to be planted out, jatropha can bear first fruits within a year (Lutz, 1992). The yield from hedges is 1kg of nuts per metre, which give 0.2l of oil. Three harvests per year are possible, depending on the rainfall; in semi-arid areas one harvest is more likely. It was reported that under very favourable conditions jatropha can yield 8t/ha of seed. Under less favourable conditions 0.2–2t/ha can be harvested (Cape Verde) (Münch and Kiefer, 1989). Yields from Jatropha hedges can reach 200l/km of purified oil. Extension work from Udaipur in Rajasthan indicated an average seed yield of respectively 0.125, 0.250, 0.550 and 0.750t/ha in rainfed 2, 4, 6 and 8 year old plantations, and 0.30, 0.60, 0.90, 1.05t/ha under irrigation (Henning, 1995). Maximum seed and oil yield in Rajasthan are 2t/ha/year of seed and 688kg/ha/year of oil at 75% field capacity and 1.9t/ha/year of seed and 632kg/ha/year of oil at 40% field capacity; 1.26t/ha/year of seed and 404kg/ha/year of oil can be achieved under rainfed conditions. These results indicate that jatropha does not need excessive irrigation for optimum yield (Jones and Miller, 1991).

In comparison to other oil producing species, jatropha has a low productivity. If appropriate conditions are provided, economic production may become feasible – for instance, in remote areas. In addition, if jatropha is grown in hedges, as windbreak or to stop erosion, the oil is a welcome by-product.

The more oil that is produced, the more press cake is available as fertilizer to improve the soil and stabilize the field crop yield without expensive chemicals. The CO_2 balance by using the oil in engines is neutral, which helps to combat the greenhouse effect.

Processing and utilization

The ripe fruits are harvested using a picking device; fallen fruits are collected. The dry pericarp is removed. The kernels can be stored. The efficiency of the harvest is a crucial point. It is estimated that a single person may harvest up to 5kg/hour.

The oil of jatropha is bitter and poisonous (Lutz, 1992). For the extraction of the oil, the nuts are milled and pressed. Therefore the nuts must be clean, dry, properly stored and preheated for pressing. Heating increases the yield as much as 25%, but the seed must not be overheated. Typically, the nuts will be spread out on a black sheet of plastic, and solar heat will suffice for good drying.

The purified oil is used as a fuel for stationary engines - for example, for generating electricity or driving water pumps and grist mills. The oil is also in demand for the production of soap. Jatropha oil and its constituents (phorbol esters) have insecticidal effects. Simple methods of extracting the active substances have been developed. The press cakes make a valuable organic fertilizer.

The phorbol esters were found to have molluscicidal effects on the water snails that carry bilharzia and on snails like *Lymnaeaamicularia rubigniosa*, an intermediate host of the river fluke. The effect on water snails is very pronounced. The pure oil is toxic, even more than the extracted esters (0.01% of the esters have an insecticidal effect). For use as an insecticide, the oil is mixed with water in the ratio 1:2 and used at 40l/ha. The esters can be extracted from the oil with methanol, which is distilled off in vacuum to obtain the pure esters.

The energy content in comparison with diesel is about 3% lower, related to the volume. Before use it is only necessary to filter the cold pressed oil. The oil can be stored in vessels, if possible airtight, to avoid polymerization (Henning and Mitzlaff, 1995).

The oil does not need to be processed before use in special diesel engines (Elsbett engines); careful filtering is enough. Diesel engines can run with transesterified jatropha oil. Attempts to produce an ester of the oil have been undertaken, but production on a large scale would necessitate major investments in equipment and personnel. A 2000t methyl ester pilot plant was to be built in Nicaragua in 1996.

The by-products can be used for energy production. The shells may be transformed into biogas in a fermenter (Foidl *et al.*, 1996). The press cake is used as organic fertilizer, and contains more nitrogen and organic matter than that provided by conventional mineral fertilizer. Turned under, it mineralizes more quickly than cattle dung. It contains more nitrogen and phosphorus than poultry and cattle dung. After drying, it can be used as a solid fuel, which makes energetical use of the remaining oil in the press cakes.

The ratio between fuel and soap production depends on the price of diesel fuel. In countries like Mali, where diesel fuel is expensive, it is economic to use the jatropha oil as fuel. Tables 7.30 and 7.31 show, respectively, the glyceride composition of the oil, and other physical properties.

The human toxicity of the oil is caused by curcin, a toxic albumin. In some investigations it was discovered that jatropha oil has a cancerous effect after contact with the skin (Heller, 1992), but other investigations gave no indications of a mutagen or cancerous effects (Henning and Mitzlaff, 1995).

The oil's insecticidal effects resulting from the 2% phorbol esters content mean that it is a source of insecticides that can be locally produced and applied. The phorbol esters are toxic, even in a concentration of 250ppm.

Table 7.30. The glycerides of jatropha oil's fatty acids (Münch and Kiefer, 1989).

Glyceride	Number of C atoms and double bonds	Content (%)
Myristic acid	(C 15/0)	1.4
Palmitic acid	(C 16/0)	10–17
Stearic acid	(C 18/0)	5–10
Oleic acid	(C 18/1)	36–64
Linoleic acid	(C 18/2)	18–45

Table 7.31. Some physical properties of jatropha oil (Münch and Kiefer, 1989).

Property	Value
Density (g/cm^3)	0.92
Flash point (°C)	340
Melting point (°C)	–5
Kinematic viscosity ($10^{-6}m^2/s$)	75.7
Cetane number	23/51
Iodine number	103
Heat of combustion (MJ/kg)	39.628

Selected references

Foidl, N., Foidl, G. and Sanchez, M. (1996) *Jatropha curcas* L. as a source for the production of biofuel in Nicaragua. *Bioresource Technology*.

Gliese, J. and Eitner, A. (1995) *Jatropha Oil As Fuel*. DNHE-GTZ, Project Pourghère.

Heller, J. (1992) *Untersuchungen uber genotypische Eigenschaften und Vermehrungs- und Anbauverfahren bei der Purgiernuß* (Jatropha curcas *L.)*. Verlag Dr. Kovaç, Hamburg.

Henning, R.K. (1995) Combating desertification: fuel oil from jatropha plants in Africa. Information packet from the GTZ Projet Pourghère, Bamako, Mali.

Henning, R.K. and Mitzlaff, K.V. (1995) Nachwachsende Rohstoffe, *Der Tropenlandwirt*, Beiheft 53, Universität Gesamthochschule Kassel.

Jones, L. and Miller, J.H. (1991) Jatropha curcas: *A Multipurpose Species for Problematic Sites*, Astag Technical Paper, Land Resource Series No. 1, The World Bank Asia Technical Department Agriculture Division, 12 pp.
Lutz, A. (1992) Vegetable oil as fuel. *GTZ Gate* 4/92.
Münch, E. and Kiefer, J. (1989) Die Purgiernuß, *Schriftenreihe der GTZ*, No. 209.
Rehm, S. and Espig, G. (1991) *The Cultivated Plants of the Tropics and Subtropics*, Verlag Joseph Margraf, Preise GmbH, Berlin.
Sujatha, M. and Mukta, N. (1996) Morphogenesis and plant regeneration from tissue cultures of *Jatropha curcas*. *Plant Cell, Tissue and Organ Culture* 44(2): 125–41.

JOJOBA (*Simmondsia chinensis* (Link) Schneid.)

Description

Jojoba is a dioecious shrub from the south-western USA and the neighbouring regions of Mexico (Sonoran desert). It is cultivated mainly in Arizona, northern Mexico, Argentina and Israel; there is some small scale planting in Australia, Chile and India (Benzioni, 1997). The plant is evergreen, and the leaves are covered with wax to prevent excessive loss of water. The trunk is branched directly over the ground. Various types are observed: some grow like shrubs, others grow up to a height of 5m. Some types bear fruits regularly, others do not. The roots reach depths of as much as 10m (Huber, 1984). The fruits normally contain one seed, but up to three are possible. The seeds contain 47–62% wax, which becomes liquid at 7°C (Rehm and Espig, 1991).

Ecological requirements

Jojoba tolerates drought. It thrives under soil and moisture conditions that are not suitable for most other agricultural crops. The shrub has been known to survive as long as a year with no rainfall at all. Jojoba requires water during winter and spring months to set its flowers and seeds. Its summer requirements are low (National Academy of Sciences, 1975). An annual rainfall of 250mm is enough for the plant to survive, but 370–450mm allow a quicker and luxuriant growth; 750mm precipitation or irrigation is adequate; 2000 mm is acceptable as maximum (Huber, 1984). The plant needs a cool period with a mean temperature of about 15°C for about 2 months to break the dormancy of the flower buds. At constantly high temperatures it never flowers. The optimum temperatures are 21–36°C, but it tolerates extreme desert temperatures (National Academy of Sciences, 1975). Seedlings may tolerate frost of –5°C, older plants –9°C. In addition, it seems that irrigated plants are less tolerant to frost than non-irrigated plants.

The pH value of the soil can range from 5 to 8, the optimum pH value is 6–7. Jojoba is not very demanding on soil fertility and is to a certain degree tolerant to salt. It needs a well drained, deep and light, sandy soil, and will not tolerate undrained wet soils. The plant is adapted to intense sunshine and the changes associated with high day temperatures and low temperatures at night. At first the young plants grow rather slowly, because they are endeavouring to develop long roots to ensure a good supply of water.

Propagation

Propagation is possible by seed or with cuttings. The seed may be planted at a depth of 2–5cm in wet, well drained soil. A seeded plantation of jojoba has genetic heterogeneity, and low average yields. Half of the seedlings are males that should be rooted out, because only 8–10% males are sufficient for pollination. All new plantations are from vegetatively propagated plants originating from cuttings from selected clones. Rooted cuttings are produced in a greenhouse and after hardening planted out (Benzioni, 1997). Breeding is in progress for high yielding monoecious and early bearing cultivars and for types adapted to local climates. In the meantime, vegetative propagation of highly productive genotypes is the most promising approach. Different varieties in growth and regularly nut producing varieties are selected. The aim of breeding is to obtain cultivars that are suited to mechanized picking and that are resistant to diseases. Jojoba can also be propagated by tissue culture (Huber, 1984).

Crop management

Jojoba thrives best in areas of low humidity and is drought hardy. Wind affects the shape and habit of growth and the plant often only reaches a height of 0.6–0.9m because of the harshness of the environment. Jojoba can be grown at altitudes from sea level to 1500m. In plantations the ratio between male and female plants is 1 to 5–10.

Plantation cultivation is possible with or without irrigation, but in regions with an annual rainfall of less than 500mm intensive cultivation is not feasible. The low yields per hectare are compensated by using a bigger area. The decision depends on the price of land and irrigation water. In plantations the jojoba is planted in hedges with rows 3m apart. The spacings between the plants range from 23cm to 75cm or more (Johnson and Hinman, 1980).

Jojoba is said not to be very susceptible to diseases and pests. In Mexico the drilling insect *Saperda candida* damages the wild living jojoba shrubs. Some fungus diseases (*Phytophthora parasitica*, *Phytium aphanidermatum*, *Verticillium dahliaae* and *Macophomia phaseolina*) can be dangerous to the plant (Huber, 1984). The most serious problems in the first two years of a plantation are weeds. Phytopathological problems in the nursery are caused by *Fusarium solani* and *F. oxysporum* (Benzioni, 1997).

Production

The first yields are obtained after 4–5 years, and full yields after about 10 years. Jojoba bears for 100 to 200 years. The fruits mature 90–180 days after fertilization. Jojoba has a high yielding potential: up to 4.5t/ha of seed has been reported (Rehm and Espig, 1991), though for 7–10 year old plants a yield of 3t/ha is a realistic estimation. The oil achieves a high price – about 2–4 times higher than that of other vegetative oils. The world's demand is estimated at 20,000–200,000t. The seed yields can vary from 0.5–15kg per tree. The average yield obtained from planted stands is 2.25t/ha, or about 1.5kg per plant.

Processing and utilization

The high costs of collecting the small fruits may be a limiting factor. The harvest of the nuts can be mechanized. One principle is to use 'over the row' machines like those used for mechanized grape harvesting. The fruits are shaken off and fall into collectors. As the nuts do not all ripen at the same time, the process has to be repeated. Another option is to wait for the ripe fruits to fall off and later collect them from the ground with a kind of vacuum collector (Huber, 1984).

Jojoba oil has special properties that distinguish it from other vegetable oils. It is mainly used for technical purposes. The oil is obtained by pressing the seeds or by extraction with solvents. In the raw state it has a light yellow colour. Strictly speaking, jojoba oil is strictly speaking a wax: it is not a triglyceride like other vegetable oils, but it is composed mainly of straight chain monoesters of C-20 and C-22 fatty acids and alcohols. The ester molecule has two double bonds, one at each side of the ester bond. Table 7.32 shows the make up of the oil with respect to wax esters. Other esters occur in amounts lower than 1%, and small amounts of free acids and alcohols can be found. (Additional chemical and physical properties of jojoba oil can be found in Table 7.34.) The almost complete absence of glycerine means that jojoba oil differs radically from all other known oils (Ayerza, 1990).

The oil is stable against oxidation and does not become rancid, and is a high quality lubricant that withstands high pressures and temperatures. Jojoba oil is a high quality

Table 7.32. The main wax ester composition of jojoba oil.

Wax ester	Amount (%)[a]
C-40	30.56
C-42	49.50
C-44	8.12
C-38	6.23

[a] Other esters occur in amounts lower than 1%.

Table 7.33. Viscosity of jojoba oil (Johnson and Hinman, 1980).

Temperature (°C)	Viscosity[a]	
	Pure oil	Sulphurized oil (9.88% sulphur)
37.8	127	3518
99	48	491

[a] Sayboldt universal seconds.

Table 7.34. Chemical and physical properties of the jojoba oil.

Property	Value
Freezing point	7–10.6°C
Melting point	3.8–7°C
Boiling point (757mm, N_2)	398°C
Smoke point[a]	195°C
Flash point[a]	295°C
Fire point	338°C
Heat of fusion	21cal/g
Refractive index (25°C)	1.4650
Specific gravity	0.863
Total esters	52%
Average molecular weight of wax esters	606
Hardness (Brinell number) of hydrogenated oil	1.9

[a] Determined according to Cc9A-48 of the American Oil Chemists' Society.

substitute for sperm oil. Because of its chemical structure (double bonds) and natural purity it is processed to waxes, creams, polishes and cosmetics. It can be used as a source of long chain alcohols.

The oil is not damaged by repeated heating, and the viscosity does not change after repeated temperature changes (National Academy of Sciences, 1975). Table 7.33 shows the viscosity of pure and sulphurized jojoba oil. The hydrogenated oil is a hard crystalline wax, which is comparable to carnauba wax (Franke, 1986). Because of its manifold practical uses in chemistry and industry and its high price, jojoba oil is not yet used for energy purposes.

The defatted meal has about 30% protein. However, the seeds also contain about 11% antinutritional compounds like simmondsin, simmondsin 2´-ferulate, 5-desmethylsimmondsin, and didesmethylsimmondsin. Several methods are under development to eliminate these materials in order to use the meal as animal feed (Kleiman, 1990). The glycosides are found in leaves, stems and roots too (Benzioni, 1997).

Selected references

Ayerza, R (1990) The potential of jojoba (*Simmondsia chinensis*) in the arid chaco: Rooting capacity and seed and wax yield. In: Naqvi, H.H., Estilai, A. and Ting, I.P (eds), *Proceedings of the First International Conference on New Industrial Crops and Products*, Riverside, California.

Benzioni, A. (1997) Jojoba; via Internet, http://www.hort.purdue.edu/newcrop/crops/cropfactsheets/jojoba.html

Franke, W. (1985) *Nutzpflanzenkunde*, 3rd edn, Georg Thieme Verlag Stuttgart and New York.

Huber, H. P. (1984) *Jojoba*, Fortuna Finanz Verlag, CH-8172 Niederglatt ZH.

Johnson, D.J. and Hinman, C.W. (1980) Oils and rubber from arid land plants. *Science*, vol. 208, May.

Kleiman, R. (1990) Chemistry of new industrial oilseed crops. In: Janick, J. and Simon, J.E. (eds), *Advances in New Crops*, Timber Press, Portland, OR, pp. 196–203.

National Academy of Sciences (1975) *Underexploited Tropical Plants with Economic Value*, Washington DC.

Rehm, S. and Espig, G. (1991) *The Cultivated Plants of the Tropics and Subtropics*, Verlag Joseph Margraf, Preise GmbH, Berlin.

KENAF (*Hibiscus cannabinus* L.)

Description

Kenaf is a short day, annual, herbaceous plant possessing high quality cellulose. It is a member of the Malvaceae family along with cotton and okra, and is endemic to Africa. In Switzerland the plant grows to a height of between 1.5m and 3m (Ott *et al.*, 1995). The rough-haired stem, most having small spines or thorns, consists of a central pith with short wood fibres and a bark with long fibres. Lower quality paper can be made from the short wood fibres, while high quality paper can be made from the long fibres (Manzanares *et al.*, 1995). At medium height the plant possesses heart shaped to deeply lobed leaves (Rehm and Espig, 1991).

Ecological requirements

Kenaf is able to adapt to a wide range of soil types and climates. The limits for cultivation are roughly 45°N to 30°S, which consists mainly of the warm temperate zone to the equator, with a maximum elevation of approximately 500m. To avoid the reduced growth associated with flowering and fruit formation, it is recommended that the day length be greater than 12.5 hours during the growing season (Rehm and Espig, 1991).

Kenaf is very cold sensitive. The optimal temperature range is from 15 to 27°C. The optimal pH ranges from 6 to 7. Kenaf grows well on light to middle weight quickly warming soils, with sandy soils also showing very good growth. Precipitation requirements are 500–700mm rainfall. Very wet soils are not suitable and kenaf cannot tolerate waterlogging. Fields with high weed levels should be avoided (Rehm and Espig, 1991). A test plot on sandy soil in Sardinia showed that kenaf needs frequent irrigation and more attention than sorghum (Bombelli *et al.*, 1995).

Propagation

Kenaf is most commonly propagated by seed.

Crop management

Sowing takes place in the middle of May, though later sowing dates may benefit from the higher ground temperature. Test plots in Spain resulted in the recommendation of crop sowing in June. This later date provides warmer temperatures that lead to quicker seed germination and seedling growth (Manzanares *et al.*, 1995).

The ground temperature should be at least 15°C, with warmer temperatures resulting in an increase in growth rate. The dry ground should be ploughed to a depth of 20cm, then left sitting for 10–15 days if possible (Ott *et al.*, 1995). Seeds are sown at 45–50 seeds per m^2 and a depth of 2–3cm, which amounts to 13–15kg/ha of seed. A spacing of 45–50cm between rows is recommended. A 70% success rate results in a crop density of 30–35 plants per m^2. The duration of germination is 4 to 7 days.

Seeds should be disinfected against anthracnose (*Colletotrichum hibisci* Poll.) and fungal diseases including *Rhizoctonia solani* Kühn and *Macrophomina phaseolina* Tassi.

In the early stages a pre- or post-emergence herbicide can be used, or a single hoeing after germination may prove sufficient for combating weeds. If a more persistent weed problem is present, then hoeing twice may be necessary. This would be performed after the kenaf is at least 15cm high and the weeds are in the germinating leaf to two leaf stage (Ott *et al.*, 1995). Because of kenaf's fast growth, weeds are not much of a problem once the plant is established.

Nematodes (*Meloidogyne* spp.) are significant kenaf pests and *Sclerotinia* is one of the few diseases affecting kenaf. *Botrytis* fungi may occur, but do not need combating because they do not affect fibre quality. During the vegetation period, which can be between 70 and 140 days, violet discoloration and leaf necrosis may occur after a cold period.

The following herbicides are acceptable for use according to conditions in Switzerland. Herbicides containing trifluralin (Trifluralin PSO) are acceptable for pre-sowing, herbicides containing alachlor (Lasso, Micro-Tech), dimethenamid (Wing), pendimethalin (Wing, Stomp) or metolachlor (Dual 960) are acceptable for pre-emergence and herbicides containing cycloxydim (Focus Ultra), haloxyfop-(R)-methylester (Gallant 535), or quizalfopethyl (Targa Super) are acceptable for post-emergence (Ott *et al.*, 1995).

Adequate fertilization is needed for large yields. Before ploughing, an initial soil fertilization using dung or liquid manure is possible. The recommended levels for crop fertilization as used in Germany and Switzerland are shown in Table 7.35.

Table 7.35. Recommend fertilizer rates for kenaf (Ott et al., *1995).*

Fertilizer	Amount (kg/ha)
Nitrogen (N) at time of sowing	20–30
Nitrogen (N) at approx. 20cm plant height (June)	50–60
Potassium (as K_2O)	120
Phosphorus (as P_2O_5)	60

Kenaf has proved to be very resistant to wilting and if the plant has been broken or blown over it will continue to root and grow.

Kenaf can be placed into a normal crop rotation. Because the harvest is in winter, summer crops with sowing dates starting on or after the middle of March can follow kenaf, but avoid preceding kenaf with crops that carry *Sclerotinia* spp. A cultivation break of 3 years is recommended for kenaf (Ott *et al.*, 1995).

Production

Although a strong market for kenaf has yet to be established, price and cost estimations have been made. For example, with an estimate of 5t/ha dry matter at a value of at least 290DM per dry tonne, this gives 1450DM/ha. After the subtraction of estimated costs of 1250DM and the addition of a set aside land subsidy, this leaves approximately 950DM/ha (H.W., 1995).

Test plots in northern Italy, using 10 May sowing and 10 October harvesting dates, have provided the following cultivar fresh biomass yields: 'Khon Kaen-60' 164t/ha, 'Kenalf' 149t/ha, 'Tainung-1' 140t/ha, and 'Everglades-41' 137t/ha. These yields correspond to dry matter yields of approximately 30t/ha (Petrini and Belletti, 1991).

A study in Spain was conducted to determine the appropriate cultivar under Spanish climatic conditions (Manzanares *et al.*, 1995). From the three cultivars tested – 'El Salvador' (long cycle), 'Everglades-71' (intermediate cycle) and 'PI-343129' (short cycle) – 'Everglades-71' proved to be most appropriate for the test plot's latitude and climate. The mean dry matter stem yields for 'Everglades-71' for 1992 and 1993 were 8.9t/ha and 8.03t/ha, respectively.

Processing and utilization

After the plant has died due to frost and achieved its highest dry matter percentage (60–70%), it is ready to be harvested. Kenaf can be harvested with a row independent maize harvester. Only the stems are harvested; the leaves will have died and fallen because of the frost. The harvest takes place in the winter between December and February. At this time the fibre and wood are still easily separable. Fresh stems consist of 5–6% fibre, while the dry weight is 18–22% fibre. Under average conditions a fibre yield of 1–2t/ha can be expected, while under favourable conditions yield reaches 3–3.5t/ha (Rehm and Espig, 1991). After being harvested, the kenaf is dried for 12 hours in a gas dryer to prepare it for transport.

The entire plant can be used to produce a pulp yield of 20t/ha for the paper industry. The fibre is mechanically or manually removed and used for weaving such items as sacks. After the fibre has been removed the remaining stems can be used as firewood or by the paper industry. The paper industry uses kenaf to make mulch paper for gardening and agriculture, for corrugated board, and to improve the tensile strength of recycled paper (Rehm and Espig, 1991). Kenaf as a high yielding plant species is a potential energy crop when used as a whole crop. Also, the residues from different industrial processes can be utilized as energy sources.

Selected references

Bombelli, V., Piccolo, A., Marceddu, E., Parrini, F. and Schenone, G. (1995) Biomass for energy in Sardinia (*Eucalyptus globulus*, *Arundo donax*, *Cynara cardunculus*, *Sorghum bicolor*, *Hibiscus cannabinus*). In: *Biomass for Energy, Environment, Agriculture and Industry, Proceedings of the 8th EC Conference*, Vienna, 1994. Pergamon, Vol 1, pp. 423–30.

H.W. (1995) Neuer Fixstern oder nur Sternschnuppe? DLG-Mitteilungen 4/1995, p. 50.

Manzanares, M., Tenorio, J.L. and Ayerbe, L. (1994). Some aspects of kenaf (*Hibiscus cannabinus* L.) agronomy in the central area of the Iberian peninsula. In: *Biomass for Energy, Environment, Agriculture and Industry, Proceedings of the 8th EC Conference*, Vienna, 1994. Pergamon, Vol 1, pp. 717–22.

Ott, A., Ammon, H.U., Mediaville, V., Zürcher, N. and Grether, T. (1995) *Anbau und Verwertung von Kenaf*. Landwirtschaftliche Forschung Beratung, Merkblatt, UFA-Revue, Bern.

Petrini, C. and Belletti, A. (1991) Kenaf: adaptability and productive potentialities in the north centre of Italy. In: *Proceedings, Biomass for Energy, Industry and Environment, 6th EC Conference*, Athens, pp. 292–6.

Rehm, S. and Espig, G. (1991) *The Cultivated Plants of the Tropics and Subtropics*, Verlag Joseph Margraf, Priese GmbH, Berlin.

LEUCAENA (HORSE TAMARIND) (*Leucaena leucocephala* (Lam.) de Wit)

Description

Leucaena originates in Mexico and Central America, though its origin is obscured by wide distribution by humans. Today, leucaena is grown in many parts of the tropics. It grows naturally as a weed in many regions (Brewbaker, 1995). It is an arborescent deciduous small tree, up to 20m high, with 10–25cm trunk diameter. The alternating leaves are evergreen, 10–25cm long and bipinnate with 3–10 pairs of pinnae. The numerous flowers are white, sitting axillary on long stalks, occurring in dense global heads. The fruit pods are 10–15cm long, 1.6–2.5cm wide and thin. They become dark brown and hard and contain 15–30 seeds. Flowers and fruits are formed nearly throughout the year. The plant develops a long, strong taproot.

Leucaena is grown for many purposes. It can be grown as a protein rich fodder plant, for land reclamation, erosion control, water conversation, reforestation, soil improvement (Duke, 1983). Leucaena is one of the few woody tropical legumes that is highly digestible and relatively non-toxic. It is planted for forage in many sites. All animals favour leucaena. It is a supplement to grass pastures. The foliage carries a mild toxin, the amino acid mimosin, which causes depilation in non-ruminant animals. The use of the wood includes fuelwood, lumber, pulpwood, craftwood and charcoal. Improved varieties allow the harvesting of lumber and paper pulp on a large plantation basis, with clear 30cm boles on 14m high trees in 8 years. Leucaena coppices well and withstands almost any type or frequency of pruning or coppicing (Brewbaker, 1995). Seedlings less than one year old will produce viable seeds (Duke, 1983).

Today, the species used commercially outside the New World is almost exclusively *L. leucocephala* (Lam.) de Wit (2n=104). However, hybrids of this and other species are grown to an increasing extent: notably, *L. pallida* Britton and Rox. (2n=104) and *L. diversifolia* (Schlecht) Benth. (2n=104) are used for hybridization. These three polyploid species intercross with ease; the last two provide cold tolerance, and *L. pallida* provides resistance to psyllid insect damage. The common is a self-pollinating *L. leucocephala* with virtually no genetic diversity outside the centre of its origin. New cultivars of arboreal type outyield the common cultivars and are less weedy. Lesser known species with some commercial interest include *L. collinsii*, *L. diversifolia* (2n), *L. diversifolia* (4n), *L. esculenta*, *L. pallida*, and *L. pulverulenta*. All *Leucaena* species are woody perennials of the New World, ranging from Texas south to Ecuador, in dry and mesic secondary forests (Brewbaker, 1995).

The leaves are used for mulching around other crops. They are said to increase the yields. The plant can fix nitrogen at 500kg/ha/year. Leucaena wood is hard and heavy; the specific gravity is 0.7g/cm^3 (Duke, 1983)

Ecological requirements

The tolerated climatic conditions range from the warm temperate dry to moist, through tropical very dry to wet forest life zones. Leucaena requires long, warm wet growing seasons. It does best under full sun. In the wild it is mostly found below 500m, though in Indonesia and Java it is grown at altitudes of 1000m and above. The growth rate is slower at high altitudes (Duke, 1983). Interspecific hybrids greatly extend this range. An annual precipitation of 650–1500mm is typical, but leucaena can be found on drier or wetter sites depending on drainage or competing vegetation (Brewbaker, 1995).

The plant accepts a wide range of soils, but does best where the soil is a deep, fertile, moist and alkaline clay (Duke, 1983). Leucaena does not like poor drainage and waterlogging or acid soil that is low in calcium and/or high in aluminum and manganese (Middleton, 1995). The low tolerance to aluminum and the high calcium demand contribute to the poor growth on soils with a pH below 5. The performance is excellent on coralline or other calcareous sites up to pH 8. It can be found on alkaline soils, but the tolerance to salinity is low. Growth is poor when the mean annual temperature is below 20°C (Brewbaker, 1995). Low temperatures cause a decline in growth. Leucaena is susceptible to frost, which will kill the top growth. Regrowth occurs quickly from aerial stems (slight frost) or from the base (heavy frost). Cold temperatures depress the germination of the seeds. Young seedlings will be killed by heavy frost (Middleton, 1995). The plant is known for drought tolerance and endures a wide range of heat and desiccation.

Propagation and crop management

Leucaena can be propagated by seed or with cuttings. Seedlings are usually transplanted from the nursery at an age of 3–4 months. They may also be transplanted bare-root, rolled in mud, if the soil moisture is adequate. Typical growth results in canopy closure in 4–6 months (Brewbaker, 1995). Some seedlings than less one year old will produce viable seed. The seeds are covered with a hard, waxy seed coat, so must be scarified before planting to ensure rapid and even germination. The seeds are either immersed in water of 80°C for 3–5 minutes, or in boiling water for 4–5 seconds (Middleton, 1995).

Inoculation with rhizobia is recommended for many soils, though rhizobia are almost universal in tropical soils. For forage, seeds should be sown 2.5–7.5cm deep. Planting should take place at the onset of the wet season. The crop soon produces a dense stand (Duke, 1983). The space between the rows depends on the annual rainfall. In areas with little precipitation, the spacings increase.

Production

Yield depends on climatic conditions. It takes several years before a plant is ready for full yield harvests (Prine *et al.*, 1997). On 3–8 year old trees the annual wood increment varies from 24–100 m^3, averaging 30–40m^3 (Duke, 1983). Oven dry stem wood produced by leucaena harvested annually averaged 31t/ha/year over four growing seasons for 12 selected accessions. The average total biomass yields of 10 leucaena accessions in Gainesville, Florida, planted in 1979 and harvested January 1994 after 4 seasons growth from 1990 through 1993, was 76.9t/ha, or 19.2t/ha/year. The yields ranged from 43.1 to 102.2t/ha (Prine *et al.*, 1997). Rahmani *et al.* (1997) report an annual dry matter yield of 35 t/ha.

Processing and utilization

Fuelwood and lumber harvests optimize at 3–8 year rotations when growth is not severely limited in some way. Trees coppice rapidly to 8m in one year (Figure 7.7) (Brewbaker, 1995). In regions with frost, leucaena will be harvested annually or every second year. New growth will coppice from stumps if the plant is cut or frozen. If frost kills the top growth, the dead stems will stand for one season while the next growth occurs and both seasons' growth can be harvested at the same time during the following winter (Prine *et al.*, 1997). The wood can serve as a biofuel for heat and electricity production using different conversion technologies.

Selected references

Brewbaker, J.L. (1995) *Leucaena*, New Crop Fact Sheet, via Internet.
http://www.hort.purdue.edu/newcrop/crops/cropfactsheets/leucaena.html

Figure 7.7. One year old leucaena logs, USA (Photo: Prine).

Duke, J.A. (1983) *Handbook of Energy Crops*, published via Internet. http://www.hort.purdue.edu/newcrop/duke_energy/leucaena_leucocephala.html

Middleton, C. (1995) *Leucaena for Beef Production in Queensland*, via Internet. http://leaky.rock.lap.csiro.au/facts/leucext2_txt.html

Prine, G.M., Stricker, J.A. and McConnel, W.V. (1997) Opportunities for bioenergy development in the lower south USA. In: *Making Business From Biomass, Proceedings of the Third Biomass Conference of the Americas*, Montreal, Quebec, Canada, 24–29 August, pp. 227–35.

Rahmani, M., Hodges, A.W., Stricker, J.A. and Kiker, C.F. (1997) Economic analysis of biomass crop production in Florida. In: *Making Business From Biomass, Proceedings of The Third Biomass Conference of the Americas*, Montreal, Quebec, Canada, 24–29 August, pp. 91–9.

LUPINS (*Lupinus* spp.)

Description

Lupins originated from the Mediterranean region and South America. Although the genus *Lupinus* is quite large, with over 300 species and a large genetic variability, only a few species are found to be agriculturally important. The domesticated species considered here are the 'Old World' species *L. albus* (white lupin), *L. luteus* (yellow lupin) and *L. angustifolius* (blue lupin), which are of Mediterranean origin, and the 'New World' species *L. mutabilis* (pearl lupin), which is of South American origin

(Hondelmann, 1984). One of the largest differences between the Old and New World species is seed size. Old World plants have larger seeds, with a diameter greater than 5mm and a weight of more than 60mg. New World plants have smaller seeds, with a diameter less than 5mm and a weight of less than 60mg.

L. albus grows to a height of up to 1.2m. The stalk and leaf stems are slightly silky. The stems have 5–9 egg-shaped leaves with a ciliated edge and a tip that comes to a fine point. The racemes are 5–30cm long and almost stemless. The corollas are white with a blue–violet tint. The plant's pods contain 3–6 seeds each. The seeds are smooth and white with a salmon/pink tint or with dark brown speckles.

L. angustifolius grows to a height of up to 1.5m. The stalk is slightly silky. The stems have 5–9 linear spatula shaped leaves. The racemes are 0.5–2.0cm long and are almost sessile. The corollas are light to dark blue with a violet tint. The plant's pods contain 4–7 seeds each. The seeds are smooth and variably coloured.

L. luteus grows to a height of up to 0.8m. The stalk is hairy. The leaves are linear to ovate and pointed. The racemes are 0.5–2.5cm long. The corollas are bright yellow and sweet smelling. The plant's pods contain 4–6 seeds each. The seeds are smooth and white or white with brown to black speckles.

L. mutabilis grows to a height of up to 2m. The stalk is more or less smooth to adjacently haired. The stems have 5–9 leaves which are long, ovate and pointed with a ciliated edge. The racemes are 5–30 cm long. The corollas are blue and/or pink and white with distinct yellow flecks. The plant's pods contain 2–6 seeds each. The seeds are smooth and black, brown-black, red-brown or white.

Elaboration on these descriptions of the plants is found in Hondelmann, 1996. All four of the *Lupinus* species mentioned here are annual plants. Lupins are able to fix nitrogen from the air and to dissolve phosphorus and other minerals in the soil.

The value of lupins as a food source is seen in the high protein contents of their seeds, with averages ranging from 31% for *L. angustifolius* up to 44% for *L. mutabilis*. *L. albus* and *L. mutabilis* are considered rich in fat/oil due to their 10 to 20% dry matter lipid contents (Hondelmann, 1996).

World production of lupin legumes is estimated at 1.5 million tonnes, coming mainly from Australia and the Commonwealth of Independent States (Rehm and Espig, 1991).

In 1872, a disease was first described that occurred in Germany with animals that were fed the alkaloid containing seeds from lupins as food. The disease, referred to as 'lupinose', was detected by the presence of one of two effects, mainly in sheep. There was either irreversible central nervous system damage or liver inflammation caused by Mycotoxicose from *Phomopsis leptostromiformis*. Both effects were often deadly. This alkaloid poisoning was linked to the species *L. luteus* and *L. angustifolius* (Hondelmann, 1996). The first species that were low in alkaloid content began to appear at the beginning of the 1930s.

Ecological requirements

L. luteus has the smallest number of ecological demands. It prefers a slightly acidic to acidic sandy soil or sandy loam soil, has a low water demand and can tolerate light frost.

L. angustifolius, on the other hand, is sensitive to drought and requires a high humidity, but has been shown to tolerate light to medium frost. It prefers a weakly acidic to neutral sandy loam or loamy sand soil.

L. albus prefers a slightly acidic loamy sand to loam soil. The plant is lime sensitive and has been shown to have the best growth of the four lupins on acidic soils. It is also slightly frost tolerant except during seed ripening.

Both *L. mutabilis* and *L. albus* have their highest seed yields in climates with warm summers. An advantage of lupin cultivation is that the plant can be grown on marginal land. In general, the recommended soil types are sandy to sandy loam soils which are in the slightly acidic to acidic pH range. The climate should be warm/mild and very sunny. Summers should not be too wet (Sator, 1979; Bramm and Bätz, 1988; Plarre, 1989).

Propagation

Lupins are most commonly propagated by seed. Here it is important that the seeds are of high quality and that the appropriate seed variety is selected according to the proposed use of the crop. It should also be ensured that the seeds are free of contaminants.

Crop management

To help maximize the yield and because lupins use nitrogen fixing bacteria to obtain nitrogen it may be necessary to inoculate the lupin seeds with a strain of nitrogen fixing bacteria if the bacteria isn't already present in the soil. Although the exact strain of bacteria depends on which type of lupin is being planted, it has been found that lupins are most commonly the host plants for *Bradyrhizobium* spp. and *Rhizobium loti* (Rehm and Espig, 1991). Before inoculation the seeds should be disinfected.

When preparing the field, although ploughing and harrowing are normal, it may be sufficient just to loosen the earth. Seed sowing/drilling in subtropical Mediterranean regions takes place in the autumn, from the end of August to November, with the beginning of the rainy season. Seed sowing/drilling in temperate regions takes place in the spring. The suggested crop density for *L. luteus* and *L. angustifolius* is 80 to 90 plants per m^2 and for *L. albus* 50 to 60 plants per m^2. The distance between rows should be 18 to 25cm with a seed planting depth of 2 to 4cm (Bramm and Bätz, 1988).

In Germany, the use of organic fertilizers and nitrogen fertilizers is not recommended. Suggested fertilizers and amounts are shown in Table 7.36.

Table 7.36. Recommended fertilizer rates for lupins.

Fertilizer	Amount (kg/ha)
Phosphorus (P_2O_5)	60–90
Potassium (K_2O)	90–120
Magnesium (MgO)	25–35

Keeping weeds under control is very important, especially during the early phases of plant growth. Because of this a herbicide and mechanical weeding may be necessary. Pre-emergence herbicides have shown to be appropriate (Bellido, 1984).

The diseases and pests which attack lupin crops usually vary from region to region. Lupins in general have been found to be attacked by fungi and rusts such as *Fusarium* spp. and *Uromyces lupini*. *L. mutabilis* is often afflicted with anthracnose caused by *Colletotrichum gloeosporioides*. Viruses also play a role with lupin crops, in particular cucumber mosaic virus, which can cause significant damage to *L. angustifolius* crops.

Pests that are known to attack lupin crops include, among others, insect larvae from butterflies and flies, lice, root parasites, snails and rodents.

It is recommended that lupins be cultivated after cereal crops and before crops such as potatoes, turnips and cabbage. Legume crops should not be planted directly before or after a lupin crop. A lupin crop cannot follow another lupin crop in the crop rotation; in addition a cultivation break of 4 years for lupins should be adhered to (Bramm and Bätz, 1988).

Because of the presence of nitrogen fixing bacteria from the lupins, the following crops will benefit from a significant amount of nitrogen remaining in the soil (Hondelmann, 1996).

Production

Lupins have a wide range of seed yields that are affected by the location and species of cultivation along with the form of crop management. This wide yield range is shown in Table 7.37.

Table 7.37. Lupin seed yield with respect to species (Bellido, 1984).

Species	Yield (t/ha)
L. albus	0.5–4
L. angustifolius	0.5–4
L. luteus	0.8–2.6
L. mutabilis	0.3–2

Since the beginning of the 1990s, average annual Australian seed yields have been in the 1.0–1.2t/ha range (Nelson, 1993). Germany has had *L. luteus* seed yields ranging from 1.71t/ha to 1.93t/ha, depending on sowing density (Sator, 1979).

Processing and utilization

Harvesting in Mediterranean regions takes place in June and July with dead ripeness of the crop. The entire plant is harvested using a harvester (Hondelmann, 1996).

The above ground portion of the plant can be used as fodder or hay and for green manuring. The seeds can also be directly used as animal feed and for human food (Hondelmann, 1996). Sweet lupins such as *L. luteus* and *L. angustifolius*, which have no or low alkaloid contents, are planted if the crop is to be used as feed or food (Hoffmann *et al.*, 1985).

The similarities in the amounts of raw nutrients of lupins and soybean as fodder can be seen in Table 7.38. Table 7.39 shows the raw protein and fat contents of seed dry matter from lupin crops. The characteristic that makes *L. albus* and *L. luteus* of interest is the percentage of fatty acids with respect to the total fat content of the seed oil (Table 7.40).

Table 7.38. Nutrient contents of lupins and soybean (Kellner and Becker, 1966).

Raw material (%)	*L. albus*	*L. angustifolius*	*L. luteus*	Soybean
Dry matter	14.1	15.0	13.4	16.2
Raw protein (N x 6.25)	3.4	2.7	2.2	2.9
Raw fibre	3.5	3.7	4.4	4.4
Raw fat	0.3	0.5	0.4	0.6
Raw ash	1.0	2.0	1.2	1.8
N-free extract materials	5.9	6.1	5.2	6.5

Table 7.39. Raw protein and fat contents of lupin seed dry matter (Plarre 1989).

Dry material	*L. albus*	*L. angustifolius*	*L. luteus*	*L. mutabilis*
Raw protein (%)	34–45	28–40	36–49	32–49
Raw fat (%)	10–16	5–9	4–9	13–23

Table 7.40. Fatty acid contents of L. albus *and* L. luteus *(Sator, 1979).*

Fatty acid	*L. albus*	*L. luteus*
Oleic acid	52.6–60.6	23.8–39.1
Linoleic acid	16.5–23.4	45.0–48.8
Linolenic acid	2.5–8.1	0.9–7.6

Research has shown that grain yields can achieve 2.5–6.0t/ha with an oil content from 8–20%. This can be used as a biofuel source. The total biomass can reach 5 to 15t/ha. The energy balance is very adequate because lupin is a legume crop and does not need substantial nitrogen fertilization, and it is a very convenient crop for improving the physical characteristics of the soil.

Selected references

Bellido, L.P. (1984) World report on lupin. *Proceedings of the 3rd International Lupine Conference*, La Rochelle, France, pp. 466–87.

Bramm, A. and Bätz, W. (eds), (1988) Topinambur. In: *Industriepflanzenbau: Produktions- und Verwendungsalternativen*, Bundesminister für Ernährung, Landwirtschaft und Forsten, Bonn, pp. 28–9.

Hoffmann, W., Mudra, A. and Plarre, W. (1985) Lupinen (*Lupinus* spec.) In: *Lehrbuch der Züchtung landwirtschaftlicher Kulturpflanzen*, vol. 2, Parey, Berlin, pp. 185–96

Hondelmann, W. (1984) Lupin – ancient and modern crop plant. *Theor. Appl. Genet.* 68: 1–9.

Hondelmann, W. (1996) Die Lupine – Geschichte und Evolution einer Kulturpflanze. *Landbauforschung Völkenrode*, Sonderheft 162.

Kellner, O. and Becker, M. (1966) *Grundzüge der Fütterungslehre*, 14th edn, Parey, Berlin.

Nelson, P. (1993) The development of the lupin industry in Western Australia and its role in farming systems, In: *Proceedings VII International Conference*, Evora, Portugal.
Plarre, W. (1989) Lupinus spp. In: Rehm, S. (ed.), *Spezieller Pflanzenbau in den Tropen und Subtropen. Handbuch der Landwirtschaft und Ernährung in den Entwicklungs-ländern*, 2nd edn, vol. 4, Ulmer, Stuttgart.
Rehm, S. and Espig, G. (1991) *The Cultivated Plants of the Tropics and Subtropics*, Verlag Joseph Margraf, Priese GmbH, Berlin.
Sator, C. (1979) Lupinenanbau zur Körnerproduktion unter besonderer Berücksichtigung der Gelben Lupine (*Lupinus luteus* L.). *Garten Organisch* 3(79): 74–7.

MEADOW FOXTAIL (*Alopecurus pratensis* L.)

Description

Meadow foxtail is a loose growing perennial grass most commonly found on damp to moderately wet meadows and other similar locations. The plant grows to a height of up to 1m. The flat, bare leaf blades are grooved and mid-green to dark green, with a glossy underside. They grow to a length of over 110mm and a width of over 8mm. The inflorescence is a 5–6cm long and 8mm wide panicle with 4–6 or more spikelets on each lateral branch. The flowering period begins in the middle of May (Siebert, 1975; Kaltofen and Schrader, 1991).

Ecological requirements

Meadow foxtail has a relatively high demand for nutrients and water. Even pastures with an insufficient water supply are not very suitable for foxtail. Most appropriate are heavy to medium heavy, humus and nutrient rich soils with good aeration. The plant can tolerate occasional flooding, snow cover, cold and late frosts (Kaltofen and Schrader, 1991).

Crop management

Meadow foxtail is propagated using seed. The seed is sown some time around the middle of May. A nitrogen fertilizer rate of three times 30kg/ha of liquid manure (NH_4–N) or three times 30kg/ha mineral fertilizer (N) can be applied.

A recommendation for a crop rotation including meadow foxtail has not been established, partly because the plant is perennial.

Production

Meadow foxtail dry matter yields consist of approximately 56% leaves, 37% stems and 7% inflorescences (Siebert, 1975). Mediavilla *et al.* (1993, 1994, 1995) provide the dry matter yields from meadow foxtail field tests that were conducted using liquid manure fertilizer at two locations in Switzerland: Reckenholz (R) and Anwil (A) (Table 7.41).

Table 7.41. Three years of meadow foxtail yields (t/ha) from field tests at two locations in Switzerland.

	1993		1994		1995	
	(R)	(A)	(R)	(A)	(R)	(A)
Dry matter yield	10.3	13.3	7.1	9.5	8.7	7.5

The crop was harvested three times in 1993 and twice in each of 1994 and 1995. At Reckenholz between 70 and 90kg/ha/year NH_4–N liquid manure was used, while Anwil used up to 180kg/ha/year NH_4–N.

Processing and utilization

The crop is harvested by mowing. Success has been achieved with harvests of one, two and three times per year. The crop is then dried in the field to prepare it for storage or for economic utilization.

Meadow foxtail is among the grasses that are seen as suitable crops for the production of biomass for use as a biofuel. This involves burning the biomass for the production of heat (directly) and electricity (indirectly). The primary utilization of the crop is as fodder, especially in the early growth stages in the springtime. If it is cut later, the crop will have a higher lignocellulose content. This will negatively influence the quality of the crop as fodder, but improve the possibility of its utilization as an energy crop.

Selected references

Kaltofen, H. and Schrader, A. (1991) *Gräser*, 3rd edn, Deutscher Landwirtschaftsverlag, Berlin.

Mediavilla, V., Lehmann, J. and Meister, E. (1993) *Energiegras/Feldholz – Teilprojekt A: Energiegras, Jahresbericht 1993*, Bundesamt für Energiewirtschaft, Berne.

Mediavilla, V., Lehmann, J., Meister, E., Stünzi, H. and Serafin, F. (1994) *Energiegras/Feldholz – Energiegras, Jahresbericht 1994*, Bundesamt für Energiewirtschaft, Berne.

Mediavilla, V., Lehmann, J., Meister, E., Stünzi, H. and Serafin, F. (1995) *Energiegras/Feldholz – Energiegras, Jahresbericht 1995*, Bundesamt für Energiewirtschaft, Berne.

Siebert, K. (1975) *Kriterien der Futterpflanzen einschließlich Rasengräser und ihre Bewertung zur Sortenidentifizierung*, Bundesverband Deutscher Pflanzenzüchter e.V., Bonn.

MISCANTHUS (*Miscanthus* spp.)

Description

Miscanthus is a genus of woody, perennial, tufted or rhizomatous grasses that originated in South-east Asia and is related to sugarcane. Miscanthus was originally introduced to Europe as an ornamental garden grass. It is high in lignin and lignocellulose fibre and uses the C_4 photosynthetic pathway. The erect stems are slim yet vigorous and are not usually branched. The solid pith stems are approximately 10mm in diameter and can reach a height of a little over 2m the first year and up to 4m each consecutive year in Europe and 7–10m in China (Figures 7.8 and 7.9) (El Bassam, 1994). The lower height in the first year is a result of a great amount of the plant's energy being used to establish its extensive root and rhizome system. The herbaceous leaves are usually cauline, flat and linear with a basal tuft that is only weakly developed. The flowers are arranged on the stem in abundant racemes that lie along a central axis (Rutherford and Heath, 1992). The rhizomes make up a highly branched storage system. The roots usually penetrate down well over a metre into the soil. Even though much of the below ground growth takes place in the first year, the crop usually does not reach maturity until after 2–3 years.

Ecological requirements

The growing season temperature has a large effect on crop yield for miscanthus. This is because *Miscanthus* spp. possess the C_4 photosynthetic pathway, which is more efficient at high temperatures and light intensities. Although the crop prefers warmer climates it has been shown that miscanthus can be grown with favourable results throughout Europe. No optimum growing temperature or range has been determined yet, but the guidelines used for maize appear to be the most appropriate.

Miscanthus has evolved in regions of the world that have large temperature fluctuations between summer and winter. This evolution has led to the development of characteristics that make the plant resistant to frost. With the presence of a thick bud around the growing point, and the latent buds on the rhizomes still being under the soil, miscanthus is able to tolerate light spring frosts (Rutherford and Heath, 1992). At temperatures below –5°C the developing shoots and leaves are destroyed, though the crop has proved able to do well in the Great Lakes region of Canada where

Figure 7.8. Miscanthus sacchariflorus, *China (Photo: Deuter).*

for several weeks during the winter the mean temperature can be as low as –40°C (Green, 1991).

The first occurrence of frost represents the end of the growing season for maize and it is also believed for miscanthus (Bunting, 1978). It is at this time that crop senescence is accelerated, nutrients are sequestered into the rhizomes and the plant begins to dry out.

Wind also plays a role in the success of miscanthus crops. Again using maize observations to characterize miscanthus, before the extended stems can lignify there is the chance of strong winds leading to lodging and/or damage to the leaves. Sheltering the crop from wind also reduces crop cooling, which can be responsible for hindering plant growth (Bunting, 1978).

Miscanthus does not make many demands on the soil, a fact demonstrated by its ability to grow on many types of arable land. Sands and sandy loams consisting of up to 10% of clay have shown to be the preferred soil types in Denmark (Knoblauch *et al.*, 1991). Success on sands and very stony soil is dependent on sufficient rainfall. A good yield has also been experienced on well drained soils with high humus.

In Japan it was observed that *Miscanthus sacchariflorus* produced more dry matter in wet areas where the water is 50mm from the surface than in dry areas where water is 150mm from the surface. Although the wet plot crops produced more shoots, they also showed a lower number of rhizomes as compared to the dry plots. The lower number of rhizomes was attributed to flooding (Yamasaki, 1981). In the long run, the insufficiency of rhizomes will probably lead to the crop's inability to maintain itself.

In order for the crop to establish itself in April and May it is necessary that the soil be sufficiently aerated and have a fine tilth, thus making soils of greater than 25% clay probably unsuitable. The soil must also be able to support farm machinery during the harvest. This makes wet, peaty soils inappropriate for miscanthus crops. Soil with 70 to 95mm of available water per 500mm depth of soil is recommended for miscanthus (El Bassam *et al.*, 1994).

Figure 7.9. Miscanthus giganteus, *Germany.*

Miscanthus is a deep rooting crop with roots going down more than 1.0m. This leads to the crop's preference for deeper soils such as on plains and in valleys. Results in Japan have also shown that *M. sinensis* grows better on the deep humus of a concave slope (Numata, 1975).

Soil texture, colour and pH can also affect miscanthus growth rate. Maize has displayed an increased growth rate as result of light textured, dark soil's ability to warm up rapidly (Bunting, 1978). A darker soil colour and lighter texture is also believed to assist in a faster miscanthus growth rate. Data from Denmark and the UK suggest that the optimum pH range is between 5.5 and 7.5 for growing miscanthus (Knoblauch *et al.*, 1991). This excludes soils that are very acidic or very chalky as potential supporters of optimum miscanthus growth.

Propagation

Seed: Although seed may prove to be the cheapest form of plant propagation, it has the limitations of seed heterozygosity and availability. In addition, some species do not produce viable seeds. *M. sinensis* 'Giganteus' in Bavaria has displayed very limited flowering. Chilling and long term storage do not seem to have negative effects on the viability of fertile seeds. Studies testing the effects of gibberellic acid, pH and temperature on the germination of seeds revealed no substantial preferences. But the results suggested that pH extremes should be avoided and that the temperature range 20–30°C may be best (Aso, 1977). Increasing water stress has been proved to decrease the growth rate of seedlings (Hsu, 1988).

Division: Propagation by division is a slow process, but it is simple and quickly leads to a large plant. *M. Sinensis* 'Gracillimus', *M. sinensis* 'Variegatus' and *M. sinensis* 'zebrinus' do not produce viable seeds, so plant division is used for their propagation. Stock plants are grown, from which 5–7.5cm long sections are cut in the spring. The sections are given a covering of slow release fertilizer and are ready for sale after three to six months (Gibson, 1986).

Stem cuttings: Mature, non-flowering stems are use to obtain tip and basal cuttings with two or three nodes. The cuttings are then planted. The use of stem cuttings has gained success with many perennial grass species, but little success has been achieved with miscanthus (Corley, 1989).

Rhizome cuttings: Rhizomes form an intertwined matrix 10–15cm below ground. This form of propagation has the advantage that more potential plants can be produced from a single plant, but it takes longer to achieve a full grown plant. It is presently recommended that in November, stock plants two to three years old be used for dividing up the rhizomes into lengths of 8–10cm. Shorter rhizome lengths can be cut, but they are more susceptible to winter kill. The cuttings can be stored between −1°C and 1°C. Tests have had success with cuttings planted in the beginning of May at a depth of 3–6cm and a density of 10,000 rhizomes per hectare (Nielsen, 1987). Data from the UK point to *M. sacchariflorus* as having the best success rate for propagation from rhizome cuttings, though plantations from rhizome cuttings in Denmark have suggested a 65–75% success rate. With respect to *M. sinensis* 'Giganteus', Huisman and Kortleve (1994) explain that farmers can grow and harvest rhizomes using a rotary cultivator to chop up the above ground plant, then a flower bulb harvester to harvest the rhizomes. Pot-grown transplants, 30–35cm high, possessing at least one shoot and noticeable robust roots, have repeatedly provided 100% establishment (Knoblauch *et al.*, 1991).

Micropropagation and tissue culture: This method of propagation has the advantage that a large number of plantlets can be produced in a short period of time. In addition, this method can be easily converted into large scale production (Rutherford and Heath, 1992). At Piccoplant Laboratories in Germany, *M. sinensis* has been used to develop a micropropagation system in which meristem tissue is removed from selected plants, then grown in culture. Field tests have indicated that the micropropagated plants are stronger than other propagated plants. Piccoplant has suggested they could produce 2 million plantlets per year. In the USA, immature inflorescences have displayed positive enough results to suggest that they could be used for tissue cultures (Gawel *et al.*, 1987).

Crop management

So far, crop management studies have concentrated on *M. sinensis* 'Giganteus' in Denmark (Knoblauch *et al.*, 1991) and Germany (Sutor *et al.*, 1991). It is recommended that young plants and stored rhizomes be planted when the planting depth temperature of the soil is 10°C or greater. This corresponds to approximately late April to early June. The important factors are sufficient soil moisture, a fine tilth, and the avoidance of young plant destruction from frost.

Rhizome cuttings have been successfully planted using machines designed for planting seed potatoes, while forestry and vegetable transplanting machinery have demonstrated positive results when planting young miscanthus plants.

The results of experiments in Denmark recommend that for transplanting, young plants should have a height of at least 30–35cm. The plants should also have vigorous roots and a minimum of one shoot. For planting rhizomes, it is recommended that lengths of at least 10cm be used for rhizome cuttings, though under favourable conditions lengths as small as 5cm have been successful. Here a planting depth of 5–7cm has been recommended. For planting pot grown plants it is suggested that the tops of the pots be covered with 2cm of soil.

Germany and Denmark have developed slightly different recommendations concerning plant spacing. The suggested spacings from Germany are 0.7–1.0m between plants and 0.8–1.0m between rows. Tests using spacings of one, two and three plants per m^2 have also shown that higher crop density results in higher dry matter yield. One exception was found, in that *M. sinensis* 'Giganteus' provided the highest second year yield at a spacing of two plants per m^2 (Kolb *et al.*, 1990). The suggested spacings from Denmark are 0.75m between two rows and 1.75m between these groupings of two rows, with 0.8–1.0 plants per m^2.

In the UK, planting of miscanthus takes place in March or April using hand planted micropropagated plantlets or rhizome pieces. Although new shoots are quick to emerge after planting, the period until May or June is marked by slow growth. With the more favourable temperatures of summer comes rapid growth and the development of stalks of over 2m in the first year and 4m in the following years (Heath *et al.*, 1994).

Weed control is an important factor, especially during the establishment and first two years of the crop. It is recommended that before planting the field be completely cleared of all perennial weeds. Because miscanthus is a perennial crop that can live for

15 years or more, a thorough preplanting weeding is important, because once the crop is in place weed clearing will be greatly hindered. The predominant problem weeds are wild oat (*Avena* spp.), creeping thistle (*Cirsium arvense*) and couch grass (*Elymus repens*). In Germany and Denmark, a number of herbicides have proved successful, including atrazine, fluroxypyr, mecoprop, propyzamide and some sulfonyl ureas.

In its native habitats miscanthus falls victim to a number of diseases, but in the areas where it has been relocated it has not yet shown much sign of being attacked by pathogens. Miscanthus has already proven itself to be quite resistant against diseases, but the more miscanthus is bred and grown throughout Europe and the Mediterranean region, the greater the chance of it being attacked by diseases already common to these areas. Diseases that attack or are likely candidates for attacking miscanthus in its native habitats can be grouped into Uredinales (rusts), Ustilaginales (smuts), Sphaeropsidales, Clavicipitales, Hyphomycetes, Peronosporaceae (downy mildews) and Pythiaceae. Miscanthus has been shown to be attacked by one virus disease, miscanthus streak virus (Rutherford and Heath, 1992).

Little research has yet been completed on the nutrient requirements of miscanthus, but some calculations have been made in order to determine the amounts of nutrients that would need to be replaced in order to maintain a proper soil balance (Table 7.42).

Table 7.42. Nutrients needed to maintain a proper soil balance for miscanthus (Rutherford and Heath, 1992).

Nutrient	Amount (kg/ha)
Nitrogen (N)	50
Potassium (K_2O)	45
Phosphorus (P_2O_5)	21
Sulphur	25
Magnesium	13
Calcium	25

Because of the low nutrient requirements of miscanthus it is believed that the soil and atmosphere will be able to supply/return much of the nutrients, but that the addition of nitrogen, phosphorus and potassium may at times be necessary. It may also be necessary to add magnesium on sandy soils with less than 3–4mg/l magnesium concentration. It seems that 50kg N, 20kg P_2O_5, and 100kg K_2O are sufficient to support adequate yields.

The best time for nutrient application is in the spring, before the new growth season but after the previous harvest. Applicators for liquid or granular fertilizers are believed to be adequate.

Water plays a large role in crop yield. During the crop's growing season it is estimated that 600mm of precipitation would be necessary to produce 20t dry matter yields. Greater yields become possible with increased irrigation. To provide for adequate establishment and first year growth of young plants, irrigation has been deemed necessary in Germany and Denmark (El Bassam *et al.*, 1994).

Production

Harvesting of miscanthus is performed in February/March, roughly late winter/early spring. This is the time at which the highest dry matter content, up to 80%, can be achieved and much of the nutrients has been sequestered in the rhizomes. Machinery for harvesting must be used carefully so as to limit compaction and damage to the underlying rhizome bed. Currently, mechanical harvesters are being studied to determine what form of machinery will work best. One promising option is a modern sugarcane harvester that does not remove the leaves as conventional harvesters do.

The yields of miscanthus are found to vary considerably according to site and climate, with the highest yields being recorded in Southern European sites when water was not a limiting factor. Yields of over 24t/ha dry matter were recorded in Portugal, Greece and Italy in the third year (all with irrigation) in comparison with third year yields of 11t/ha in Ireland, 16t/ha in Britain and 18.3t/ha in northern Germany. In China, yields up to 40t/ha and more could be achieved. The yields in Ireland and Britain compare very favourably with the annual dry matter production from short rotation coppice, which is grown as an energy crop in these countries.

Processing and utilization

At present, modifications to machinery are necessary for the handling of the miscanthus crop. Several general handling methods are presently seen as possible: pelleting, briquetting, baling and wafering. For the production of pellets and briquettes the crop is cut, then the biomass is immediately processed in the field or taken to a processing facility where it is highly compressed into pellets or briquettes. Baling consists of cutting and windrow drying, then using a large baler to produce large dense crop bales. Conventional baling produces a bale compressed to 120kg/m^3 while in Germany a new technique can produce a density of up to 450kg/m^3. Wafering takes the windrowed miscanthus and compresses it into a wafer. Wafering has the advantage of creating a denser final product than conventional baling, but at a significantly higher operational cost.

Drying and storage are possible, but so far have been the focus of little research. It is believed that after harvest the product can be covered and left outside or moved to a storage facility, or the excess heat from combustion of already dry crop matter can be used for crop drying, leading to long term storage. The important factor with storage is the moisture content. Jørgensen and Kjeldsen (1992) explain that big bales can be stored at 25% moisture content, but that denser bales must have a moisture content below 18%.

To minimize costs it is best to process and utilize the crop as much as possible on the farm where it is grown. The main goal for the utilization of miscanthus is whole plant compaction to produce a solid fuel, with gasification as a large scale option. Compaction can involve in-field or stationary pelleting or briquetting of the crop, or further compaction of bales. Compaction makes the product both easy to transport and ready for combustion for heat or electricity generation. These are possible directly on the farm, thus cutting costs. The solid fuel could also be transported to larger user facilities. The compact biofuel can be burned in individual homes, communities, or industries to produce heat, or regionally and nationally to generate electricity. The energy density of miscanthus is 18.2MJ/kg.

Selected references

Aso, T. (1977) Studies on the germination of seeds of *Miscanthus sinensis* Anderss. Scientific Report, Yokohama National University, Section II, Biological & Geological Science 23: 27–37.

Bullard, M.J., Heath, M.C., Nixon, P.M.I., Speller, C.S. and Kilpatrick, J.B. (1994) The comparative physiology of *Miscanthus sinensis*, *Triticum aestivum* and *Zea mays* grown under UK conditions. COST 814 Workshop.

Bunting, E.S. (1978) Agronomic and physiological factors affecting forage maize production. In: Bunting, E.S, Pain, B.F., Phipps, R.H., Wilkinson, J.M. and Gunn, R.E. (eds), *Forage Maize, Production and Utilisation*, Agricultural Research Council, London, pp. 57–65.

Corley, W.L. (1989) Propagation of ornamental grasses adapted to Georgia and the US southeast. *International Propagators' Society Combined Proceedings* 39: 322–37.

Deuter, M. (1996) Personal communication.

El Bassam, N. (1994) Miscanthus – Stand und Perspektiven in Europa. Forum für Zukunftsenergien e.V. – Energetische Nutzung von Biomasse im Konsenz mit Osteuropa, International Meeting, March 1994, Jena, pp. 201–12.

El Bassam, N., Dambroth, M. and Jacks, I. (1994) The utilization of *Miscanthus sinensis* (china grass) as an energy and an industrial raw material. *Natural Resources and Development* 39: 85–93.

Gawel, N.J., Robacker, C.D. and Corley, W.L. (1987) Propagation of *Miscanthus sinensis* through tissue culture. *Hortscience* 22: 1137 (abstract).

Gibson, J.G. (1986) Hints for propagating ornamental grasses, azaleas and junipers. *American Nurseryman* 164: 90–7.

Green, D. (1991) The new North American garden. *The Garden*, 116 (Part I), January, pp. 18–22. Royal Horticultural Society, London.

Heath, M.C., Bullard, M.J., Kilpatrick, J.B. and Speller, C.S. (1994) A comparison of the production and economics of biomass crops for use in agricultural or set-aside land. *Aspects of Applied Biology* 40: 505–15.

Hsu, F.H. (1988) Effects of water stress on germination and seedling growth of *Miscanthus*. *Journal of Taiwan Livestock Research* 21: 37–52.

Huisman, W. and Kortleve, W.J. (1994) Mechanization of harvest and conservation of *Miscanthus sinensis giganteus*. COST 814 Workshop.

Jørgensen, U. and Kjeldsen, J.B. (1992) *Production of elefantgræs*. Danish Research Service for Plant and Soil Science, Grøn Viden 110, 6 pp.

Kilpatrick, J.B., Heath, M.C., Speller, C.S., Nixon, P.M.I., Bullard, M.J., Spink, J.G. and Cromack, H.T.H. (1994) Establishment, growth and dry matter yield of *Miscanthus sacchariflorus* over two years under UK conditions. COST 814 Workshop.

Knoblauch, F., Tychsen, K. and Kjeldsen, J.B. (1991) *Miscanthus sinensis* 'giganteus' (elefantgræs). *Landbrug Grøn Viden 85*, 6 pp. [English version: *Manual for Growing* Miscanthus sinensis *'Giganteus'*. Danish Research Service for Plant and Soil Science, Institute of Landscape Plants, Hornum, Denmark.]

Kolb, W., Hotz, A. and Kuhn, W. (1990) Untersuchungen zur Leistungsfähigkeit ausdauernder Gräser für die Energie- und Rohstoffgewinnung (Investigations relating to the productivity of perennial grasses for the production of energy and raw materials), Institute of Viticulture and Horticulture, Würzburg. *Rasen-Turf-Gazon* 4: 75–9.

Nielson, P.N. (1987) Vegetative propagation of *Miscanthus sinensis* 'giganteus'. *Tidssk Planteavl* 91: 361–8.

Numata, M. (ed.) (1975) *Ecological Studies in Japanese Grasslands with Special Reference to the IBP Areas – Productivity of Terrestrial Communities*, Japanese Committee for the International Biological Program (JIBP Synthesis), 13, University of Tokyo Press, 268 pp.

Rutherford, L. and Heath, M.C. (eds) (1992) *The Potential of Miscanthus as a Fuel Crop*. Energy Technology Support Unit (ETSU) B1354, Harwell, UK, 125 pp.

Sutor, P., Sturm, M., Hotz, A., Kolb, W. and Kuhn, W. (1991) Anbau von *Miscanthus sinensis* 'giganteus'. Bayerische Landesanstalt für Bodenkultur und Pflanzenbau, Technical University, Munich, 8/91, pp. III 5–III 10.

Yamasaki, S. (1981) Effect of water level on the development of rhizomes of three hygrophytes. *Japanese Journal of Ecology* 31: 353–9.

NEEM TREE (*Azadirachta indica* Juss.; syn. *Antelaea azadirachta* (L.) Adelbert)

Description

The neem tree is a small to medium sized, usually evergreen, tree reaching a height of 5–25m and a trunk diameter of 30–80cm. It branches low and forms a wide spreading, dense ovoid crown. The seeds and leaves contain the insect repellent azadirachtin. The neem tree is native to Africa, South-east Asia and India (Gliese and Eitner, 1995) It can be found at altitudes up to 1500m and it occurs naturally within the latitudinal range 10–25°N.

Ecological requirements

The optimum temperature for growing the neem tree is 26–40°C, but a temperature range of 14–46°C is acceptable. Frost will kill the tree. An annual rainfall between 450–1200mm is adequate. The best pH value of the soil is 5.5–7. The neem tree needs bright light and short day conditions. It accepts a wide range of soil texture, but the soil should be well drained and deep. The tree can be successfully grown in soils with low fertility, but it will not grow on seasonally waterlogged soils (FAO, 1996).

Propagation

Modern techniques used for propagation include *in vitro* culture of leaf discs: 12–15 plantlets can be obtained from a single leaf disc within 6 months (Ramesh and Padhya, 1990). It is also possible to culture the embryos after dissection from the fruits. Embryos of 3–5mm were ready for transplanting within 45 days (Thiagarajan and Murali, 1994).

Crop management

The neem tree is planted as a windbreak and for shade and shelter. The growing period is 150–210 days per year. The tree can reach a height of 4–7m after 3 years and 5–11m after 8 years. In western Africa it is grown on a fuelwood rotation of 8 years. It withstands a dry season of 5–7 months. The tree is drought hardy, it regenerates rapidly and is shade tolerant in youth. The plantations need control because the neem may aggressively invade neighbouring areas and become a weed. It will not grow on seasonally waterlogged soils or where the dry season water table lies below 18m. Apart from the deep root system the tree also has lateral roots that may extent up to 15m away from the trunk (FAO, 1996).

Production

The yield of neem oil can reach 600l/ha. The optimum wood yield of 8 year old trees is 169m^3/ha. The annual wood production is 5–18m^3/ha.

Processing and utilization

The neem tree can be used as an energy plant in two ways. It can be grown as a source of fuelwood, or its oil can be used as fuel. The seeds contain about 45% oil, used primarily in the manufacture of soap, insecticides and insect repellents and for various medicinal purposes (Rehm and Espig, 1991). Neem tree oil is opaque, bitter and unpalatable, though it can be transformed into an edible oil containing 50% oleic acid and 15% linoleic acid (Gliese and Eitner, 1995). The neem tree is not only used as an oil producing plant, but also as a pesticide producing plant. The seeds and leaves contain the insect repellant azadirachtin. The oil, bark extracts, leaves and fruits have medicinal properties. The oil can be used in soaps, disinfectants or as a lubricant.

Selected references

FAO (1996) Ecocrop 1 Database. Rome.
Gliese, J. and Eitner, A. (1995) *Jatropha Oil as Fuel.* DNHE-GTZ, Project Pourghère.
Rehm, S. and Espig, G. (1991) *The Cultivated Plants of the Tropics and Subtropics*, Verlag Joseph Margraf, Preise GmbH, Berlin.
Ramesh, K. and Padhya, A. (1990) *In vitro* propagation of neem, *Azadirachta indica* (A. Juss), from leaf discs. *Indian Journal of Experimental Biology* 28(10): 932–5.
Thiagarajan, M. and Murali, P. M. (1994) Optimum conditions for embryo culture of *Azadirachta indica* (A. Juss). *Indian Forester* 120(6): 500–3.

OIL PALM (*Elaeis guineensis* Jacq.)

Description

The oil palm is indigenous to the humid tropics of West Africa. It occurs wild along the banks of rivers in the transition zone between the rain forest and the savanna. Today, the most common oil palm is *E. guineensis*, which makes up the largest part of the stocks in the plantations. Only a little is known about the origin of the oil palm, which is found in Africa and America. It is assumed either that the oil palm was brought from its habitat in Africa to America by (slave) traders, or that the oil palm existed already in the territory when Africa and South America were connected by a land bridge (Lennerts, 1984).

The oil palm is an unbranched evergreen tree, reaching a height of 18–30m and having a stout trunk with the thickness of 22–75cm and covered with leaf bases. The tree sprouts adventitious roots, growing from the bottom 1m of the trunk. A few deeply penetrating roots anchor the tree, but most of the roots grow horizontally in the top 1m of the soil, and as far away as 20m from the trunk. The roots lignify in the early states of growth; the intake of nutrients takes place through the tips of the roots. This explains the importance of the soil quality for the growth of the oil palm. The stem terminates in a crown of 70–100 leaves at the very top. The leaves are up to 7m long. The photosynthetic pathway is C_3 II (FAO, 1996).

The palm is monoecious, and male and female flowers do not appear on the plant at the same time. The African oil palm (*Elaeis guineensis*) and the American oil palm (*Elaeis oleifera*) can be crossed to obtain fertile hybrids. Each leaf axil produces a blossom that develops into an infructescence weighing 15–25kg, of which 60–65% is fruit. An infructescence contains 1000–4000 oval fruits measuring 3–5cm in length. The largest proportion of oil is in the fruit pulp (mesocarp), and about an eighth is in the endosperm. Fruit flesh and kernel deliver different types of oil. As Table 7.43 shows, three types of palms are distinguished, depending on the thickness of the nutshell of the palm kernels (endocarp).

Table 7.43. Palm type based on nutshell thickness (Rehm and Espig, 1991).

Palm type	Nutshell thickness	Nutshell as % of total fruit weight
Dura	2–8mm	35–55
Tenera	0.5–3mm	1–32
Pisifera	No shell	

Each tree changes periodically between male and female inflorescences. In the first years of a newly founded plantation, it may be necessary to pollinate the female flowers artificially if the palms do not produce enough male flowers. The pulp of the fruit contains an average of 56% oil. The ratio between the pulp and the kernel depends on the climatic conditions and on the variety of the tree. The oil content of the whole fruit can vary widely, depending on the cultivar and the growth conditions. The dura, grown in dry regions, has an oil content of 14%, the wild and semi-wild dura fruit 20–27%, and the tenera 36% (Gliese and Eitner, 1995).

The inflorescences begin to develop 33 to 34 months before flowering; the sex is determined 24 months before flowering. The development of the fruit takes 5–9 months from pollination to ripening (Rehm and Espig, 1991). One month before the final ripening, delicate oil drops and carotenes begin to form in the fruit. The fruits change colour from green to orange during ripening.

Ecological requirements

The oil palm requires mean temperatures between 24 and 28°C. The cultivation zone is restricted to the humid tropics between the latitudes 10°N and 10°S with an elevation under 500m above MSL. The yields are low in cool conditions. The annual precipitation should be in the 1500–3000mm range. Dry periods lasting not longer than 3 months are tolerated (Rehm and Espig, 1991). The oil palm needs a large amount of sunshine – 5–6 hours per day are optimal. The soil should be deep and well drained. Brief and occasional flooding does not cause any damage, but undrained soil is incompatible.

The best pH range is 5.5–7, though the palm thrives even at pH 4 if provided with the correct fertilizers. The majority of nutrient absorbing roots lie 0–30cm deep, in a circle with a radius of 5m. The nutrient intake is high: a 1t fruit bunch contains 6kg N, 1kg P, 8kg K and 0.6kg Mg; B and Cl are important trace elements (Rehm and Espig, 1991). In climates like that of Malaysia and the Congo where no distinct dry periods occur, harvesting is possible through the whole year. In areas like Cameroon, Nigeria and Sierra Leone, with dry and rainy periods, the main flowering takes place in the former and the harvest in the latter (Lennerts, 1984).

Propagation

The natural germination is slow and irregular. In modern cultivation the seeds are treated in a special way. First the seeds are kept in dry heat for 40 days at a temperature of 38–40°C. Then follows a water absorption period lasting 7 days. The moist kernels are then kept at 28–30°C. The germination begins after 10 days and is complete after 30–40 days. The germinated seeds are transplanted into nursery beds or containers. After 4 or 5 leaves have sprouted, they are transferred into a growing field or bigger containers. They can be planted out into the plantation at an age of 10–12 months. At first the trunk does not grow in height, but it grows in width until the diameter of the normal trunk is reached. Thereafter the palm begins to grow in height. High temperatures caused by solar radiation are good for the development of the seedlings.

Breeding: The aim is the selection of high yielding strains and production of hybrid seeds, such as tenera (= dura x pisifera). Apart from the major aims of a thin endocarp and a thick mesocarp, breeders are striving to achieve earlier fruiting and slower stem elongation, higher fruit weight in the fruit bunch and resistance against fusarium and dry basal rot (Rehm and Espig, 1991). Propagation by tissue culture technique is practised on large scale and can speed up the breeding of improved palm cultivars, but there are problems because of somaclonal variations. The crossing of *E. guineensis* and *E. oleifera* has not yet resulted in commercially usable cultivars (Rehm and Espig, 1991).

Crop management

If there are not enough male inflorescences, artificial pollination is possible. The pollen can be stored for a month without special treatment. Hybrid seeds of tenera varieties are produced in specialized breeding stations (Rehm and Espig, 1991). The distance between the palms in the plantation is 8–9m (130–150 palms per hectare). Between the plants, legumes can be sown as a ground cover and to provide the oil palm with nitrogen. In the first years intercropping is possible.

Most diseases and pests can be avoided by:

- raising healthy plants (treatment of the nuts with TMTD and streptomycin, weeding out weak plants in the nursery, applying pesticides in the nursery beds);
- choosing an appropriate site for the plantation;
- breeding cultivars resistant against main diseases.

The main diseases of the oil palm are spear and bud rot (*Fusarium oxysporum* Schlecht. f. *elaeidis* Toovey, and other microorganisms), leaf spot disease (*Cercospora elaeidis* Stey.), and dry basal rot (*Ceratocystis paradoxa*). The cause of tip rot or fatal yellowing has not yet been discovered. It is suspected that it is a physiological disorder, perhaps associated with micronutrient deficiency (Lieberei *et al.*, 1996). With good care of the plants, disease control measures in the stand are not usually necessary. Insect pests like *Oryctes* spp. and *Rhynochophorus* spp. are held within acceptable limits by sanitary measures. The use of insecticides is only occasionally necessary (Rehm and Espig, 1991).

After 30 years, replanting is necessary because the palm is then so tall that harvesting becomes too difficult. Since the import of *Elaeidobius* beetles from West Africa, where they improve pollination, yields in Malaysia have been increased. The ratio between kernel and fruit flesh was increased from 0.2:1, to 0.27:1, without decreasing the oil content of the flesh (Schuhmacher, 1991).

Production

The palms begin to bear fruit after 4–5 years, while still in the rosette stage. The yield potential has increased because of breeding and selecting high yielding strains. Unimproved African palms yield about 0.6t/ha, and bred varieties produce 6t/ha. Currently, annual yields in Malaysia are 7t/ha of mesocarp oil and 1.25t/ha of kernel oil (Rehm and Espig, 1991).

Processing and utilization

The fruits are harvested when they change colour (black to orange), or after the fall of individual fruits. The fruit bunches are cut off or knocked down.

The fruits must be processed within 24 hours after harvesting because some fruits are always damaged and the oil is cracked by lipase into fatty acids and glycerin, reducing the value of the oil (Rehm and Espig, 1991). The red raw palm oil has a high carotene content (Gliese and Eitner, 1995). The oil obtained from the fruit flesh differs significantly from the kernel oil (Table 7.44).

Table 7.44. Composition and properties of palm oil (mesocarp oil) and kernel oil (Diercke, 1981).

Composition of the palm oil (mesocarp oil)		Composition of the kernel oil	
Component	Amount (%)	Component	Amount (%)
Palmitic acid	35–40	Palmitic acid	7
Oleic acid	34	Oleic acid	18.5
Myristic acid	1.5	Myristic acid	12
Stearic acid	5	Stearic acid	4
Linoleic acid	10	Capric acid	6
		Lauric acid	55
Property			
Melting point	27–42°C		23–26°C (fresh) 27–28°C (old)
Iodine number	53.6–57.9		10–18[a]
Saponification number	200–205		245–255[b]

[a] Lennerts, 1984; [b]Römpp, 1974.

The fruit bunches are treated in an autoclave to destroy the lipases and to facilitate threshing. The separated fruits are treated in a digester and stirred to a pulp at a temperature of 95–100°C for 20–75 minutes. Then the fruit flesh is pressed. The palm nuts are separated from the pulp, cleansed and dried. After the shells are broken and the shells and kernels separated, the kernel oil is extracted by pressing or solvent extraction.

The oil palm is a rich source of vegetable oil. It can be used as liquid fuel and in industry as a lubricant or for soap and candle production. Palm oil can be used as a vegetable substitute for diesel, and not only in agricultural machinery: in Malaysia processed (transesterified) palm kernel oil is used to run buses, and plants have been built to produce the transesterified oil on an industrial scale (Lutz, 1992).

The press cakes from the fruit have a high fibre content. After drying they can be used as solid fuel. The ashes are used as fertilizer. The press cakes from the kernel have a high protein content and are used as fodder (Franke, 1985).

The fruit bunches can be used as a high value energy source; they are dried to a water content below 50% and chopped for use as a fuel. The ashes contain 41% K_2O, 3.7% P_2O_5, 5.8% MgO and 4.9% CaO, and are used as fertilizer. Increasing quantities of fruit bunches are used for mulching to provide the soil with organic matter.

The fibres and nut shells are suitable for energy provision because of their low water content. They are usually burned at the oil mill to provide it with heat and power through turbine powered generators.

During processing in the oil mill, for each tonne of palm oil 2.5t of sewage is produced. One tonne of sewage can produce 28m^3 of biogas. After the concentration of hydrogen sulphide has been reduced to below 1000ppm, the gas is suitable to operate special gas units that can drive generators and supply 1.8kWh per m^3 of biogas. Conventional treatment of the sewage in sewage treatment plants releases climate affecting gases into the atmosphere during the process. It is possible to get 110kWh from the by-products of 1t of fruit bunches. If the energy needed for the production of palm oil, palm kernel oil, sewage treatment and threshed fruit bunches is subtracted from this amount of energy, the oil mills could provide 370kWh from each tonne of palm oil. The use of by-products could reduce the emission of CO_2 equivalents during the production of palm oil, especially by avoiding methane emission from the sewage. Using the by-products as an energy source is economically feasible and has no negative impacts on the environment (Germer and Sauerborn, 1997).

Selected references

Diercke (1981) *Weltwirtschaftsatlas* 1, Deutscher Taschenbuchverlag/Westermann, Germany.

FAO (1996) Ecocrop 1 Database, Rome.

Franke, W. (1985) *Nutzpflanzenkunde*, 3rd edn, Georg Thieme Verlag Stuttgart and New York.

Germer, J. and Sauerborn, J. (1997) The oil palm (*Elais guinesis*) as a source of renewable energy. In: *Proceedings of the International Conference on Sustainable Agriculture for Food, Energy and Industry*, in press.

Gliese, J. and Eitner, A. (1995) *Jatropha Oil as Fuel*. DNHE-GTZ, Project Pourghère.

Lennerts, L. (1984) *Ölschrote, Ölkuchen, pflanzliche Öle und Fette*, Verlag Alfred Strothe Hannover, Germany.

Lieberei, R., Reisdorff, C. and Machado, D.M. (1996) *Interdisciplinary Research on the Conservation and Sustainable Use of the Amazonian Rain Forest and its Information Requirements*, GKSS-Forschungszentrum Geesthacht, GmbH, Germany.

Lutz, A. (1992) Vegetable oil as fuel. *Gate* 4/92.

Rehm, S. and Espig, G. (1991) *The Cultivated Plants of the Tropics and Subtropics*. Verlag Joseph Margraf, Preise GmbH, Berlin.

Römpp (1974) *Römpps Chemie Lexikon*, 7th edn, Franckh'sche Verlagshandlung, Stuttgart.

Schuhmacher, K.-D. (1991) Analyse des Weltmarktes für Ölsaaten, Ölfrüchte und pflanzliche Öle, Dissertation, Institut für Agrarpolitik und Marktforschung der Justus Liebig Universität, Gießen.

OLIVE TREE (*Olea europaea* L.)

Description

The olive is native to the Mediterranean area; there and in the Middle East 98% of olives are produced. It is an evergreen tree with small broad leaves. The leaves have a grey-green colour on their upper surface and are silver-white beneath. They are dropped after 2 years. The tree can reach a height of 10m, and can live for several hundred years. The buds arise from the leaf axils. The tree blooms in early June with yellow-white flowers. The olives ripen between October and December (Gliese and Eitner, 1995). At first they are green, then later they become dark green to black. The oil is formed in the flesh during the change of colour. The tree requires 210–300 days from flowering to harvest. The roots reach up to 12m away from the trunk and grow up to 6m deep (Rehm and Espig, 1991). Its vitality allows the tree to live for several hundred years and form new branches even if the trunk has become hollow (Franke, 1985).

The fruit flesh (mesocarp) contains 23–60% oil (40–60% from oil cultivars), and the seed (endosperm and embryo) contains 12–15% oil. Oil from fruit flesh and seed both belong to the oleic acid group (Rehm and Espig, 1991).

Ecological requirements

A temperature of 12–15°C is sufficient for blooming and fructification, while the olives require 18–22°C to develop well. The tree tolerates temperatures as high as 40°C. An annual rainfall of 200mm is adequate (Gliese and Eitner, 1995). Where the annual rainfall is 400–600mm 100 trees per hectare can be planted, but dry regions can support only 17–20 trees per hectare. The olive's extended root system means that the tree can take in water from a large volume of soil, so it is rather drought tolerant (Rehm and Espig, 1991). The trees may survive a temperature of −10°C, but it is better not to plant olive trees where the temperature regularly falls below −5°C in the dormancy period. Late spring or early autumn frosts are undesirable (FAO, 1996).

Propagation

The olive tree is reproduced vegetatively by grafting on seedlings, cuttings or wood of the trunk (Rehm and Espig, 1991). Breeding aims are to obtain varieties with special properties like larger fruits, early ripening and different oil yields (Parlati *et al.*, 1994).

Crop management

The olive fruit fly *Dacus oleae* Gmel., scale insects, the olive moth *Prays oleae* and other pests are difficult to control (Rehm and Espig, 1991). New plantings of olive trees for oil production are economically sound only in dry regions or where the soil is too stony for other forms of cultivation. The tree can be grown up to an altitude of 1220m and is well adapted to dryness. Most, but not all, cultivars require a chilling period through the last two months in the winter with temperatures between 1.5 and 15.5°C to stimulate the initiation of the flowers. Late spring or early autumn frosts are undesirable, as are hot dry winds or cool wet weather during flowering.

Production

The tree starts bearing at an age of 8–10 years. The maximum yields are obtained between 60 and 100 years of age. An average tree produces 60–65kg fruit per year. The oil yield is rather low: in the Mediterranean region the average is 400kg/ha/year. On the best sites of California, yields of 12.5t/ha of table olives have been achieved; 14–16l of oil can be obtained from 100kg of fruits (Gliese and Eitner, 1995). The bearing alternates: a full yield can be expected only in every second year. There are high labour costs during harvesting (Rehm and Espig, 1991).

Processing and utilization

Harvesting takes place from December to February. The oil olives are stripped off using machines, shaken down (or knocked down) from the tree, and then collected from the ground. The fruits are harvested before they are fully mature, at a time when the oil content of the flesh has reached its maximum at 50% (Gliese and Eitner, 1995). The oil content does not increase after this, but the proclivity of the oil to become rancid does. Manually picked olives provide the best oil. Damaged olives may go mouldy, and therefore the oil quality decreases (Franke, 1985).

The main part of the olive oil is used for nutrition; only a small part is used as a source of energy. To obtain the oil, the whole olives are crushed into a paste and then pressed (Rehm and Espig, 1991). First (cold) and second (warm) pressings give a high quality edible oil. A third (hot) pressing gives oil for fuel and industrial purposes (lubricant, soap production). The press cake contains 8–10% oil. By treatment with hexane all but about 2% of the remaining oil can be extracted (Gliese and Eitner, 1995). The oil is yellow to greenish-yellow and non-drying. The extraction remains are used as fuel or as fertilizer.

Olive oil contains 99.6% fat and 0.2% each of water and carbohydrates. Table 7.45 shows the acid composition of the oil, and Table 7.46 some of olive oil's properties.

Table 7.45. Acid composition of olive oil (Diercke, 1981).

Acid component	Amount (%)
Oleic acid	66.3
Linoleic acid	12.3
Palmitic acid	8.9
Eicosenoic acid	4.9
Palmitoleic acid	4.7
Arachidic acid	0.3
Myristic acid	0.2

Table 7.46. Some properties of olive oil.

Property	Value
Density (g/ml)	0.91–0.92
Melting point	6°C
Cloud point	10°C
Iodine number	75–94
Saponification number	185–203
Acid number	0.3–1

Selected references

Diercke (1981) *Weltwirtschaftsatlas* 1, Deutscher Taschenbuchverlag/Westermann, Germany.
Franke, W. (1985) *Nutzpflanzenkunde*, 3rd edn, Georg Thieme Verlag, Stuttgart and New York.
Gliese, J. and Eitner, A. (1995) *Jatropha Oil as Fuel*. DNHE-GTZ, Project Pourghère.
Parlati, M.V., Bellini, E., Pandolfi, S., Giordani, E. and Martelli, S. (1994) Genetic improvement of olive: Initial observations on selections made in Florence. *Acta Horticulturae* 356: 87–90.
Rehm, S. and Espig, G. (1991) *The Cultivated Plants of the Tropics and Subtropics*, Verlag Joseph Margraf, Preise GmbH, Berlin.
Römpp (1974) *Römpps Chemie Lexikon*, 7th edn, Franckh'sche Verlagshandlung, Stuttgart.

PERENNIAL RYEGRASS (*Lolium perenne* L.)

Description

Perennial ryegrass is a perennial plant whose greatest importance has been its utilization as a feed. The plant produces above ground stolons and grows to heights ranging from 30 to 70cm or more. The growth of the mature crop varies between semi-erect and semi-prostrate. The leaf blades are small, with a width of no more than 5mm. They are dark green with a grooved top side and a keeled, smooth and glossy underside. The inflorescence is a two rowed ear that grows to a length of up to 20cm. The plant flowers in May or June.

Perennial ryegrass is commonly infected with the endophytic fungus *Acremonium lolii*. In this mutualistic relationship the fungus produces alkaloids that have been shown to lead to disorders in livestock, but confer insect resistance upon the plant.

Ecological requirements

Temperate zones, subtropics and tropical highlands are regions with the most suitable climate for perennial ryegrass cultivation. The plant does not tolerate drought, wet soil and salt. Most important for the crop's growth are firm, nutrient rich soil and a mild, humid climate. Depending on the variety, it is also sensitive to snow cover and late frosts, though it has been shown to have good regenerative capabilities (Villax, 1963).

Propagation

Perennial ryegrass is propagated using seed.

Crop management

The crop is sown at the end of summer or the beginning of autumn using about 25–35kg/ha of seed.

Perennial ryegrass has a high demand for fertilizers, especially nitrogen. In Germany ryegrass field tests at several locations have used fertilizer rates that varied as shown in Table 7.47. A large percentage of these quantities was supplied in the form of liquid manure. Fertilizer rates vary with soil composition.

Table 7.47. Fertilizer rates for perennial ryegrass, as used in field tests in Germany.

Fertilizer	Amount (kg/ha)
Nitrogen (N)	284–380
Phosphorus (P_2O_5)	93–160
Potassium (K_2O)	176–463
Magnesium (MgO)	49

Among the diseases most damaging to annual ryegrass crops are mould (*Fusarium nivale*) and rusts (*Puccinia* spp.).

Little research has been undertaken to determine an appropriate crop rotation for perennial ryegrass. It is a perennial plant with several years of significant yields once the crop has reached maturity.

Production

Field tests at four locations in Germany where perennial ryegrass was harvested once a year at the onset of flowering provided yields averaging between 7.4 and 12.7t/ha of dry matter. The plant's dry matter content averaged around 32–33%. The heating value averaged approximately 17MJ/kg.

Additional research in Germany that tested different varieties of perennial ryegrass resulted in late ripening plants producing the highest fresh and dry matter yields. In 1994 the average fresh yield was 78.1t/ha and dry yield 13t/ha. The highest fresh matter yields were produced by the varieties 'Morenne', 'Phoenix t' and 'Sambin', and the highest dry matter yields were produced by the varieties 'Parcour' and 'Tivoli t'. Late ripening plants produce a consistent yield throughout the entire summer.

Processing and utilization

The perennial ryegrass crop is harvested by mowing. Research is under way into harvests varying from 1–5 times per year. If the crop is harvested five times per year, the first harvest takes place when the crop is approximately 40–60cm high. This corresponds to harvests taking place about every 3–4 weeks from around the end of May through until the middle of October. The crop should be carefully harvested and quickly dried when it is to be used as fodder. When the grass is grown simply for biomass production the feed quality of the harvest is not significant. It is important that the harvested material be sufficiently dried for storage and/or economic utilization as a biofuel.

Traditionally, perennial ryegrass has been an important feed crop; now it is being examined as a producer of biomass for conversion into biofuel. New technologies are under development for the utilization of biomass and biofuels to produce energy in the form of heat and electricity, and existing technologies have been improved.

Selected references

Feuerstein, U. (1995) *Erzeugung standortgerechter zur Ganzpflanzenverbrennung geeigneter Gräser für die Nutzung als nachwachsende Rohstoffe*. GFP-Project F 46/91 NR-90 NR 026, Deutsche Saatveredelung Lippstadt – Bremen GmbH.

Kaltofen, H. and Schrader, A. (1991) *Gräser*, 3rd edn, Deutscher Landwirtschaftsverlag, Berlin.

Kusterer, G. and Wurth, W. (1992) Ergebnisse der Landessortenversuche mit ausdauern-den Gräsern 1985–1991. *Information für die Pflanzenproduktion*, 1. Landesanstalt für Pflanzenbau Forchheim.

Kusterer, B. and Wurth, W. (1995) Ergebnisse der Landessortenversuche mit ausdauern-den Gräsern 1994. *Information für die Pflanzenproduktion*, 3. Landesanstalt für Pflanzenbau Forchheim.

Rehm, S. and Espig, G. (1991) *The Cultivated Plants of the Tropics and Subtropics*, Verlag Joseph Margraf, Priese GmbH, Berlin.

Villax, E. J. (1963) *La Culture des Plantes Fourragères dans la Région Méditerranéenne Occidentale*. Inst. Nat. Rech. Agron., Rabat.

Wurth, W. (1994) Ergebnisse der Landessortenversuche mit ausdauernden Gräsern 1993. *Information für die Pflanzenproduktion*, 2. Landesanstalt für Pflanzenbau Forchheim.

PIGEONPEA (*Cajanus cajan* (L.) Millspaugh)

Contributed by: A. K. Gupta

Description

Pigeonpea is a C_3 plant that belongs to the family Phaseoleae and subfamily Cajaninae under suborder Papilionaceae of the order Leguminosae. The pigeonpea plant is a woody, short lived perennial shrub growing up to 4m high but is most often cultivated as an annual reaching up to 2m in height. Pods are usually short (5–6cm) and contain four to six seeds. The cultivated species show a wide range of diversity of various morphological characters and also of maturity periods. Peninsular India or eastern Africa is credited with its homeland but pigeonpea has already been acclimatized for multiple uses in more than 60 countries in the world. It is produced commercially in India, Myanmar, Kenya, Malawi, Uganda, the Dominican Republic, Haiti and Puerto Rico. In India, the crop is grown on more than 3.6 million hectares (FAI, 1993).

Pigeonpea is an ecologically highly desirable plant species because it is endowed with several unique features that give it many excellent qualities and make it an important crop for drier regions and waste lands of poor fertility. Its prolific root system allows the optimum moisture and nutrient utilization. Pigeonpea ameliorates poor soils through its deep, strong rooting system, leaf drop at maturity and addition of nitrogen (Nene and Sheila, 1990). Its characteristic root exudates give pigeonpea a high ability to take up phosphorus even from normally unavailable forms (Ae *et al.*, 1990; Otani and Ae, 1996). Pigeonpea has tremendous potential for use in agroforestry systems.

With the increasing shortage of energy, pigeonpea sticks are likely to be in demand for fuel and many farmers will be tempted to grow more pigeonpea (Nene and Sheila, 1990). A number of benefits of growing pigeonpea, when weighed against its low input requirements, justify its expansion to non-traditional areas such as the hilly lands of tropical Asia and marginal lands in Australia and the Americas.

Ecological requirements

Pigeonpea is predominantly a crop of the tropics and subtropics, but a wide range of maturity groups enables the crop to adapt to diverse agroclimatic areas and cropping systems. The range of maturity varies from 90 to 300 days, depending mainly on genotype and time of planting. Its new dwarf, short duration cultivars and hybrids can be cultivated up to 45° latitude on both sides of the equator. Although it is highly drought and heat resistant (Gooding, 1962) the crop performs better when annual precipitation exceeds 500mm. The plant also grows in subhumid ecologies where ripening can occur during the dry season (Rachie and Roberts, 1974).

The traditional landraces and cultivars are highly photosensitive (short day) plants, whereas new lines have been shown to be highly insensitive. Regardless of plant type, pigeonpea as a sole crop is relatively inefficient because of its low initial growth rate (Willey *et al.*, 1981), and is frequently mixed with other crops – mainly short term, hot weather cereals or other grain legumes to make efficient use of growth resources. Short term and intermittent waterlogging severely impairs plant growth, so areas of high rainfall and/or impermeable soils are unsuitable for pigeonpea. Furthermore, the plant is very susceptible to frost in the winter.

Pigeonpea is often reported to be a crop well adapted to marginal conditions (Whiteman *et al.*, 1985) and is grown on a wide range of soils types from gravel to heavy clay loam. It can grow in the pH range 5–8. In highly acidic soils, crop growth is adversely affected by aluminium toxicity or calcium deficiency. Farmers in India often grow pigeonpea on poor soils where no other crop can be used so profitably. However, well drained, deep fertile loam with neutral pH is ideal.

Crop management

Pigeonpea is propagated using seed. The optimum temperature range for germination is wide (19–43°C), with the most rapid seedling growth occurring between 29 and 36°C (de Jabrun *et al.*, 1981). Traditionally, the crop is sown in India at the beginning of the monsoon or rainy season, which is from the middle of June to the end of July. In eastern Africa, pigeonpea is sown in October/November at the onset of short rains. Pigeonpea does not require special land preparation. Ploughing to a depth of 15cm is sufficient to obtain a good crop. The optimum depth of seeding is 4–5cm. The seed shape is very suitable for machine sowing. The inoculation of seeds with rhizobium culture ensures good performance.

Once established, pigeonpea roots are capable of penetrating hard pan layers. Kampen (1982) found the broad bed and furrow system of sowing to be more useful. Dry sowing is usually preferred as it is difficult to work on vertisols when they are wet. Ratoon cropping – a multiple harvest system in which stubbles of the first sown crop are allowed to regenerate for subsequent production – is also becoming popular using some short duration (100–140 days) varieties in parts of India where the winter is mild. This system minimizes the cost of cultivation, avoids the risks associated with sowing a second crop in rainfed conditions, and provides high returns (Ali, 1990).

Historically, pigeonpea has been grown in mixtures without a distinct row arrangement, but the modern method is to sow in rows for inter-row cultivation and mechanical harvesting. A plant population of 60, 80, and 80–100 thousand plants per hectare has been recommended for summer sown long, medium, and short duration varieties respectively; winter planting as practised in the north-eastern hill regions of India required a higher plant population of 250–300 thousand plants per hectare (Chandra *et al.*, 1983). The most popular spacings are 60cm × 18–20cm.

To produce 1t of pigeonpea grain requires about 50kg N, 5kg P and 22kg K (Kanwar and Rego, 1983). Under intense management systems more nutrients will be required to produce higher biomass. Rao (1974) estimated that to produce 2t of grain and 6t of stalks from 1ha, the short duration variety removed 132, 20 and 53kg of N, P and K respectively. It may be stated that 20kg N, 60kg P and 5–6t of farmyard manure per hectare will be necessary to obtain a good yield.

With regard to water management, more than 90% of the pigeonpea growing area is rainfed (Singh and Das, 1985), but both biomass and seed yield may be enhanced in pigeonpeas of all maturity groups by the application of irrigation (Venkataratnam and Sheldrake, 1985). Flowering and pod setting stages are most crucial with respect to water stress. Pigeonpea, being widespread and slow growing, suffers heavily from weed infestations; these can be tackled biologically by inserting an vigorously establishing intercrop or chemically by the pre-emergence incorporation of Basalin at 1kg AI per hectare. Pigeonpea when intercropped with a cereal ensures higher biomass production and a measure of income stability.

Production

Pigeonpea is grown in a multitude of contrasting production systems aimed at obtaining higher grain yield of better quality. The composition of seed is presented in Table 7.48.

Table 7.48. General composition of pigeonpea seed (%).

Crude protein	Starch	Soluble sugars	Fat	Crude fibre	Ash	Gross energy (MJ/kg)
21–25	50–60	3–5	1–2	1–5	3–4	16–18

Favourable growing conditions can result in a seed yield of up to 2.5t/ha; a high yield of 5t/ha of dry seeds was reported from India (Anonymous, 1967). Akinola and Whiteman (1972) recorded the highest yield of 7.6t/ha of dry seed from experimental plots in Queensland, Australia. The production potential in India, especially in the farmer's field, is severely reduced by a moisture deficit during critical growth stages, the use of traditional low yielding varieties, or by poor management including the lack of any plant protection measures (Chandra *et al.*, 1983). A moderately well managed crop routinely yields about 2t/ha of grain and 7–8t/ha of stalks. Figures given in Table 7.49 project the average breakdown of dry matter in new genotypes of pigeonpea.

Production of biomass can be further enhanced by introducing a compatible intercrop. When pigeonpea is to be planted for energy, seed quality is not important. In India, growers have recently been encouraged to advance the sowing of short duration pigeonpea to April in order to produce more stalks (up to 15–17t/ha) and up

Table 7.49. Partitioning of dry matter (% weight) in pigeonpea (Gupta and Rai, unpublished).

Roots	Stem sticks	Branches and leaves	Husk	Seed
11.0±0.4	38.6±3.7	29.0±4.0	7.2±1.8	14.2±1.7

to 50% more seed yield compared to conventional June sowings. Besides this, the succeeding wheat crop is reported to produce a 25% higher yield than that following the June sown crops (Panwar and Yadav, 1981). However, the April sowing is feasible only where there are ample irrigation facilities.

Processing and utilization

Pigeonpea has traditionally been a valuable pulse crop. It can simultaneously satisfy the need for the '4 Fs' (food, feed, fodder and fuel) and thus ensures a self sustaining system. Its *dal* (a thick soup cooked from dry decorticated split cotyledons) is a rich source of protein for vegetarian folk in India; the tender green pods serve as a vegetable; crushed dry seed forms an excellent feed for cattle, pigs and poultry; green or dry leaves provide palatable fodder; and the woody stem serves as fuel and raw material for making huts, brooms and baskets. Recent studies in Bangladesh (Akhtaruzzaman *et al.*, 1986) and in Pakistan (Shah, 1997) indicate the possibility of using pigeonpea to produce paper pulp. This is a crop that can be used in many diverse ways. Presently, little is known about the efficient utilization of the entire plant as a biofuel. Pigeonpea can be regarded as a potential source of bioenergy because of its high biomass yield in comparison to the amount of off-farm inputs used.

Selected references

Ae, N., Arihara, J., Okada, K., Yoshihara, T. and Johansen, C. (1990) Phosphorus uptake by pigeonpea and its role in cropping systems of the Indian subcontinent. *Science* 248 (4954): 477–80.

Akhtaruzzaman, A.F.M., Siddique, A.B. and Chowdhury, A.R. (1986) Potentiality of pigeonpea (arhar) plant for pulping. *Bano Biggyan Patrika* 15(1–2): 31–6.

Akinola, J.O. and Whiteman, P.C. (1972) A numerical classification of *Cajanus cajan* (L.) Millsp. accessions based on morphological and agronomic attributes. *Australian Journal of Agricultural Research* 23(23): 995–1005.

Ali, M. (1990) Pigeonpea: Cropping systems. In: Nene, Y.L. *et al.* (eds), *The Pigeonpea*, CAB International, UK, pp. 279–301.

Anonymous (1967) Regional Pulse Improvement Project. Annual Report, No. 5 RPIP. New Delhi, India

Chandra, S., Asthana, A.N., Ali, M., Sachan, J.N., Lal, S.S., Singh, R.A. and Gangal, L.K. (1983) *Pigeonpea Cultivation in India: 'A Package of Practices'*, All India Coordinated Pulse Improvement Project, Kanpur: Extension Bulletin no. 2, 36 pp.

de Jabrun, P.L.M., Byth, D.E. and Wallis, E.S. (1981) Imbibition by and effects of temperature on germination of mature seed of pigeonpea. In: *Proceedings of the International Workshop on Pigeonpeas*, 15–19 December 1980, Patancheru (India), vol. 2, pp. 181–7.

FAI (1993) *Fertiliser Statistics*, Fertiliser Association of India (FAI), New Delhi, 270 pp.

Gooding, H.J. (1962) The agronomic aspects of pigeonpeas. *Field Crop Abstracts* 15(1): 1–5.

Kampen, J. (1982) *An Approach to Improved Productivity on Deep Vertisols*, ICRISAT Bulletin no. 11, p. 14.
Kanwar, J.S. and Rego, T.J. (1983) Fertiliser use and watershed management in rainfed areas for increasing crop production. *Fertiliser News* 28(9): 33–43.
Nene, Y.L. and Sheila, V.K. (1990) Pigeonpea: geography and importance. In: Nene, Y.L. *et al.* (eds), *The Pigeonpea*, CAB International, UK, pp. 1–14.
Otani, T. and Ae, N. (1996) Phosphorus (P) uptake mechanisms of crops grown in soils with low P status. I: Screening crops for efficient P uptake. *Soil Sci. Plant Nutr.* 42(1): 155–63.
Panwar, K.S. and Yadav, H.L. (1980) Response of short-duration pigeonpea to early planting and phosphorus levels in different cropping systems. In: *Proceedings of the International Workshop on Pigeonpeas*, 15–19 December 1980, Patancheru (India), vol. 1, pp. 37–44.
Rachie, K.O. and Roberts, L.M. (1974) Grain legumes of the lowland tropics. In: N.C. Brady (ed.), *Advances in Agronomy*, Academic Press, New York, pp. 1–132.
Rao, J.V. (1974) Studies on fertiliser management of wheat in maize-wheat and arhar-wheat cropping systems. Ph.D. thesis, Indian Agricultural Research Institute, New Delhi, India, 117 pp.
RPIP (1967) *Regional Pulse Improvement Project*. Annual Progress Report no. 5, RPIP, New Delhi, India.
Shah, P. (1997) Biomass for industry: Primary productivity of pigeonpea as affected by plant population and phosphorus. *Abstract, International Conference on Sustainable Agriculture for Food, Energy, and Industry*, Braunschweig, 22–28 June 1997, p. 288.
Singh, R.P. and Das, S.K. (1985) Management of chickpea and pigeonpea under stress conditions, with particular reference to drought. In: *Proceedings of the Consultants' Workshop on Adaptation of Chickpea and Pigeonpea to Abiotic Stresses*, 19–21 December 1984, pp. 51–61.
Venkataratnam, N. and Sheldrake, A.R. (1985) Second harvest yields of medium-duration pigeonpea (*Cajanus cajan*) in peninsular India. *Field Crops Research* 10: 323–32.
Whiteman, P.C., Byth, D.E. and Wallis, E.S. (1985) Pigeonpea (*Cajanus cajan* (L.) Millsp). In: Summerfield, R.J. and Roberts, E.H. (eds), *Grain Legume Crops*, Collins Professional and Technical Books, London, p. 658.
Willey, R.W., Rao, M.R. and Natarajan, M. (1981) Traditional cropping systems with pigeonpea and their improvement. In: *Proceedings of the International Workshop on Pigeonpeas*, Patancheru (India), 15–19 December 1980, vol. 1, pp. 11–25.

POPLAR (*Populus* spp.)

Contributed by: W. Bacher

Description

Different *Populus* species (family Saliaceae) exist in temperate climate zones. *Populus alba* is grown mainly in southern and central Europe, and *P. tremula* L. in Europe and Asia (Franke, 1981), while *P. tremuloides* Mic. is grown mainly in North America, Canada and in parts of Alaska (Schütt *et al.*, 1994). The hybrid *Populus x canadensis* (Canadian poplar) is cultivated in Europe (Franke, 1981). *Populus* spp. are pioneer trees with a capacity for fast growth. In Europe, cultivated poplars grow to heights of between 30 and 35m, and the hybrid *Populus x canadensis* is able to reach 50m. Poplar wood is long grained, relatively soft, and easy to divide. Poplar is being used as a short rotation coppice crop (SRC), and is cultivated on biomass plantations for use for direct combustion, making briquettes or gasification.

Ecological requirements

Like willow (*Salix* spp.), poplar grows successfully on a wide range of soil types – on sandy soils as well as on loamy-clay soils with a pH range of 6.0 to 8.0 and an optimum pH of 6.5 (FAO, 1996). Irrigation is beneficial to yield if the yearly rainfall is lower than 600mm (Parfitt and Royle, 1996). Poplar grows well on fine sandy–loamy soils with organic matter and a good supply of water. This species is more dependent on water than other agricultural crops, so extremely dry land should be avoided (Johansson *et al.*, 1992). The tree's consumption of water is as much as 4.8mm per day. The minimum growing temperature of the different *Populus* species has a range between 5°C and 10°C and the maximum growing temperature between 30°C and 40°C; the optimal growing temperature range is between 15°C and 25°C. Poplar species cannot survive if temperature reaches –30°C or below.

Propagation

Many poplar clones could be used as short rotation crops and are planted as plantlets or cuttings. Poplar cuttings incorporating primary and undeveloped secondary buds should be planted as vertically as possible with no more than 2.5 to 3cm remaining above ground (Parfitt and Royle, 1996). For ease of planting, successful rooting and subsequent management, the site should be subsoiled if necessary, deep ploughed (25–30cm) in autumn and power harrowed to produce a level and uncompacted tilth. Planting is practicable in the spring. About 12,000–18,000 of 20cm long and more than 0.8cm thick cuttings can be planted per hectare (Johansson *et al.*, 1992).

The *Populus* cuttings are placed in double rows with a row spacing of 75cm and 125cm between the double rows. The plant spacing in the row is 55cm (Johansson *et al.*, 1992).

First year growing shoots on each cutting are cut back to the ground level in the first winter and can be used for making cuttings to establish new plantations. The regrowth of cut back poplar induces the generation of a large number of new shoots.

Crop management

Uncontrolled weeds in the first season will reduce growth up to 50% and dry matter yield up to 20% of that from poplars in weed free conditions (Parfitt *et al.*, 1991). Therefore, land for growing poplar must always be cleaned of perennial weed rhizomes before planting, and weed control is necessary during the first few years. Poplar plantations are able to suppress weeds from the second growing year on.

No fertilizer is applied in the planting year because of the risk that the weeds benefit more than the poplars. An average application of 60–80kg N, 10–20kg P and 35–70kg K per hectare per year is suitable in the following years, depending on soil fertility. The fertilizer regime should be adapted to the soil type and the natural mineralization, and to the provision of nitrogen via rainwater.

Poplar can be attacked by numerous leaf consuming, stem sucking and wood boring insects (Hunter *et al.*, 1988). In most cases, none of these pests represents a serious problem. A poplar plantation is used by a great number of different insects, birds and animals. The insects' natural enemies and the high number of shoots and leaves on 1–3 year parts of the rootstocks should ensure that normal levels of insect attack are not seriously damaging. Diseases are more serious, and numerous pathogens are described, especially different species of rust. Different clones have varying susceptibilities to fungi and insects and can be selected for better resistance. Generally, there is little reason to introduce inputs for insect and disease control. Considerable damage may be caused by mice, rabbits, deer and farm animals during the establishing phase of poplar. Established cultivars may be at less risk.

Production

Populus plantations grown as SRC can be harvested for the first time 3–4 years after cut back. This holds true for all subsequent harvest periods. The best harvest time is in winter with a dry matter percentage of 50%. Current expectations are that there are eight or more cycles of harvests possible during the life of a poplar plantation. Annual yields up to 12–15 ODT/ha/year and more (ODT = oven dried tonne) are obtained in Germany from the second and third ratoon on (El Bassam, 1997). Increased yields are expected when the full benefits from improved materials have been obtained and when agricultural practices have been optimized.

Processing and utilization

Harvesting of stems takes place in winter when the leaves have fallen off. The two available harvesting methods – direct chipping and whole shoot harvesting – need different techniques. Direct chipping is possible if chips can be combusted at the harvest moisture content (50%) in large heating plants. Chips with a high moisture content will quickly deteriorate through microbial activity and must be combusted directly or ventilated for storage. Drier chips are needed in smaller plants and stationary ovens on farms. In this case, shoots are harvested whole to dry in piles during the summer. This material reaches a dry matter content of about 70%.

Dry wood chips for energy production need no further processing and can be burned or gasified (Parfitt and Royle, 1996). *Populus* chips can be utilized in two ways. Direct combustion in special automated boilers will give low grade space heating of 50 to 400kWh. With high pressure steam techniques, steam turbines can be powered with up to 10MW installed capacity (Graef, 1997). Another possibility is the gasification (pyrolysis) of wood chips to drive an engine to power a generator and produce heat (a system known as combined heat and power, or CHP). A growth rate of 15ODT/ha/year is equivalent to the 7000l oil needs of two one-family villas.

Selected references

El Bassam (1997) Unpublished data.

FAO (1996) Ecocrop 1, Crop Environmental Requirements Database, FAO Software Library.

Franke, W. (1981) *Nutzpflanzenkunde – nutzbare Gewächse der gemäßigten Breiten, Subtropen und Tropen*, Thieme, Stuttgart.

Graef, M. (1997) Personal communication.

Hunter, T., Royle, D.J. and Stott, K.G. (1988) Diseases of Salix biomass plantations: Results of an international survey in 1987. *Proceedings of the International Energy Agency/Bioenergy Agreement Task II Workshop Biotechnology Development*, Uppsala, Sweden, pp. 37–48.

Johansson, H., Ledin, S. and Forsse, L.S. (1992) Practical energy forestry in Sweden: A commercial alternative for farmers. In: Hall, D.O. *et al.* (eds), *Biomass for Energy and Industry*, Ponte Press, Bochum, pp. 117–26.
Parfitt, R.J. and Royle, D.J. (1996) *Populus*, poplar. In: *Renewable Energy – Potential Energy Crops for Europe and the Mediterranean Region*, REU Technical Series, 46, FAO.
Parfitt, R.J., Clay, D.V., Arnold, G.M. and Foulkes, A. (1991) Weed control in new plantations of short rotation willow and poplar coppice. *Aspects of Applied Biology* 29: 419–23.
Schütt, P., Schuck, H.J., Aas, G. and Lang, U.M. (eds) (1994) *Enzyklopädie der Holzgewächse, Handbuch und Atlas der Dendrologie*, Ecomed Verlagsgesellschaft Landsbergam Lech, III-2, pp. 1–16.

RAPE (*Brassica napus* L.)

Description

Rape is an annual C_3 plant which originated from the Mediterranean region. It is a member of the Cruciferae family and has both winter and spring forms. The name is derived from the Latin word *rapum*, which means 'turnip', a plant to which rape is closely related. The plant germinates quickly, forming a deep growing taproot and a rosette of blue-green leaves from which emerge 7–10 lateral shoots. On the ends of the branched stems grow the gold-yellow flowered racemes. Each plant has approximately 120 long slender seed pods; 40 to 60 of these are found on the main shoot. Each seed pod contains 18–20 seeds (2000–3000 seeds per plant) (Honermeier *et al.*, 1993). The seeds are small, round and black.

Based on its seed oil, rape belongs to the erucic acid group of oil plants. Of the total fatty acid content, about 6% is saturated fatty acids and 94% is unsaturated fatty acids. Among the unsaturated fatty acids, 14% is oleic, 45% erucic, 14% linoleic and 10% linolenic (Rehm and Espig, 1991). Low erucic acid and acid free cultivars (0-rape) and cultivars also low in glucosinolate have been developed in Europe and Canada because of health problems associated with the consumption of erucic-acid-containing oils. Cultivars with no erucic acid and low glucosinolate content have also been bred (00-rape). Canola oil is the name used for low-erucic-acid rapeseed oil.

Ecological requirements

Rape is an ecologically demanding plant. A deep sandy loam rich in humus and nutrients and with an optimal lime content is the most appropriate soil for rape cultivation, followed by humic loam, loam and clayey loam in decreasing order of suitability. Humic and loamy soils can be appropriate when a sufficient water supply is guaranteed in April, though in general the success of the crop is greatly influenced by a sufficient water supply during the vegetation period. Because of spring rape's weaker root development it is more sensitive to water deficit than winter rape. Marshy or waterlogged soils are unsuitable (Honermeier *et al.*, 1993).

After sowing winter rape, the plant should have about 100 days with temperatures over 2°C to reach the 8–10 leaf stage and develop the taproots necessary for wintering. The winter can be moderately cold with a light snow cover. Temperatures below –15°C may be damaging. There should be very little spring frost, and the spring should be moderately warm. In late winter or early spring when the temperature remains constantly above 5°C the plant will begin leaf growth (Honermeier *et al.*, 1993). Summer rape is planted, usually in northern latitudes, where the winters are too severe for winter rape to survive the hibernation period.

Propagation

Rape is propagated using seed. Variety selection is based on the desired glucosinolate and erucic acid in the end product. Other factors that influence seed selection are location, climate, yield potential, date of ripening, special tolerances and planting time (that is, winter and spring forms). The most commonly used forms fall into the categories '0-rape' (elimination of erucic acid) and '00-rape' (elimination of erucic acid and greatly reduced glucosinolate content). These characteristics are necessary if the crop's seeds are to be used by the food industry (Gross, 1993). In Germany the best yields have been obtained from the winter rape varieties 'Lirajet', 'Wotan' and 'Silvia', with 'Falcon', 'Liberator' and 'Vivol' having reached middle value yields in comparison. Spring rape seeds with high yield and oil contents are 'Lisonne' and 'Evita' (Rottmann-Meyer *et al.*, 1995).

Crop management

Tests comparing conventional tillage (CT) (25cm deep plowing) and minimum tillage (MT) (10–15cm deep disk harrowing) were conducted over a 3 year period on very sandy soil in Italy. The results showed that there was no significant difference in rapeseed grain or biomass yields between the two forms of tillage, but MT was found to lead to a progressive worsening of soil conditions for plant root growth. MT was found, however, to lead to an average reduction of 55% for working time, fuel consumption, energy requirement and cost when compared with CT (Bonari *et al.*, 1995).

Winter rape is sown from the middle of August to the beginning of September. As a guide, areas in northern Germany usually plant rape between 10 and 20 August, while in southern Germany planting takes place between 20 August and the beginning of September. The desired crop density is between 60 to 80 plants/m^2, which is achieved by using 3–4kg/ha of seed. The seed should be planted at a depth of 1.5–3.0cm, with a distance between rows of 13.5cm, as with wheat. The distance between rows can be increased to 20–28cm if the soil is in optimal condition. Spring rape is sown from the end of March to the beginning of May. Up to twice as much seed as with winter rape may be necessary to obtain the same desired crop density (Rottmann-Meyer *et al.*, 1995).

Pre-emergence or post-emergence herbicides can be applied to the field before sowing. Problem weeds are: chickweed, field foxtail grass, couch grass, chamomile, blind nettle, pansy, annual wild oats and reappearing wheat, among others.

Fungicides may also be used. Among the most common fungi to attack rape crops are *Sclerotinia sclerotiorum*, *Botrytis cinerea*, *Phoma lingam* and *Alternaria brassicae*.

Insecticides can be used to combat the variety of pests that infest rape crops. These pests include fleas (*Psylloides chryocephala*), lice (*Brevicoryne brassicae*), snails (*Deroceras agreste*), pollen beetles (*Meligethes aeneus*) and weevils (*Ceutorhynchus* spp.).

Table 7.50. Fertilizer levels for winter rape as recommended in Germany.

Fertilizer	Amount (kg/ha)
Nitrogen (N)	0–50 (in the autumn) Up to 100 (at the beginning of year with start of vegetation) 80–100 (4 weeks later)
Phosphorus (P_2O_5)	80–100
Potassium (K_2O)	180–220
Magnesium (MgO)	25–30

Table 7.51. Recommended fertilizer levels for spring rape (Rottmann-Meyer et al., *1995).*

Fertilizer	Amount (kg/ha)
Nitrogen (N)	80–100 (with sowing) 60–80 (at 6–8 leaf stage)
Phosphorus (P_2O_5)	80
Potassium (K_2O)	120
Magnesium (MgO)	40

Organic fertilization using liquid manure is possible and recommended. Nitrogen fertilizer should be used according to the development of the crop. Table 7.50 gives the fertilizer levels recommended in Germany for winter rape.

Recommended levels for spring rape are shown in Table 7.51.

Crops coming before rape should be early harvested crops such as winter barley or peas. Because of the danger of nematodes infesting the crop, rape should not include sugarbeet in its crop rotation. Nor should host plants of *Sclerotinia* be used in rape's crop rotation. A break in rape cultivation of 4–5 years should be adhered to.

Studies by Christen and Sieling (1995) in northern Germany demonstrated that the best yield was obtained when rape was cultivated following peas. Rotations with cereal crops also had favourable yields. The lowest yields were obtained when rape was grown in monoculture. It was found that in general the yields of oilseed rape increased with the length of the rotation and the length of the break between two rape crops.

Production

One way of looking at the value of the biomass from a crop is to look at its coefficient of energy performance (energy balance). This is the ratio of total energy obtained from the crop to total energy used by the crop, so ratios greater than one mean that energy is generated. Rapeseed methyl ester (RME), which can be used as a substitute for diesel fuel, has an energy balance of 5.5 including straw and 2.9 not including straw; on set aside land this becomes 6.1 including straw and 3.2 not including straw. In comparison, ethanol from wheat had energy balances of 3.6, 3.9 on set aside; and ethanol from sugarbeet had energy balances of 2.46, 2.53 on set aside. Fossil fuel refining has an energy balance of 0.94, not including losses from extraction, transport, etc. (Poitrat, 1995). This shows that biofuels have a positive energy balance, while mineral fuels have a negative energy balance.

In a study of oil crops in the USA by Goering and Daugherty (1982), it was determined that the total energy input for a non-irrigated spring rape crop was 7624MJ/ha. This includes the entire range of costs from seed price to oil recovery costs. The energy output was determined to be 20,066 MJ/ha, based on a seed yield of 1233kg/ha and an oil content of 41%. The energy output/input ratio was 4.18. This was the second highest ratio of nine different crops tested under conditions of non-irrigation. Soybean had the highest output/input ratio at 4.56.

ʾrocessing and utilization

ʾhe winter rape crop is harvested in July, after a vegetation period of approximately 30 days. The seed pods and upper shoots are grey–brown; the seeds are black–brown ɔ black, hard (when fully ripe), and rattle in the seed pods when moved (Honermeier *t al.*, 1993). At this time the moisture content has sunk to under 20%. Drying may be ecessary to achieve the desired 9% moisture content that makes the seed storable and eady for further processing. The straw can be immediately chopped up to be used for ɔdder, or dried for storage or further processing. Winter rape has a yearly average seed ield of about 3t/ha and a dry matter straw yield of 10–12t/ha (Honermeier *et al.*, 1993).

The spring rape crop is ready for harvesting when the plants take on a brown colour nd the seeds have become loose in their pods. Spring rape has a seed yield of 1.5–2.5t/ha Rottmann-Meyer *et al.*, 1995).

Graef *et al.* (1994) provides a look at the components that are processed from a rape rop. A whole plant rape crop of 8650kg/ha consists of 5470kg/ha straw and 3180kg/ha eed. The seed can be processed into 1848kg/ha rape seed meal and 1332kg/ha rape eed oil. After it is refined, 1279kg/ha of the oil remains. This is mixed with 139kg/ha ıethanol to produce 1285kg/ha (1460l/ha) of rape seed oil methyl ester and 133kg/ha f glycerol. The 1460l/ha rape seed oil methyl ester has an energy content of 47.8GJ/ha.

Raw rape seed is approximately 40% oil and 60% meal. The seed is processed by ırst pressing the seed to separate the oil from the meal. The oil can be further rocessed for technical uses. Among its many uses are hydraulic oils, chain saw oils, ıbricants, biodiesel (RME), biocomponents for heating oil and diesel fuel, and onstituents in cosmetics, ointments, plaster, and for leather processing. If the oil has o or low erucic acid content then it can be used by the food industry for such products s cooking oil, margarine and mayonnaise, or as a feed ingredient for cattle, pigs and irds. The meal undergoes an extraction process to produce the remaining oil; the esidual meal product can be used as fodder (Gross, 1993; Sipilä, 1995).

The straw from the rape crop can be used in boilers or flash pyrolysis power plants ɔ obtain heat and bio-oil. The bio-oil can be used by existing oil boilers or diesel enerators to produce power and heat (Sipilä, 1995).

elected references

onari, E., Mazzoncini, M. and Peruzzi, A. (1995) Effects of conventional and minimum tillage on winter oilseed rape (*Brassica napus* L.) in a sandy soil. *Soil & Tillage Research* 33(2): 91–108.

ʾhristen, O. and Sieling, K. (1995) Effect of different preceding crops and crop rotations on yield of winter oil-seed rape (*Brassica napus* L.). *Journal of Agronomy and Crop Science [Zeitschrift für Acker und Pflanzenbau]* 174(4): 265–71.

ʾoering, C.E. and Daugherty, M.J. (1982) Energy accounting for eleven vegetable oil fuels. *Transactions of the ASAE* 25(5): 1209–15.

ʾraef, M., Vellguth, G., Krahl, J. and Munack, A. (1995) Fuel from sugar beet and rape seed oil – mass and energy balances for evaluation. In: *Biomass for Energy, Environment, Agriculture and Industry, Proceedings of the 8th EC Conference*, Vienna, Pergamon, 1159–64.

ʾross, L. (1993) *Warum der 00-raps so interessant für uns ist ... Arbeitsheft*, Verband Deutscher Oelmühlen, Bonn.

ʾonermeier, B., Adam, L., Gottwald, R., Hanff, H., Hoffmann, G., Krüger, K., Patschke, K. and Zimmermann, K.H. (1993) *Ölfrüchte – Empfehlungen zum Anbau in Brandenburg*, Druckhaus Schmergow GmbH, Potsdam.

Poitrat, E. (1995) Energy balance of bioethanol and rapeseed methyl ester. In: *Biomass fo Energy, Environment, Agriculture and Industry, Proceedings of the 8th EC Conference*, Vienna Pergamon, 1156–58.
Rehm, S. and Espig, G. (1991) *The Cultivated Plants of the Tropics and Subtropics*, Verlag Josep Margraf, Priese GmbH, Berlin.
Rottmann-Meyer, Schindler, Rose, H. and Martens, R. (eds) (1995) *Nachwachsende Rohstoff – Möglichkeiten und Chancen für den Industrie- und Energiepflanzenanbau*, 2nd edn Landwirtschaftskammer, Hannover.
Sipilä, K. (1995) Research into thermochemical conversion of biomass into fuels, chemical and fibres. In: *Biomass for Energy, Environment, Agriculture and Industry, Proceedings of th 8th EC Conference*, Vienna, Pergamon, 156–67.

REED CANARYGRASS (*Phalaris arundinacea* L.)

Description

Reed canarygrass is a perennial C_3 plant. This rhizomatous grass is most commonl found growing on creek and river banks, in ditches and on damp to wet meadows i subtropic and temperate zones throughout the Northern Hemisphere. It produces a expansive and well developed underground rhizome and root system. The stem i generally stiff and upright, with branching often present in the lower part. The plan grows to a height of between 0.5 and 2.0m or more and produces a strong leaf shoot The ungrooved leaf blades are from 8 to almost 20cm long and 0.8–2.0cm wide, wit a matte underside. The inflorescence of canarygrass is a panicle between 10 and 20cm long. Canarygrass usually flowers in June and July (Siebert, 1975; Kaltofen an Schrader, 1991).

Ecological requirements

Success of the crop depends on a sufficient supply of nutrients and oxygen rich, non stagnant water. Reed canarygrass does not tolerate drought or salt. It is very toleran to wet soil, but waterlogged soil is not suitable because of the deficiency of oxyge (Kaltofen and Schrader, 1991).

Propagation

Reed canarygrass is propagated from seed, and vegetatively.

Crop management

Canarygrass is most appropriately sown in damp to wet fields, though the extent c dampness depends on the soil's ability to support the necessary farm machinery.

In field trials conducted by Lechtenberg *et al.* (1981), reed canarygrass was give 112, 224 or 336 kg/ha of N and was harvested two, three or four times a year. Yield at 336kg/ha N averaged 11.5t/ha. Plant recovery of N applied to canarygrass average 64, 67 and 55% for the 112, 224 and 336kg/ha rates respectively, when the plant wa harvested four times a year. Recoveries averaged 82, 67 and 63% respectively when i was harvested only three times. The yield increase resulting from an application of 11 nd 224kg/ha N was 30% greater with two and three cuts than with four cuts a year. n a biomass production system where herbage quality is not important, an infrequent arvest system may yield more biomass per unit of applied N.

Some of the benefits as seen in Sweden for reed canarygrass cultivation (Olsson, 993) are:

a low investment level for establishment
the machinery and facilities already exist
there is no need for contractors
flexible land use.

Canarygrass is known to be affected by rusts and mildew, along with other fungi.

Because reed canarygrass is a perennial plant there is currently no recommended rop rotation. A single crop duration of 10 years or more may be possible.

Production

Tests in Sweden comparing the varieties 'Palaton', 'Venture', 'Vantage' and Mutterwitzer' resulted in dry matter yields (after winter losses) of about 4.8t/ha, 3.3t/ha, .2t/ha and 4.2t/ha respectively. In addition, it was shown that despite the higher ields from summer harvested crops, the spring harvested crops had lower production osts (Olsson, 1993).

Mediavilla *et al.* (1993; 1994, 1995) obtained the dry matter yields (t/ha) from reed anarygrass field tests that were conducted using liquid manure for fertilizing at two ocations in Switzerland, Reckenholz and Anwil. The yields (t/ha) for 1993, 1994 and 995 were respectively 12.4, 11.0 and 13.5 at Reckenholz, and 19.3, 17.0 and 12.3 at nwil. The crop was harvested three times in 1993 and twice in each of 1994 and 1995. t Reckenholz between 70 and 90kg/ha NH_4-N liquid manure per year was used, hile at Anwil the rate was up to 180kg/ha NH_4-N.

Processing and utilization

Canarygrass is most commonly harvested once a year by mowing and baling. Drying onsists either of swath drying or barn drying, with storage in a barn or silo or under anvas. Drying important for reducing transportation costs and for maintaining good bre quality.

In Sweden, research on crop production using a delayed harvesting method is nder way (Olsson, 1994). This method involves harvesting the canarygrass crop after ne snow has melted in the spring (May) instead of the usual summer harvesting. This ads to death of the above ground plant and, along with field drying, moisture ontents as low as 10–15% can be achieved with delayed harvest crops. In addition, uring wintering, the plant returns much of its nutrients to the soil, but maintains the nportant cellulosic matter content.

Their generally lower density means that spring harvested crops can be baled using high density baler. Spring harvested crops have been shown to have improved quality or processing into fuel and pulp raw material. A disadvantage of this method is that inter losses of 20–30% in dry matter have been experienced.

Large scale trials comparing a traditional harvest time in August and the delaye harvest time in the spring showed that for the August harvest of 10.4t of total cro yield, 35% was lost during harvesting, resulting in 6.8t net yield. For the spring harves of 10.4t of total crop yield, 29% was lost during the winter and 18% was lost durin harvesting, resulting in 5.5t net yield (Olsson, 1993).

One of the most heavily researched and promising uses of reed canarygrass is a pulp for fine paper making. Qualities that are required of short fibres in printing grad papers are a large number of fibres per unit weight, and stiff, short fibres (lo hemicellulose content, and low fibre width to cell wall thickness ratio). Canarygras has short, narrow fibres (0.72mm mean length) and produces pulp with a greate number of fibres than most hardwoods. The delayed harvest crops have been show to lead to approximately 20% more pulp per tonne of grass compared to summe harvest crops. The grass fibres are easy to pulp and can be cooked and bleached usin existing techniques.

Because of its low moisture content, spring canarygrass makes a good candidate fc fuel upgrading to pellets, briquettes and powder (with qualities similar to woo powder). In addition, the delayed harvest crop shows a reduction in undesirabl elements, making it more suitable for combustion as well as better for the environ ment. The heating value of the grass is about 4.7kWh/kg dry matter.

Table 7.52 provides the fuel characteristics of reed canarygrass as found in summe harvest and delayed harvest crops. The information was compiled from 1991 and 199 Swedish trial sites (Olsson, 1994).

Table 7.52. Fuel characteristics of reed canarygrass.

Parameter (% of dry matter)	Summer harvest	Delayed harvest
Net heating heat value (MJ/kg)	17.9	17.6
Ash	6.4	5.6
Carbon	46	46
Hydrogen	5.7	5.5
Nitrogen	1.33	0.88
Sulphur	0.17	0.09
Chlorine	0.56	0.09
Volatile matter	71	74
Initial ash deformation (°C)	1074	1404
Potassium	1.23	0.27
Calcium	0.35	0.20
Magnesium	0.13	0.05
Phosphorus	0.17	0.11
Silica	1.2	1.85
Cadmium (mg/kg)	0.04	0.06

Reed canarygrass, along with a few other plants, are currently being used as plar cover for wastewater treatment in the Czech Republic. This is part of a ne constructed wetlands technology (Vymazal, 1995).

Although canarygrass is presently part of breeding programs to make it mor advantageous for biofuel and pulp production, it is still a good fodder grass for hay an silage production, especially in flood plain meadows.

Selected references

Kaltofen, H. and Schrader, A. (1991) *Gräser*, 3rd edn, Deutscher Landwirtschaftsverlag, Berlin.

Lechtenberg, V.L., Johnson, K.D., Moore, K.J. and Hertel, J.M. (1981) Management of cool-season grasses for biomass production. *Agronomy Abstracts*, 73rd annual meeting, American Society of Agronomy. Madison, Wisconsin, USA.

Mediavilla, V., Lehmann, J. and Meister, E. (1993) *Energiegras/Feldholz – Teilprojekt A: Energiegras*, Jahresbericht 1993, Bundesamt für Energiewirtschaft, Bern.

Mediavilla, V., Lehmann, J., Meister, E., Stünzi, H. and Serafin, F. (1994) *Energiegras/Feldholz – Energiegras*, Jahresbericht 1994, Bundesamt für Energiewirtschaft, Bern.

Mediavilla, V., Lehmann, J., Meister, E., Stünzi, H. and Serafin, F. (1995) *Energiegras/Feldholz – Energiegras*, Jahresbericht 1995, Bundesamt für Energiewirtschaft, Bern.

Olsson, R. (1993) Production methods and costs for reed canary grass as an energy crop. In: *Bioenergy Research Programme*, Publication 2, Bioenergy 93 Conference, Finland.

Olsson, R. (1994) A new concept for reed canary grass production and its combined processing to energy and pulp. In: *Non-wood Fibres for Industry*, Pira Int./Silsoe Research Institute Joint Conference, Bedfordshire, UK.

Pahkala, K.A., Mela, T.J.N. and Laamanen, L. (1994) Mineral composition and pulping characteristics of several field crops cultivated in Finland. In: *Biomass for Energy, Environment, Agriculture and Industry, Proceedings of the 8th EC Conference*, Vienna, Pergamon, 395–400..

Siebert, K. (1975) *Kriterien der Futterpflanzen einschließlich Rasengräser und ihre Bewertung zur Sortenidentifizierung*, Bundesverband Deutscher Pflanzenzüchter e.V. Bonn.

Vymazal, J. (1995) Constructed wetlands for wastewater treatment in the Czech Republic – State of the art. *Water Science and Technology* 32(3): 357–64.

ROCKET (*Eruca sativa* (L.) Mill)

Contributed by: A. K. Gupta

Description

Rocket is a C_3 annual plant that grows to a height of 0.3–1m. It is a member of the rapeseed and mustard group of crops (n=11). The plant is native to the Mediterranean region but is distributed from India in the east, to the central USA in the west, and to Norway in the north. Rocket is known by different local names such as: Ölrauke or Senfkohl (Germany), roquette (France), ruchetta or ruccola (Italy), roqueta, oruga or jaramago (Spain), rocket, rocket-salad or hedge mustard (UK), taramira (in Hindi), etc.

The roots of the rocket plant are of spindle form and less branched (Schuster, 1992). The stem is branched and the leaves are compound resembling those of spinach. The flavour of young, tender leaves is pungent and they used in southern Europe as a salad. However, rocket is commercially cultivated for its oil bearing seeds, though the oil (about one-third of the seed weight) is non-edible because of the presence of erucic acid and it is used only for secondary purposes.

India, with more than 6 million hectares under oleiferous Brassicaceae, seems to be the largest producer of rocket. Here it is raised as a main crop on moisture deficient and saline areas of the mustard belt in the north-western parts of the country. More frequently, rocket is chosen as a contingent crop in the event of failure of a winter crop, or because of delayed winter rains from a receding monsoon. These factors mean that there is no trend over time with respect to the area and production of rocket in India.

Ecological requirements

The ecological requirements of rocket are similar to those of rapeseed and mustard. It is a cool season crop where the winter is mild. Nevertheless, the plant has some unique inherent features that allow it to survive under adverse soil and atmospheric conditions in which crop plants can hardly grow. These characteristics include an efficient and fast penetrating root system that imparts drought hardiness, tolerance to salinity, a wide adaptability towards sowing time, tolerance of frost, and less susceptibility to biotic stresses (Kumar and Yadav, 1992). The crop thrives well on manured loamy soils with a pH range of 6–8 and having good provision for drainage.

Propagation

Rocket is propagated using seed. Because of its small seed size, rocket requires a seed bed of fine tilth. When the crop is to be sown in summer fallow, disc harrowing after every effective shower is the best method of conserving moisture. After the cessation of rains, harrowing should invariably be followed by planking to prepare a fine and compact seed bed. In rainfed areas, germination can be improved by presoaking the seed in water, or by using 'Jalshakti' starch polymer (Gupta and Agarwal, 1997). Because of its confinement in small and isolated pockets, rocket has been a neglected crop and, consequently, high yielding seed is a great problem. A normal seed rate of 3kg/ha is adequate to obtain the desired plant population of 225–250 thousand plants per hectare, but a higher rate is usually adopted to compensate for the poor germination that is mainly due to suboptimal moisture conditions.

Seed treatment with a fungicide such as Bavistin, Thiram, etc. protects the crop from seedborne diseases and in turn ensures uniform crop emergence and an optimum stand. Spacing of 30cm × 15cm is ideal for optimizing biomass yield. Seed should be sown to a depth of 3cm in open furrows with a ridge seeder. As with any other crop, the date of sowing (mainly governed by temperature) is an important factor in determining the growth and performance of rocket. For instance, around the first week of October has been shown to be the optimum sowing time in north-west India. Delayed sowing resulted in reduced yield both of biomass and oil primarily as a result of the shortening of the reproductive phase and an increased incidence of aphids. On the other hand, high temperature in early sowing led to improper canopy development and reduced branching of stems. The attack of seedling killing insects such as *Bagarada hilaris* was also highly observed at early sowing.

Crop management

Because rocket has been grown on wastelands, it has continued to rely on natural selection for survival. As a result, the prevailing landraces and cultivars do not respond to intense input management. The problem of soil inhabiting insects can be tackled by applying methyl parathion at 25kg/ha. With regard to nutrient management, the available literature suggests that rocket is a moderate feeder. An application of 40kg/ha

N, 20kg/ha P_2O_5, 15kg/ha K_2O and 90kg S will be enough. Further, pre-emergent application of Fluchloralin (1 kg/ha AI) is effective against most of the weeds, whereas hand weeding requires 15 labourers per hectare.

Irrigation is not required if rain is well distributed. The pod formation stage is recognized to be most crucial with regard to water stress; ensuring adequate moisture at this stage significantly boosts the seed yield. Although the established crop is not prone to significant damage by insects, even here the probability of any economic injury can be prevented by spraying a systemic insecticide. Rocket has been shown to be twice as frost tolerant as Indian mustard; however, a spray of 0.1% sulphuric acid is reported in the literature to protect a crop against frost injury. Rust/blight and/or mildew can be controlled by spraying a 0.2% solution of Dithane M-45.

Its deep root system makes rocket a good preceding crop in a crop rotation. In India, a crop rotation of cowpea or clusterbean/rocket has proved to be more profitable. In multiple cropping systems, rocket /summer greengram or a fodder crop/maize or cowpea can be an appropriate option. There are no recommendations or restrictions for possible crops to follow or precede rocket.

Production

A higher production level cannot be expected from the local varieties and the poor management they receive. Seed and straw yields of 1.5t/ha and 4t/ha respectively have been obtained in various locations in India. There is considerable scope for raising the productivity level of this crop through crop improvement and proper management.

Processing and utilization

The crop is considered mature and ready to harvest when the siliquae have turned brown–yellow and the seeds inside the pods have become loose. In India, the crop matures in about 140–150 days. The crop is harvested manually using sickles. Harvesting needs 18 labourers per hectare. The harvested material is brought to an area for threshing to reduce the moisture content of the seed to about 9%, which makes it storable and ready for processing. Currently, the seed is threshed out manually by beating the plants with a long stick or by hitting small bundles of the crop against a hard surface. Seed is sold in the market, and then oil is extracted and used for various purposes.

The aerial part of a mature plant consists of about 20% seed and 80% straw. The seed can be processed into approximately 33% oil and 67% oilcake. The energy value of rocket oil is not available, but is believed to be comparable with that of rapeseed. Oilcakes serve as a concentrated feed for dairy animals or are used as a manure. Straw is used for bedding material in cattle sheds or burned to generate heat. The present method of disposing of straw cannot be called efficient, effective and environmentally sound. An alternative strategy is needed to make the best possible use of rocket biomass.

Selected references

Gupta, A.K. and Agarwal, H.R. (1997) Taramira: A potential oilseed crop for marginal lands of north-west India. In: *Proceedings of the International Conference on Sustainable agriculture for Food, Energy and Industry*, 22–28 June 1997, FAL, Braunschweig, Germany. In press.

Kumar, D. and Yadav, I.S. (1992) Taramira (*Eruca sativa* Mill) research in India – present status and future programme on yield enhancement. In: Kumar, D. and Rai, M. (eds), *Advances in Oilseeds Research*, vol. 1, Scientific Publishers, Jodhpur, India, pp. 329–58.
Schuster, W.H. (1992) *Ölpflanzen in Europa*, DLG Verlag, Frankfurt am Main, Germany, 240 pp.

ROOT CHICORY (*Cichorium intybus* L.)

Description

The best known wild plant from the genus *Cichorium* (family Cichoriaceae) is *Chicory intybus* L. var. *intybus*. It is a biennial C_3 plant with strong cylindrical to cone shaped roots. The first year is marked by the development of a flat adjacent leaf rosette and the roots. The inflorescence is developed in the second year and by the end of July the plant will produce blue flowers (Franke, 1985). One of the main values of chicory is its root content of inulin, a polyfructoside, which can be processed by the chemical, pharmaceutical and food industries for various uses. Extracts from roasted roots have also been used to improve the aroma of coffee. The traditional areas of chicory cultivation have been northern France and Belgium, though the total EU cultivated area was approximately 14,000ha in 1992.

Ecological requirements

Appropriate soils for chicory cultivation are sandy loams, loams and other soils that have less than a 30% clay content. The soil should be well drained, but not too dry, and quick to warm up in the spring. The optimal pH value is between 6 on sandy soils and 7 on loamy soils. Not appropriate are soils that are excessively wet or acidic, have too many stones, or do not permit an even, level seedbed (Dhellemmes, 1987; Baert, 1991, 1993).

Chicory is most commonly produced in a mild, moist climate such as in north-western France and Belgium. There the crop can be harvested up until December.

The nutrient content of the roots depends, to a degree, on the variety of chicory planted. In general, 100g of root dry matter consists of 0.9–1.0% of nitrogen, 1480–1986mg potassium, 34–42mg sodium, 212–237mg calcium, 88–94mg magnesium and 271–318mg phosphorus.

Propagation

Chicory is most commonly propagated using seed. The plant has already undergone extensive breeding and is still the subject of breeding. Among the traits that are being improved are better emergence, adaptability to an earlier sowing, faster growth, higher sugar yield, salt tolerance in the germinating plants, resistance to disease and damage, etc. The plant may also be propagated by planting seedlings, which has the advantage of controlling undesired crossings. When choosing the variety to be planted it is important to take note of the potential yield and date of ripening, among other things.

Crop management

Following a spring ploughing of the field to a depth of about 30cm, the sowing of the crop by seed takes place between the end of April and the beginning of May. For smaller seeds it is important that the field be flat and level and has a soil temperature of at least 10°C. A machine for drilling sugarbeet can be used. The seed can be either coated or uncoated. Uncoated seed is sown at a depth of 0.5–1.0cm. Coated seed is sown at a depth of 1.0–1.5cm. The distance between rows should be 45cm and the distance between plants should be about 9cm. An appropriate crop density is 140,000 to 160,000 plants per hectare, which is achieved with approximately 250,000 seeds.

Chicory reacts very sensitively to poor irrigation and crop management during germination, or to an encrusted soil surface after a heavy rainstorm. If after 14 days no germinating plants can be seen, the crop should be resown (Dhellemmes, 1987; Baert, 1991).

Mineral fertilizers should not be used shortly before sowing because of possible erosion of the germ roots. The soil should be analysed to determine its composition and to help determine fertilizer levels, but according to Bramm and Bätz (1988), recommended fertilizer levels in Germany are as shown in Table 7.53.

Table 7.53. Recommended fertilizer levels in Germany for chicory crops.

Fertilizer	Amount (kg/ha)
Nitrogen (N)	100–130
Phosphorus (P_2O_5)	60–130
Potassium (K_2O)	200–250
Magnesium (MgO)	40–60

Two-thirds of the nitrogen should be applied 3 weeks before sowing, and the remaining one-third at the six leaf stage.

Fields with serious weed problems should not be used for cultivating chicory. Because chicory is slow to establish itself, a pre-sowing and/or pre-emergence herbicide is recommended to help control weeds.

Chicory should not be cultivated after sugarbeet, carrot, sunflower, bean, lettuce or endive (wild chicory) crops because of the danger of *Sclerotinia*. There are no recommendations or restrictions for possible crops to follow chicory. Chicory has a deep growing and wide spreading root system, so the soil is usually left in a very good condition after the crop.

Production

With yield variations depending on the climate, location and crop management, root yields of 32 to 50t/ha (Anonymous, 1993) have been achieved, and yields as high as 56t/ha with an inulin content of 18.2% can be foreseen. This would produce an inulin-fructose yield of over 10t/ha (Baert, 1993). Root dry matter yields between 10.6 and 16.5t/ha have been observed. This resulted in inulin–fructose yields between 8 and 12.2t/ha (Meijer and Mathijssen, 1991).

Processing and utilization

Chicory seeds are usually harvested in the first half of September of the second year. Seed moisture contents of 12–15% make it appropriate to allow the seeds to dry naturally in an elevator. If the moisture content is more than 15% then the seeds will need to be dried before storing. Although the seed yield largely depends on the year's weather, yields of 600 to 1200kg/ha can be expected (Delesalle and Dhellemmes, 1984).

Chicory roots are usually harvested in the autumn, around the middle of November depending on the time of ripening for the variety used. A modified six row sugarbeet rooter can be used to harvest the roots.

According to Frese *et al.* (1991), chicory roots have a dry matter content of 22.2–26.8%, a total sugar content of 17.4–21.9%, a fructose content of 14.6–19.1% and a glucose content of 2.5–3.5%. Because the fructose fraction in syrups produced from inulin is up to 75%, fructose syrup production from chicory is possible (Guiraud, 1983). The polyfructosides, which grow in chain length throughout the vegetation period, also have the potential to be used for ethanol production.

Selected references

Anonymous (1993) Inulin Gewinnung bei Suiker-Unie. *Starch/Stärke* 45: 158.

Baert, J.R.A. (1991) Cultivation and breeding of root chicory for inulin production, *Proceedings of the 1st European Symposium on Industrial Crops and Products*, Maastricht, Netherlands.

Baert, J.R.A. (1993) The potential of inulin chicory as an alternative high income crop, *Proceedings of EC Workshop on the Production and Impact of Specialist Minor Crops in the Rural Community*, Brussels, Belgium.

Bramm, A. and Bätz, W. (1988) *Industriepflanzenbau, Produktions- und Verwendungsalternativen*, BML, Bonn

Delesalle, L. and Dhellemmes, Ch. (1984) Semences de betteraves et de chicorée industrielles. Betterave porte-graine le repiquage: une pratique exigeante. *Bulletin Semences* 86: 39–40.

Dhellemmes, Ch. (1987) Agronomy of chicory. In: Clarke, R.J. and MacRae, R. (eds), *Coffee*, vol. 5: *Related Beverages*, Elsevier, London, pp. 179–91.

Franke, W. (1985) Inulin liefernde Pflanzen. In: *Nutzpflanzenkunde*, 3rd edn, Georg Thieme Verlag, Stuttgart and New York.

Frese, L., Damroth, M. and Bramm, A. (1991) Breeding potential of root chicory. *Plant Breeding* 106: 107–113.

Guiraud, J.P., Bajou, A.M., Chautard, P. and Galzy, P. (1983) Inulin hydrolysis by an immobilized yeast-cell reactor. *Enzyme Microb. Technol.* 5: 185–90.

Meijer, W.J.M. and Mathijssen, E.W.J.M. (1991) Inulineproduktie via aardpeer of cichorei? In: Meijer, W.J.M. and M. Vertregt (eds), *Agrobiologische Thema's CABO-DLO*, vol. 4: *Gewasdiversifikatie en Agrificatie*, Wageningen, Netherlands.

ROSIN WEED (*Silphium perfoliatum* L.)

Description

Rosin weed is a perennial wild C_3 plant originating in North America that grows to a height of 2 to 3.5m. The stalk is strong and relatively square with a diameter of 20–25mm. The plant has six to eight leaf pairs opposing each other on opposite sides of the stalk. The leaves are coarse with a length of over 30cm and a width of 25cm. The leaves towards the top of the stem grow around the stem to form a cup that captures and holds precipitation.

Flowering begins sometime in July, at which time the plant produces numerous yellow flowers, each having a diameter of 5–8cm. The flowers produce nectar and pollen that attract many insects to the plant. The seeds ripen at the end of September, with each flower producing 20 to 30 seeds. This amounts to approximately 400–500 seeds per plant. Rosin weed develops a very strong root system, which consists of one main root reaching a length of up to 100cm and numerous lateral roots that spread out near the surface of the soil.

Ecological requirements

Rosin weed has no special requirements from the soil. The plant thrives on soils with a high water level and marshy soil. Only soils with a high salt content are unsuitable. Rosin weed is also open to a wide range of climates. The plant has an excellent winter hardiness, being able to withstand temperatures below –30°C. The plant also has a distinct drought resistance. A sufficiently warm summer, such as in wine grape growing areas, is recommended for rosin weed crops. Wet and cold years lead to a diminished harvest.

Propagation

Although rosin weed can be propagated using plant cuttings, the most common propagation method is by seed. The seed can be directly sown into the field, or it can be cultivated in pots first and then transplanted into the field at a height of 6–10cm.

Crop management

The field should be properly prepared to avoid weeds in the first year. The seed can be sown in the spring, after sufficient field preparation with ploughing and harrowing. Approximately 10kg/ha of seed is used, at a sowing depth of 1–1.5cm. The distance between plants is 25–50cm and between rows 50cm. The seeds germinate 3–4 weeks after being sown. Afterwards, the plants grow very slowly and the soil needs to be hoed two or three times to remove weeds. By the time the plants have reached a height of about 1m they are large enough and crowd out the weeds. However, a weed hoeing is recommended at the beginning of each additional year.

Diseases and harmful pests of rosin weed have not yet been encountered.

Rosin weed produces a large amount of above ground and below ground biomass, which is very positively influenced by application of fertilizer. Experience from plots in Germany has concluded that an organic fertilizer such as stable manure is recommended at the beginning of the growth cycle and then later liquid manure can be used. The additional inorganic fertilizer levels have also been advised (Table 7.54).

Stable manure fertilized crops such as potatoes, turnips and cabbage are good prior crops, but because rosin weed has a life of more than 10 years it may not be necessary to develop a crop rotation.

Table 7.54. Recommended fertilizer levels for rosin weed, as used in Germany.

Fertilizer	Amount (kg/ha)
Nitrogen (N)	100–125
Phosphorus (P_2O_5)	110–115
Potassium (K_2O)	125

Production

No market or market value has been established for rosin weed. Because the plant has been used mainly as feed for animals, usually directly where it is grown, it has not been sufficiently established as a crop.

In Germany, yields have been recorded in four areas: fresh cut yield in secondary mountains was approximately 70t/ha, on dry loess soil 100t/ha, on hilly land in Thuringen 130t/ha, and on the coast 140t/ha. These yields correspond to dry matter biomass approximations of 8, 12, 15 and 16t/ha respectively.

Processing and utilization

When used for animal feed the plant is harvested in September, at which time it has a dry matter content of 13.9%. It is also possible, starting in the second year, to harvest the portion of the plant above 0.8–1.2m one to three times a year. This has been shown to lead to a higher total yield than a once a year harvest. Because planted fields have been small until now, the plant has been harvested manually with a sickle.

When used as feed, rosin weed is chopped into small pieces and fed directly to the animals. The plant as well as the techniques of harvesting, handling and storage need to be researched with a view to large scale biomass production for energy production or conversion into a biofuel.

Selected references

Festerling, O. (1993) Personal communication.

VEB (1986) *Durchwachsene Silphie (*Silphium perforliatum *L.) Anbau and Verwendung*, Saat- und Pflanzgut Erfurt, Saatzuchtstation Bendeleben, 6 pp.

SAFFLOWER (*Carthamus tinctorius* L.)

Contributed by: L. Dajue

Description

The safflower is probably native to an area bounded by the eastern Mediterranean and the Persian Gulf. After the breeding of higher yielding cultivars it is now grown in other parts of the world too, for example in America or Australia (Diercke, 1981). Safflower is commercially cultivated to a large extent in India, Mexico and the USA

(Schuster, 1989). It is a herbaceous, thistle like annual or winter annual crop. The plant is branched and reaches a height of 30–150cm.

The stalkless oval-lacerate dark green leaves are usually set with spines. Leaf size varies widely among varieties and on individual plants. Typically it ranges from 2.5 to 5cm in width and 10 to 15cm in length. The leaves are coated with wax. The branches terminate in a globular flower head (capitulum). Flowers are commonly brightly coloured yellow, orange or red. Later the capitulum contains the seed. The plant develops a deep taproot, penetrating the soil 2–3m deep, that enables it to thrive in dry climates (Dajue and Mündel, 1996).

The number of spines on the outer involucral bracts varies from none to many; the diameter of the capitula ranges from 9 to 38 mm. The weight of 1000 seeds ranges from 25 to 105g. Finally, the oil content in the seeds varies from 14.90 to 47.45%. This shows a negative correlation between oil content and 1000-seed weight, and reveals that large seed is not very good for oil content. Lines that were used for the production of herbaceous medicine yielded the least oil (Dajue and Yunzhou, 1993). The photosynthesis pathway is C_3 II (FAO, 1996). Originally, the plant was grown for pigments extracted from its petals, but now it is commonly grown as an oil seed (Rehm and Espig, 1991). The evaluation of more than 2000 accessions of the World Collection of Safflower Germplasm (WCSG) showed a considerable diversity.

Ecological requirements

The production is restricted to the region between the latitudes 20°S and 40°N. It is usually grown below 900m elevation. The climate, especially the temperature, plays an important role in safflower growth. Emerging plants need cool temperatures (15–20°C) for root growth and rosette development, and high temperatures (20–30°C) during stem growth, flowering and seed formation. The seedling in the rosette leafy state is frost resistant from –7 to –14°C, depending on the variety. The mature plant is destroyed by a slight frost of –2°C.

The plant needs a light, medium deep and well drained soil. The best pH value is 6.5–7.5. Optimum precipitation ranges from 600 to 1000mm. Its deep root system makes it drought tolerant and it can survive with an annual rainfall of 300mm. The humidity should be medium to low. Bees or other insects are generally necessary for optimum fertilization and yields (FAO, 1996). Irrigation may be necessary, depending on the precipitation levels. To help ensure a good yield it is important that the crop receives sufficient water at least during the flowering stage (Zaman and Maiti, 1990). A dry period is necessary for ripening (FAO, 1988). It does not survive standing in water in warm weather, even for a few hours. Excess rainfall, especially after flowering begins, causes leaf and head diseases, which reduce the yield or even cause the loss of the crop (Dajue and Mündel, 1996). The plant is salt tolerant, which is a very useful quality if it is grown under irrigation (Rehm and Espig, 1991). Because of its tolerance to drought, it can be grown in those areas where other oil plants fail (Diercke, 1981).

Safflower is generally considered to be a day length neutral or long day plant. However, the origin of the variety is important: when summer crop varieties from temperate regions are sown during shortening days as a winter crop in tropical or subtropical regions, they show a delayed maturity (Dajue and Mündel, 1996).

Propagation

Safflower is propagated by seed. Seeding safflower into a firm, moist seedbed enhances emergence and stand and improves the vigour of the plant. Germination is followed by a slow growing rosette stage, during which numerous leaves and the strong roots are formed. In this stage the plant is endangered by fast growing weeds. After the rosette stage, the stem develops, elongates quickly and branches. Varieties that are almost free of spines have been developed for hand harvesting of the floral parts and seeds (Dajue and Mündel, 1996). Selection for high oil content reduced the pericarp thickness. Breeding aims are the improvement of resistance against diseases, and tolerance of cold and saline–alkaline conditions.

A number of wild species can be used as good sources with resistance and tolerance to several diseases and pests. Drought hardiness and disease resistance have been incorporated into cultivated types by repeated back-crossing and selection. Resistance against the fungus *Alternaria carthami* together with resistance against the bacterial pathogen *Pseudomonas syringea* is incorporated into several varieties. Other lines are resistant against rust (*Puccinia carthami*), verticillium wilt (*Verticillium albo-atrum* Reinke & Berth.), fusarium wilt and rot (*Fusarium oxysporum* Schlecht f. ssp. *carthami* Klis. & Hous.), rhizoctonia rot (*Rhizoctonia solani* Kuhn) and other diseases (Dajue and Mündel, 1996).

Crop management

The seed is sown 2–3cm deep, with planting distances of 10cm in the row and 30–60cm between the rows. The desired crop density is 40–50 plants per m^2, and up to 70 plants per m^2 on light soil (Rehm and Espig, 1991). Depending on the climatic conditions, safflower can be sown in autumn or in spring. The growing period is 200–245 days or 120–160 days, respectively. Organic fertilization is not recommended because of the non-calculable nitrogen release. Table 7.55 shows the inorganic fertilizer levels as recommended in parts of Germany.

Table 7.55. Fertilizer rates for safflower as recommended in Germany.

Fertilizer	Amount (kg/ha)
Nitrogen (N)	50–70
Phosphorus (P_2O_5)	60–80
Potassium (K_2O)	150–210

No recommendations can be made for herbicide use, as this aspect of weed control is still being tested, but herbicides appropriate for sunflower and soybean crops are believed to be appropriate for safflower too (Schuster, 1989). Mechanical weed control is still commonly practised.

A number of diseases and pests are known to attack safflower crops. The most serious disease that affects safflower in humid climates during and after the flowering is botrytis (*Botrytis cinerea*). The use of botryticides at the proper time can lessen the effect of this disease. Further diseases are: flower and leaf rot (*Pseudomonas syringae*), leaf spot (*Cerospora carthami*), root rot (*Phytophthora drechsleri*), rust (*Puccinia carthami*), seedling blight (*Phytophthora palmivora*) and wilt (*Sclerotinia sclerotiorum*). Pests include aphids (*Macrosiphum jaccae*), caterpillars (*Perigaea capensis*), fruit fly larvae (*Acanthiophilus helianthi*) and nematodes (FAO, 1988; Schuster, 1989).

Although safflower does not have any specific crop rotation, it is suggested that it may be placed between two cereal crops in the rotation. Safflower improves the soil quality by leaving behind its taproot and root system. A 5 year break in safflower cultivation should be adopted.

Production

The yields of seed can reach 1.1–1.7t/ha/year (FAO, 1996). With irrigation and good fertilization, yields of 2.8–4.5t/ha/year can be achieved, but the world average yield is 0.5t/ha/year (Rehm and Espig, 1991). The new cultivars have 36–48% oil in the fruit. The oil contains 73–79% linoleic acid, depending on the cultivar. The scrap cake portion amounts to 18–30%.

In a study of oil crops in the USA by Goering and Daugherty (1982) it was determined that the total energy input for non-irrigated safflower was 9313MJ/ha. This includes the range of costs from seed price to oil recovery costs. The energy output was determined to be 19,483MJ/ha. This was based on a seed yield of 1233kg/ha and an oil content of 40%. On irrigated safflower, irrigation was determined to cost 48,180MJ/ha, and the total energy input increased to 63,062 MJ/ha. The energy output was 44,278MJ/ha, based on a seed yield of 2802kg/ha. The energy output/input ratio decreased from 3.39 without irrigation to 1.14 with irrigation. As a comparison, soybean and rape had the highest non-irrigated output/input ratios at 4.56 and 4.18 respectively. Corn and sunflower had the highest irrigated ratios at 3.44 and 1.96 respectively.

Processing and utilization

The harvest takes place between the middle of August and the end of September. A more precise time can be determined when the stalks, leaves and flower tops dry out and turn brown. At this time the seeds should have a moisture content of 8–9%, which makes them suitable for storage. If the seeds have a higher moisture content they need to be dried. Harvesting consists of cutting the crop at ground level, drying it in the field for several days and threshing it (FAO, 1988). The use of a combine harvester should be possible. In countries with low labour costs the seed can be harvested by hand; spineless cultivars make this easier. Tests in India showed that seed production from a ratooned crop is possible (Dajue and Mündel, 1996).

The seeds are milled, and the oil is pressed out of the pulp or extracted. According to the procedure and the intended purpose of the press cake, the seeds are either shelled or left unshelled. The oil is edible, and because of its composition it is in demand in the nutritional industry. It is used for technical purposes too. The oil is quick drying and does not become darker, so it is valued for paints and lacquers (Rehm and Espig, 1991). It is also used as a fuel for lighting and as a component in soap production. It could also be used as biodiesel, as a fuel additive or as a biodegradable chainsaw bar oil.

The hulls can also be used as fuel. They are a low ash fuel that burns readily, but they are bulky. It is possible to process the hulls into a gas that can replace natural gas. Gasification systems that use agricultural wastes like rice hulls, cotton seed hulls, straw, etc. are available. These wastes can be successfully converted to clean hot gas (Bailey and Bailey, 1996). The hulls can be processed to charcoal, though volatile materials might also be produced. The pressed cakes are used as protein rich fodder. The value of the fodder is higher if the hulls are removed before pressing.

In addition to these energy uses, safflower can be grazed by cattle or stored as hay or silage. The yields and the feed value are similar to alfalfa. Safflower straw is used in a similar way to cereal straw (Dajue and Mündel, 1996).

Safflower oils differ widely in composition, depending on the cultivar. Some safflower oils are free of stearic acid or nearly free of palmitic acid (Dajue and Yunzhou, 1993). Table 7.56 shows characteristics of safflower oil as well as of the linoleic and oleic acid portions. Table 7.57 shows the cultivar dependent fatty acid contents of the oil.

Table 7.56. Some properties of safflower oil (Römpp, 1974; Peterson, 1985).

Property	Value
Density (g/ml)	0.972
Iodine number	140–150
Saponification number	188–194
Linoleic safflower oil	
Kinematic viscosity (mm^2/s)	32.3
Heat of combustion (kJ/kg)	39,226
Oleic safflower oil	
Kinematic viscosity (mm^2/s)	42.1
Heat of combustion (kJ/kg)	39,306

Table 7.57. The composition of safflower oil, depending on cultivar (% of the fatty acids) (Dajue and Mündel, 1996).

Oil type	Palmitic acid (C-16/0)	Stearic acid (C-18/0)	Oleic acid (C-18/1)	Linoleic acid (C-18/2)
Very high linoleic	3–5	1–2	5–7	87–89
High linoleic	6–8	2–3	16–20	71–75
High oleic	5–6	1–2	75–80	14–18
Intermediate oleic	5–6	1–2	41–53	39–52
High stearic	5–6	4–11	13–15	69–72

Selected references

Bailey, W.W. and Bailey, R. Jr. (1996) Clean heat, steam, and electricity from rice hull gasification. In: *Bioenergy '96, Proceedings of the Seventh National Bioenergy Conference*, 15–20 September 1996, pp. 284–7.

Dajue, L. and Mündel, H.-H. (1996) *Safflower* Carthamus tinctoris *L. Promoting the conservation and use of underutilized and neglected crops*. 7. International Plant Genetic Resources Institute, Gatersleben/International Plant Genetic Resources Institute, Rome, Italy. 83pp.

Dajue, L. and Yunzhou, H. (1993) The development and exploitation of safflower tea. Proceedings of the Third International Safflower Conference, 9–13 June 1993, Beijing. pp. 837–43
Diercke (1981) *Weltwirtschaftsatlas* 1, Deutscher Taschenbuchverlag / Westermann, Germany.
FAO (1988) *Traditional Food Plants*, FAO Food and Nutrition Paper no. 42, Rome.
FAO (1996) Ecocrop 1 Database, Rome.
Goering, C.E. and Daugherty, M.J. (1982) Energy accounting for eleven vegetable oil fuels. *Transactions of the ASAE* 25(5): 1209–15.
Peterson, C.L. (1985) *Vegetable Oil as a Diesel Fuel – Status and Research Priorities*. American Society of Agricultural Engineers, Summer Meeting 1985, Michigan State University, East Lansing, 23–26 June, Paper No. 85-3069.
Rehm, S. and Espig, G. (1991) *The Cultivated Plants of the Tropics and Subtropics*, Verlag Joseph Margraf, Preise GmbH, Berlin.
Römpp (1974) *Römpps Chemie Lexikon*, 7th edn, Franckh'sche Verlagshandlung, Stuttgart.
Schuster, W.H. (1989) Saflor. In: Rehm, S. (ed), *Spezieller Pflanzenbau in den Tropen und Subtropen. Handbuch der Landwirtschaft und Ernährung in den Entwicklungsländern*, 2nd edn, vol. 4, Ulmer, Stuttgart.
Zaman, A. and Maiti, A. (1990) Cost of safflower cultivation and benefit–cost ratio at different nitrogen levels and limited moisture supply conditions. *Environment & Ecology* 8(1): 232–5.

SAFOU (*Dacryodes edulis* H.J. Lam; syn. *Canarium edule* Hook.)

Contributed by: A. Riedacker and T. Silou

Description

The safou tree, *Dacryodes edulis* H.J. Lam, also reported under the name of *Pachylobus edulis* G. Don, *P. saphu* Engl. and *Canarium saphu* Eng., *C. edulis* Hook or *C. mubago* Fichalo, belongs to the family Burseracea (Okafor, 1983). It is also improperly called bush butter tree (Youmbi *et al.*, 1989), African pear (Omoti and Okiy, 1987), African plum and africado (Viet Meyer, 1990). The wood of its straight stem, which may reach 8 to 10m in height and up to 25 m when growing in the dense forest, is of good quality. The tree may be found in the Guinea gulf, from Sierra Leone down to Angola, and eastwards up to Uganda. It is a promising non-conventional oil tree, both for human nutrition (pulp) and oil production (pulp and kernel) (Silou, 1996), that has been domesticated for many years in countries like Cameroon and Congo and former Zaire.

Crop management

With urbanization, the trade in safou, the local name of the fruit, has recently taken off locally and even among the countries of Central Africa (Tabuna, 1993). Production starts in June and ends in November in the northern hemisphere (Cameroon, Nigeria and northern parts of Congo and Congo ex Zaire) whereas in the southern hemisphere it starts in November and ends in April.

The tree could be planted on equatorial and moist tropical grasslands, such as the 2 million hectare Batteke Plateau in Congo (Cailhiez, 1990), instead of or together with eucalyptus trees, which today cannot be further developed because of insufficient demand for that quality of wood.

Propagation

There are male and female trees; the average ratio is 1:1 but the optimum ratio would be 5 male to 95 female. However, the seeds are viable only for a few days (Kengue and Nya Ngatchou, 1991). Vegetative propagation, long considered as difficult, has recently been solved using aerial layering (Mampouya, 1991; Silou, 1996). Layers start to bear fruits after only 18 months instead after 5 to 6 years in the case of trees from seeds (Kengue and Tchio, 1994). This technique will now permit the creation of productive orchards.

Production

The harvesting of fruits on layers in Congo is easier than on large trees from seeds, and is less damaging for the fruits. The fragile fruit may thus be conserved for a longer time (Silou and Avouampo, 1997). There is a great variability in fruits: they may be 6 to 8cm long and 3.5 to 4.5cm thick, and weigh between 50 and 65g. The pulp (mesocarp) is 0.2 to 1cm thick. Some promising trees with large seedless fruits, late maturation and excellent taste have been identified (Avouampo, 1996). But there are also acid fruits. The fruit, much appreciated in the human diet, contains, on a dry weight basis, 30 to 70% lipids (average 50%), 13 to 27% proteins, carbohydrates and fibres in the pulp, and 10% oil in the kernel.

Although this species has not yet benefited from any selection, tree breeding or fertilizer programmes, high fruit yields have been recorded: 30kg on 5 year old trees, and from 100kg (in Congo) to 200kg (in Nigeria) on 10 to 15 year old mature trees (Silou, 1996). Fruit yields of 10 to 15t/ha may thus be achieved.

Oil from safou pulp has the same composition as that extracted from the kernels. Unlike most other oil producing species like palms, olives etc., the simultaneous extraction of oil from the kernel and the pulp is therefore possible. The main three fatty acids of this palmito-oleic oil are palmitic acid (C16:0; 35.6 to 58.4%), oleic acid (C18:1; 16.9 to 35.5%) and linoleic acid (C18:2; 3.9 to 31.5%) – and these usually account for 95% of the total amount of fatty acids. *R*, the ratio of unsaturated fatty acids to saturated fatty acids, is close to 1 as it is for palm oil, and thus safou oil falls between the fluid oils (R=4) and vegetable butters (R=0.25). Esterification with alcohol (either methanol or ethanol) would increase its fluidity, as is seen with palm oil.

Oil extraction has been performed in the laboratory and in preliminary experiments with a simple extrusion screw (Tchendji *et al.*, 1987; Ndamba, 1989; Kiakouama and Silou 1990; Bezard *et al.*, 1991; Silou *et al.*, 1991; Silou *et al.*, 1994; Silou, 1996; Silou and Avouampo, 1997). Oil cakes from the pulp are similar in composition to those from coconuts and oil palms: 13 to 16% proteins, and 20% of cellulose only slightly lignified and easily digestible. Oil cakes may be promoted as animal or fish food (Silou and Avouampo, 1997; Ndamba, 1989).

Processing and utilization

At present, the safou tree has oil production potentials comparable to oil palms, one of the most promising species in the world for vegetable oil (Table 7.58). Oil yields, already 2–4t/ha, are similar to oil palm fields in the early stages. Now, with vegetative propagation being possible, yields could similarly be improved with a consistent breeding programme.

Table 7.58. Biofuels for transport: energy yield for alternative feedstock/conversion technologies.

Production	Feedstock yield (dry t/ha/yr)	Transport fuel yield (GJ/ha/year)	Transport services yield (1000 vkm/ha/year)	Ratio, km from safou from rapeseed in Netherlands
Rapeseed methyl ester, Netherlands	3.7t of rapeseed (1369 litres of oil)	47	21,000 (ICEV)[a]	1
Palm oil and Safou oil, Africa	2–4t of oil (2000–4000 litres)	67 to 137	30,000 to 61,000	1.4 to 2.85
Hardnut (*Jatropha curcas*), Mali	1t of seeds 270 litres/ha	9.3	4200	0.2
Ethanol from maize, USA	7.2t	76	27,000 (ICEV)	1.3
Ethanol from wheat, Netherlands	6.6t of grain	72	26,000 (ICEV)	1.2
Ethanol from sugarbeet, Netherlands	15.1t of beet	132	48,000 (ICEV)	2.3
Ethanol from sugarcane, Brazil	38.5t of cane	111	40,000 (ICEV)	1.9
Ethanol, enzymatic hydrolysis of wood (present technology)	15t of wood	122	44,000 (ICEV)	2.1
Ethanol, enzymatic hydrolysis of wood (improved technology)	15t of wood	179	64,000 (ICEV)	3
Methanol, thermochemical gasification of wood	15t of wood	177	64,000 (ICEV), 133,000 (FCV)[b]	3 (ICEV), 6.3 (FCV)
Hydrogen, thermochemical gasification of wood	15t of wood	213	84,000 (ICEV), 189,000 (FCV)	4 (ICEV), 9 (FCV)

[a] ICEV: internal combustion engine vehicle (standard).
[b] FCV: fuel cell vehicle.
Consumption (in litres of gasoline equivalent per 100 km) is assumed to be 6.30 for vegetable esterified oil, 7.97 for ethanol, 7.90 for methanol and 7.31 for hydrogen used in ICEVs, and 3.81 for methanol and 3.24 for hydrogen used in FCVs; 1 litre of gasoline equivalent = 0.0348GJ HHV.
Source: Second Assessment Report of IPCC 1995 (ch. 19, p. 608) completed by A. Riedacker.

However, there are other reasons for promoting the species:

- It increases the diversity of candidates for highly productive vegetable oil species in the tropics.
- Oil can be extracted with a simple and low cost extrusion screw, and requires much less time and effort than palm oil extraction on a village level.
- As a fruit tree, it enhances the security of food supplies.
- It fits into agroforestry systems (Hladick, 1993), and can be planted with other species – for instance, with coffee (Tiki Manga and Kengue, 1994);
- Oil esterification with locally produced ethanol could be performed either in relatively small rustic installations or in the more traditional facilities developed for rapeseed oil.

The esterification of oil is a readily available technique. Diesel engines can use the resulting fuel without any problem, whereas gasification and methanol producing plants from lignocellulosic material, and cars or trucks with internal combustion engines or methanol reforming devices and fuel cells, are still under development. More emphasis should therefore be laid on improving both the species and agronomic techniques related to this crop.

A small pilot project, with both oil production and extraction and alcohol production for esterification, incorporating a microdistillery and a small esterification facility, should be considered as a high priority by donors. Such a development would further increase incomes and would improve the potential for replacing fossil fuels by vegetable oil to reduce greenhouse gas emissions.

Selected references

Avouampo, E. (1996) Contribution à l'étude de la valorisation alimentaire du safou: transport, stockage, procédés de séchage des pulpes et d'extraction d'huile, Séminaire d'Information sur la Recherche en Alimentation et Nutrition, 6–8 October, Brazzaville, Congo.

Bezard, J., Silou, Th., Kiakouama, S. and Sempore, G. (1991) Variation de la fraction glycéridique de l'huile de la pulpe de safou avec l'état de maturité du fruit. *Rev. franç. Corps Gras* 38(7/8): 233–41.

Cailhiez (1990) Plantations industrielles d'Eucalyptus en République populaire du Congo, International Workshop on Large Scale Reforestation, Corvallis, Oregon, 8–10 May 1990.

Hladick, A. (1993) Perspectives de développement pour l'Agroforesterie In: *Antropologie alimentaire et développement en afrique intertropicale: du biologique au social*, Proceedings of conference, Yaoundé, 27–29 April, pp. 483–92.

Kengue, J. and Nya Ngatchou, J. (1991) Perspectives d'amélioration du safoutier. In: *Actes du séminaire sous régional sur la valorisation du safoutier*, 25–28 November, Brazzaville, Congo.

Kengue, J. and Tchio, F. (1994) Essai de bouturage et de marcottage de safoutier (*Dacryoides edulis*). In: Nya Ngatcchou, J. and Kengué, J. (eds), *Seminaire sur la valorisation du Safoutier*, 4–6 October, Douala, Cameroun, pp. 80–98.

Kiakouama, S. and Silou, Th. (1990) Evolution des lipides de la pulpe de safou *Dacryodes edulis* en fonction de l'état de maturité du fruit. *Fruits* 45(4): 403–8.

Mampouya, P.C. (1991) Marcottage du safoutier. In: *Actes du séminaire sous régional sur la valorisation du safoutier*, 25–28 November, Brazzaville, Congo, pp. 32–9.

Ndamba, J. (1989) Analyse bromatologique du tourteau de safou en vue de son utilisation en alimentation animale. Résultats préliminaires. Dakar, Thèse de Doctorat Vétérinaire EISMV, 80pp.

Okafor J.C. (1983) Varietal delimitation in *Dacryodes edulis* (G. Don) H.J. Lam (Burseraceae). *Int. Tree Crop. J.* 2: 255–65.

Omoti U., Okiy A.D. (1987) Characteristics and composition of pulp oil and cake of African pear *Dacryodes edulis* (G. Don) H.J. Lam. *J. Sci. Food Agric.* 38: 67–72.

Silou, Th. (1996) Le safoutier *(Dacryodes edulis)*: un arbre mal connu *Fruits* 51: 47–60.

Silou, Th. and Avouampo, E. (1997) Perspectives de production de l'huile de safou. In: *Bulletin Africain bioressources et energies pour le développement et l'environnement*. Enda tm Dakar no. 8, in press.

Silou, Th., Kiakouama, S., Bezard, J. and Sempore, G. (1991) Note sur la composition en acides gras et en triglycérides de l'huile de safou en relation avec la solidification partielle de cette huile. *Fruits* 46(3): 26–7.

Silou, Th., Kiakouama, S. and Koucka-Gokana, B. (1994) Evaluation de la production et étude de la variabilité morphologique et physico-chimique du safou, *Dacryodes edulis*. Mise au point méthodologique et résultats préliminaires. In: Nya Ngatcchou, J. and Kengué, J. (eds), *Séminaire sur la valorisation du safoutier*, 4–6 October, Douala (Cameroun), pp. 30–44.

Tabuna, H. (1993) *La commercialisation du Safou à Brazzaville*, Montpellier, France CIRAD-SAR Internal report no. 82/93, 28 pp.

Tchendji, C., Severin, M., Wathelet, J.P. and Deroanne, C. (1987) Composition de la graisse de *Dacryodes edulis*. *Rev. Franc. Corps Gras* 28(3): 123.
Tiki Manga, T. and Kengue, J. (1994) Strategie d'amélioration du safoutier (*Dacryoides edulis*). In: Nya Ngatcchou, J. and Kengué, J. (eds), *Séminaire sur la valorisation du safoutier*, 4–6 October, Douala (Cameroun), pp. 12–16.
Viet Meyer (1990) *Butter Fruit, 'africado'*, US National Academy of Sciences, Washington DC, draft, 13 pp.
Youmbi E., Clair-Maczulajtys, D. and Bory G. (1989) Variation de la composition des fruits de *Dacryodes edulis* (Don) Lam. *Fruits* 44(3): 149–57.

SALICORNIA (*Salicornia bigelovii* Torr.)

Description

Salicornia bigelovii Torr. is an annual C_3 vascular plant originating in the Americas that is most commonly found in coastal estuaries and salt marshes. The plant has succulent, erect shoots with articulated and apparently leafless stems that are completely photosynthetically active and take on the appearance of green, jointed pencils. Seed spikes are found on the top third of the plant. Flowering takes place from July to November (Wiggins, 1980).

Salicornia bigelovii is a halophyte, or salt tolerant plant. Halophytes are unique in that they expend energy to maintain a higher salt concentration in their vacuoles than is found in the soil. By having the higher salt concentration in the vacuoles of the leaves the flow of water into the plant is ensured. The vacuoles are able to achieve a concentration of more than 6% salt, while pure seawater is only 3.2% salt (Douglas, 1994).

Approximately 30% of the seed's total weight is oil and about 30% protein, in comparison to the soybean which is 17–20% oil. Salicornia's oil is about 72% linoleic acid, which is a healthy polyunsaturated fat. For over a decade, salicornia has been 'selectively developed' to produce higher oilseed yields. Test plots in the United Arab Emirates, Egypt, Kuwait, Saudi Arabia and Mexico have all had success (Clark, 1994). It is estimated that there are approximately 130 million hectares worldwide that would be appropriate for salicornia cultivation. In addition, salicornia could help prevent erosion, lead to the return of nutrients to the soil, sequester carbon dioxide, remove salt and heavy metals from power plant wastewater, and provide a new non-fossil biofuel (Clark, 1994; Douglas, 1994).

Ecological requirements

According to Glenn and Watson (1993), there are three types of areas that could be used for the development of large scale farms for halophytes: coastal deserts (seawater irrigation), inland salt deserts (underground or surface water irrigation) and existing arid zone irrigation districts (brackish drainage water irrigation). Although sand or sandy soil will support the development of *S. bigelovii*, heavier soils with a larger water holding capacity would be preferred in that this reduces the amount of irrigation, which would normally be in the form of flood irrigation.

Salicornia's incredible salt tolerance is seen in the fact that it can endure salt concentrations of up to 50,000ppm (5% salt) without blighting. Although the salt build up can be harmful, it can be avoided by overwatering. This pushes the salt below root level. On a test farm in Mexico this was accomplished by estimating the crop's water needs, then irrigating with an additional 25% water (Clark, 1994). From this it is estimated that *S. bigelovii* needs yearly 1–3m^3 of seawater per m^2 of soil (Douglas, 1994).

Research has been conducted that has shown that euhalophytes, more extreme halophytes, have optimal growth at salinities in the range of 100–200mol/m^3. Salinities above or below this range lead to decreases in growth (Greenway and Munns, 1980; Munns *et al.*, 1983).

Propagation

More than a decade of breeding and selection has led to the seed as the most prominent form of salicornia propagation (Clark, 1994). Seeds can also be used for the mass propagation of selected plants. This involves the germination of seeds in a greenhouse followed by the selection and growth of shoot tips in a tissue culture medium. Research has shown that it is possible to produce 12 to 30 new shoots per culture every 8 weeks (Lee *et al.*, 1992).

Crop management

S. bigelovii has its growing season from March to August. According to Clark (1994), new strategies that were planned to be implemented in the 1994–95 season in Saudi Arabia included 'lowering seed density to produce fewer but bigger plants; applying phosphorus before planting to promote general crop growth; using "socks" or tubes to carry water from the irrigation sprinkler heads directly to the ground when plants begin to pollinate; and cutting off irrigation when the largest plants reach full size, instead of waiting for the entire crop to mature'.

S. bigelovii grows in its native habitats and on test farms without being significantly affected by disease or pest, though Stanghellini *et al.* (1988; 1992) reported instances of *S. bigelovii* being attacked by *Metachroma* larvae and *Macrophomina phaseolina*. Test plots in Sonora, Mexico, showed stand losses varying from 0 to 35% in 1988. It was determined that the small beetle larvae of the *Metachroma* genus were responsible for severing the plants' roots 3–5cm below the surface. Treatment of the soil with diazinon stopped plant damage by the larvae.

Macrophomina phaseoline, which is a soilborne fungus of arid and semi-arid regions, has also been reported to attack *S. bigelovii* (Stanghellini *et al.*, 1992). On test plots in Sonora, Mexico, it was found that the root pathogen *M. phaseoline* caused rotting of the plants' roots and led to stand mortality rates as high as 80% in 1989 and 30% in 1990. The occurrence of the disease is attributed to the plant undergoing environmental and/or physiological stresses.

Because salicornia primarily grows in the wild and is only now starting to be cultivated there is no established crop rotation.

Production

The most significant cost of growing halophytes is the expense of irrigating/pumping with seawater (Douglas, 1994). The direct cost of raising the crop, which includes diesel fuel for irrigation and tilling, is estimated at US$44–53 per tonne of fresh matter (Douglas, 1994).

Using planting dates in October and November 1993 and harvesting in September, several areas on a farm in Saudi Arabia have achieved their goals of 10t/ha of forage and 1t/ha of seed, while in a field 60km away, oilseed yields have reached as much as 3.5t/ha (Clark, 1994).

On survey plots in Mexico, *S. bigelovii* finished in the top five of halophytes for dry weight (dw) productivity in a one year trial (Table 7.59).

Table 7.59. Productivity of five halophytes in Mexico (Glenn and O'Leary, 1985).

Plant	Productivity (dw g/m²/year)
Atriplex lentiformis	1794
Batis maritima	1738
Atriplex linearis	1723
Salicornia bigelovii	1539
Distichlis palmeri	1364

At the same location in Mexico after six years of trials, *S. bigelovii* showed a mean biomass yield of 2.02kg/m²/year and a mean seed yield of 203g/m²/year (Glenn *et al.*, 1991).

Processing and utilization

Salicornia is fit for human consumption as a vegetable and salad greens, and also represents a source of vegetable oil, fodder and meal. The meal is approximately 40% protein and has had success as a poultry feed additive once the 'anti-feedant' saponin has been neutralized. The leftover straw, though high in salt and low in protein, may be used as feed (Clark, 1994).

Based on the calculations that halophytes have heat contents that fall in the range of lignite coal, research conducted by the Environmental Research Laboratory (ERL) at the University of Arizona has suggested that 'using a 2-to-1 ratio of coal to halophyte biomass in a 500MW power plant would produce about 10% less heat per unit weight than coal alone but would result in 25% less carbon being contributed to the atmosphere (because of the recycling of carbon back to the biomass)'. The ERL also indicates that it is possible to use refined halophyte oil to blend with diesel fuel (Douglas, 1994).

Selected references

Ayala, F. and O'Leary, J.W. (1995) Growth and physiology of *Salicornia bigelovii* Torr. at suboptimal salinity. *Int. J. Plant Sci.* 156(2): 197–205.

Clark, A. (1994) *Aramco World*, November/December.

Douglas, J. (1994) A rich harvest from halophytes. *Resource*, May 1994, pp. 15–18.

Glenn, E.P. and Watson, M.C. (1993) Halophyte crops for direct salt water irrigation. In: Lieth, H. and Al Masoom, A. (eds), *Towards the Rational Use of High Salinity Tolerant Plants*, vol. 1, pp. 379–85.

Glenn, E.P., O'Leary, J.W., Watson, M.C., Thompson, T.L. and Khuel, R. (1991) *Salicornia bigelovii*: an oilseed halophyte for seawater irrigation. *Science* 251: 1065–7.
Glenn, E.P. and O'Leary, J.W. (1985) Productivity and irrigation requirements of halophytes grown with hypersaline seawater in the Sonoran Desert. *J. Arid Environ.* 9: 81–91.
Greenway, H. and Munns, R. (1980) Mechanisms of salt tolerance in nonhalophytes. *Ann. Rev. Plant Physiol.* 31: 149–90.
Lee, C.W., Glenn, E.P. and O'Leary, J.W. (1992) In vitro propagation of *Salicornia bigelovii* by shoot-tip cultures. *HortScience* 27(5): 472.
Munns, R., Greenway, H. and Kirst, G.O. (1983) Halotolerant eukaryotes. In: Lange, O.L., Nobel, P.S., Osmond, C.B. and Ziegler, H. (eds.), *Physiological Plant Ecology*, vol. 3: *Responses to the Chemical and Biological Environment*, Springer, Berlin.
Stanghellini, E., Mihail, J.D., Rasmussen, S.L. and Turner, B.C. (1992) *Macrophomina phaseolina*: A soilborne pathogen of *Salicornia bigelovii* in a marine habitat. *Plant Disease* 76: 751–2.
Stanghellini, E., Werner, F.G., Turner, B.C. and Watson, M.C. (1988) Seedling death of *Salicornia* attributed to *Metachroma* larvae. *The Southwest Entomologist* 13(4): 305.
Wiggins, I.L. (1980) *Flora of Baja California*, Stanford University Press, Stanford, CA, 1025 pp.

SHEABUTTER TREE (*Vitellaria paradoxa* Gaertn. f.; syn. *Butyrospermum paradoxum* Gaertn. f.; *Butyrospermum parkii* (G. Don) Kotschy)

Description

The sheabutter tree is indigenous to the Sahel and northern Guinea zones, where it grows wild or cultivated. It has local importance as a source of fat. The tree usually reaches a height of 7–15m, but it may reach up to 25m with a trunk diameter of up to 2m and a dense crown. The bark is corky; the leaves are oblong and grow at the end of the twigs. In dry seasons the leaves are dropped.

The tree blossoms from December to March; later, green to yellow fruits are formed (Franke, 1985). The sheabutter fruits are ellipsoidal, 4–5cm long with fleshy pulp and usually with a single large seed. The fruits require 4–6 months to ripen (Gliese and Eitner, 1995). The soft pericarp is edible and very sweet. The seed contains 45–48% fat, 10% protein, and 25–60% carbohydrates. The sheabutter tree is a main resource of fat in regions where neither the olive tree, cultivated in northern regions, nor the oil palm can be grown. The perennial, drought resistant trees provide the inhabitants of the growth area with nutrients if the crops fail or during long dry seasons (ICRAF, 1997).

Ecological requirements

The sheabutter tree is usually found at elevations from sea level to 1000m, but in Cameroon it reaches 1200m. It occurs in areas with a rainfall of 600–1000mm and a marked dry season of 6–8 months, or 900–1800mm and a shorter dry season of 4–5 months, but in both cases subjected to annual burning (the tree is fire resistant). The absolute minimum of precipitation is 300mm. It needs a well drained, deep soil, but its soil fertility requirements are low. The optimum temperature lies between 29 and 38°C. The plant does not withstand frost. Sheabutter occurs naturally within the latitudes 10°N and 20°N (FAO, 1996).

Production

The sheabutter tree occurs both wild and cultivated. The tree bears the first fruits after 10–15 years, but takes 20–25 years to reach full productivity. In a given year, roughly every third tree bears fruit. An average tree bears 15–20kg fresh fruits; 50kg fresh nuts yield 20kg of dry kernels, from which approximately 4kg sheabutter can be made. The kernels contain 32–45% fat (Gliese and Eitner, 1995).

Processing and utilization

The seeds are milled and pressed. An older method is to cook the mortared seeds with water and to skim off the oil. The compounds in the oil are glycerides of stearic acid (41%) and oleic acid (48%). The oil is not only used as edible oil (sheabutter), but also for technical purposes such as the traditional production of soap and candles (Franke, 1985).

Selected references

FAO (1996) Ecocrop 1, Database, Rome.

Franke, W. (1985) *Nutzpflanzenkunde*, 3rd edn, Georg Thieme Verlag, Stuttgart and New York.

Gliese, J. and Eitner, A. (1995) *Jatropha Oil as Fuel*, DNHE-GTZ, Project Pourghère.

ICRAF (International Centre for Research in Agroforestry) (1997) via Internet: World Bank Newsletter, June 1997.

SORGHUM (FIBRE SORGHUM) (*Sorghum bicolor* L. Moench)

Contributed by: W. Bacher

Description

Fibre sorghum is a new hybrid between grain and broomcorn sorghums, making it a subspecies of *Sorghum bicolor* in the sub-tribe *Sorghastrae* of the tribe *Andropogoneae* (Jvanjukovic, 1981). Fibre sorghum's stalks are fibrous, not juicy. It is a annual C_4 plant species with high water efficiency, high biomass yield and one of the highest dry matter accumulation rates on a daily basis. The plant can grow to a height of 3.5–4m (Figure 7.10). Its habitat is similar to that of maize. Although sorghum is of tropical origin, the plant is well adapted in subtropical and temperate regions, especially the fibre sorghum hybrids, which are able to grow in northwest European countries. Temperature is the most critical growth factor of all climatic parameters (GEIE Eurosorgho, 1996).

Ecological requirements

Germination of fibre sorghum takes place between 8 and 10°C. The most productive growth stage is in the temperature range of 27–30° C and when annual precipitation is from 400–600mm. Sorghum's ability to tolerate drought is due to the high efficiency and growth of its root system, which is twice as high as with maize. This, and the closer spacing between the rows, reduces the danger of soil erosion.

Figure 7.10. Fibre sorghum, Germany.

Fibre sorghum has been successfully grown in temperate climates, such as in northern Germany. The new hybrids are very productive at higher latitudes (El Bassam, 1994).

Fibre sorghum has a great ability to mobilize the natural nitrogen reserves and other minerals of the soil because of its well developing root system (GEIE Eurosorgho, 1996). Soil requirements of sorghum are very modest. Sorghum tolerates salts and alkalis and can grow in a wide pH range between 5.0 and more than 8.0. Water-logged soils and acidic soils should be avoided.

Crop management

Sowing of fibre sorghum should not be done before or after maize, as this could lead to crop rotation problems. Sorghum is propagated by seed. To prepare the seedbed a ploughing depth of 10–30 cm is suggested. The seed is sown with a drilling machine to a depth of 2–4 cm into the soil (El Bassam, 1994). Pre-emergence herbicide application is recommended as similar with maize. 'Safer' can be used as a seed dressing to minimize pre-emergence herbicide damage. The sowing date is dependent on the climate of the region, because sorghum is sensitive to frost. Therefore in northern parts of Europe, like Germany, fibre sorghum varieties with a shorter growth season than six months should be chosen. The most suitable plant density ranges between 200,000 and 300,000 plants (20-70 kg seed/ha) with a row spacing of 40–50 cm (GEIE Eurosorgho, 1996). Because of the great ability of fibre sorghum to mobilise soil nitrogen, nitrogen fertilization does not need to exceed 80–100 kg/ha and should be adjusted with regards to the soil contribution possibilities and the accessible yield. Recommended fertilization rates are 70kg/ha P_2O_5 and 140kg/ha K_2O (GEIE Eurosorgho, 1996).

Production

Total dry matter fibre sorghum yields (stem and leaves) of up to 40 t/ha are possible. Dry matter contents range from 20% to 30% (GEIE Eurosorgho, 1996). In northern parts of Germany, the highest total dry matter yield of 23t/ha was ascertained during a test of different fibre sorghum genotypes. The highest yield of stems was 14.5t/ha (El Bassam, 1994). The analysis for lignin and cellulose of selected genotypes using the ADF (Acid Detergent–Fibre Method) showed 3–10% lignin and 24–34% cellulose.

Processing and utilization

Fibre sorghum is an annual plant which has to be harvested in autumn before winter begins. Because fibre sorghum varieties are sterile hybrids, no seed production is possible. The greatest interest in fibre sorghum is for biomass production. This involves using fibre sorghum as an energy plant or as pulp for the paper industry. The harvest and the raw material preparation are decisive elements for using sorghum for the paper industry. Important for all harvesting systems is the protection of breakdowns and the jamming of the biomass, which is extremely dependent on the physiological stage of the harvested material and the weather conditions. In addition, protection against sand and earth during the harvest must be guaranteed. Chopping and baling, as well as ensiling, are possible harvesting and storing methods. For the utilization of fibre sorghum as a solid fuel, the biomass should be mowed and dried, and then afterwards baled with a high density cubic baler or with a round baler. In humid regions, fibre sorghum should be ensiled and pressed mechanically before being used as a fuel feedstock for combustion (El Bassam, 1994).

Selected references

El Bassam, N. (1994) *New hybrid plant: fiber sorghum*. Annual Report, FAIR 1-CT92-0071.

GEIE Eurosorgho, (1996) *New hybrid plant (Fiber Sorghum) as a component in mixtures of pulps standardly used in the paper industry*. Periodic Progress Report 1995/96, FAIR 1-CT92-0071.

Jvanjukovic, L.K. (1981) Specific and intraspecific classification of sorghum species. *Kulturpflanze* XXIX: 273–80.

SORGHUM (SWEET SORGHUM) (*Sorghum bicolor* L. Moench)

Contributed by: C. D. Dalianis

Description

Cultivated sorghums, under the scientific name *Sorghum bicolor* L. Moench, are classified into four major groups.

- Grain sorghum, grown for grain production, is the most important. It is extensively cultivated in several African countries, certain regions of India and in the USA. In Africa and India its grain is used for human consumption,

playing the same role as wheat in the European Union (EU), while in the USA it is used as animal feed.
- Broomcorn is grown for the brooms that are made out of the long branches of its panicles. Fibre sorghums is a hybrid between grain and broomcorn sorghums.
- Sudangrass is grown for fodder production.
- Sweet sorghum is grown for its sweet stems, which are used mainly for syrup production (Figure 7.11).

Sweet sorghum is a C_4 crop that belongs to the grass family. It has a fibrous root system that branches profusely. Under favourable conditions, the above ground nodes may produce strong adventitious roots that may help to anchor the plant and thereby reduce lodging.

Sweet sorghum plants attain a height of up to 4m. The stems are similar to those of corn, being grooved and nearly oval. The top internode is not grooved. Sweet sorghum has sweet juicy pith in the stems. Crown buds give rise to tillers. Each tiller soon develops an independent root system, though it remains attached to the main stem.

Ecological requirements

Sweet sorghum, although a native to the tropics, is also well adapted to temperate regions. Recently, sweet sorghum has been considered in the EU as a potential energy crop, mainly for ethanol production. It is expected to play a similar role in temperate climates to that sugarcane plays in tropical regions, especially Brazil. In recent years there have been many studies in EU countries of sweet sorghum's adaptability, productivity and ethanol potential under various environmental and soil conditions as well as under various cultural practices.

Figure 7.11. Sweet sorghum, China (Photo: Dajue).

Sweet sorghum is well adapted to the warm southern regions of the EU and moderately well adapted to several central EU regions with mild climates. Because sweet sorghum originated in the warm region of central Africa it is a cold sensitive plant, so its adaptation in northern cooler climates is poor. Attempts to grow sweet sorghum in Ireland have failed as a result of problems with establishment and poor growth.

However, it should be mentioned that cool tolerant sorghum genotypes are encountered at high altitudes in Africa and in temperate regions of America (Dogget, 1988). In the USA, cool tolerant grain sorghum cultivars have been bred for extending grain sorghum cultivation in cool northern areas.

The minimum temperatures are 7–10°C for germination and 15°C for growth. The optimum temperature for growth is 27–30°C (Quinby *et al.*, 1958). Sorghum withstands high temperatures better than many other grain crops.

Sorghum can be grown successfully on a wide range of soils, such as heavy clays, medium loams, calcareous soils and organic soils. It tolerates a pH range from 5.5 to 8.5, and also some degree of salinity, alkalinity and poor drainage (Hayward and Bernstein, 1958).

Sorghums are considered to be one of the most drought resistant agricultural crops. They have the ability to remain dormant during drought and resume growth when favourable conditions reappear. For this reason sorghum is frequently called the 'camel' of field crops.

One of the important factors affecting its drought endurance is the effectiveness of its large fibrous root system, mainly the secondary roots, which can extend to a distance of up to 1m laterally and to a depth of 1.8m. The root system is about twice as active as that of corn in taking up water from the soil (Martin, 1941).

Sorghum, beyond its ability to absorb water more effectively than many other crops, has a better ability to regulate water loss to the atmosphere. With the onset of water stress, sorghum showed the lowest rate of decline in relative turgidity and the least diurnal depression as compared with other crops, because it reduced its transpiration rate to a much greater extent than did other crops (Slatyer, 1955). Sorghums also have a relatively impervious waxy epidermis that retards the desiccation of stalks and leaves. The leaf blades are covered with a waxy coating that reduces water loss.

Propagation

At present, sweet sorghum is being extensively bred to increase sugar content and biomass, and to improve its resistance to lodging, frost and disease. Sweet sorghum is propagated by seed, of which the cultivars 'Keller' and 'Korall' have had significant success in Europe.

Crop management

A warm seedbed free from weed growth is essential to good seed germination of sweet sorghum. Planting takes place in the spring and may start when soil temperature remains above 15°C. Seed germination occurs within 24 hours in warm and moist soil. Pneumatic sowing machines may be used for sweet sorghum. Recommended plant distances are 70cm between rows and 10–20cm within rows. The early growth of the plants is very slow, and places them at a disadvantage in competition with weeds.

Although sweet sorghum is one of the most drought resistant crops, in southern EU regions it needs to be irrigated, otherwise the plants remain stunted and biomass yields are very low. Depending on irrigation rates, annual dry matter biomass yields of sweet sorghum in central Greece ranged from 22 to 33t/ha and in southern Italy from 18.5 to 47t/ha.

Compared with many other widely grown agricultural crops, sorghums have the highest water use efficiency. In a very old but classic work, Shantz and Piemessel (1927) found that alfalfa requires 844kg of water for the production of 1kg of dry matter, oat 583kg, common wheat 557kg, barley 518kg and corn 349kg, but grain sorghum requires only 304kg of water.

Sweet sorghum, having very high water use efficiency, is very well suited for southern EU conditions. In central Greece, water use efficiencies, depending upon irrigation rates applied throughout the growing period by the drip method, ranged from 206l/kg of dry matter, produced under highly irrigated conditions (458mm), to 181l/kg of dry matter, produced when irrigation was restricted to only 157mm – 34% of the highly irrigated rate (Dercas *et al.*, 1996).

In a comparative study in southern Italy (Mastrorilli *et al.*, 1996) the water use efficiency of sweet sorghum has been estimated at 192l/kg of dry matter produced, whereas the comparable figure for grain sorghum was 270l/kg, for sunflower 278l/kg and for soybean 357l/kg.

Several nitrogen experiments have been carried out in various EU countries and sites. So far, no significant nitrogen responses have been observed in the various experiments (Dalianis *et al.*, 1995; Dercas *et al.*, 1995; 1996; Cosentino, 1996), except in Portugal, where significant differences have been observed between the 75 and 100kg/ha nitrogen rates (Oliviera, 1996).

Beyond its low nitrogen requirements leading to less nitrogen fertilization inputs, sweet sorghum is characterized by a large and widespread root system. Therefore, even if some nitrate leaching occurs during the growing period there are increased chances for nitrates to be trapped during their downward movement by the plant's richer root system compared to the poorer systems of many other agricultural crops.

The potassium requirements of sweet sorghum are very high. It is reported (Curt *et al.*, 1996) that potassium uptake amounts to 617kg/ha for a dry matter yield of 25t/ha. The fact that all the above ground biomass is removed from the field – compared, for example, with a corn grain crop where only one-third of the potassium in the above ground part is removed with the grains – indicates that sweet sorghum is a heavily potassium depleting crop.

There are certain indications that plant density affects biomass yields from sweet sorghum. With a stable between row distance of 70cm and within row distances of 5, 10, 15 and 20cm it was found (Dalianis *et al.*, 1996) that the 15 and 20cm within row distances were superior in terms of total fresh and dry matter stem yields. Significant also is the effect of plant density on sugar yields and on final stem diameter. Sugar yields and stem diameter were higher with the 20cm within row plant density.

Production

Sweet sorghum adaptation and productivity have been studied around the world. Fresh biomass yields ranged from 22.4 to 44.8t/ha in Ohio (Lipinski *et al.*, 1978), 40 to 45t/ha in southern USA (Bryan *et al.*, 1981), and up to 64t/ha in Alabama (Soileau and Brandford, 1985). It is reported from China (Nam and Ma, 1989) that for the varieties 'Wray' and 'Keller' fresh stem yields were around 49.5t/ha, with a simultaneous grain yield of 1.8t/ha for 'Wray' and 2.1t/ha for 'Keller'. Recent work in China (Dajue and Yonggang, 1996) reports fresh biomass yields, depending on variety, from 82.4 to 128.1t/ha with a simultaneous seed production of 1.43 to 6.67t/ha.

Under certain EU environmental conditions, sweet sorghum biomass yields are very high. From previous work, fresh biomass yields of sweet sorghum, depending upon site, weather conditions, soil fertility, varieties and cultural practices, were 80–100t/ha in Germany (Bludau, 1987), 92.6t/ha in Spain (Curt *et al.*, 1994), and up to 129t/ha in Greece (Dalianis *et al.*, 1995). It should be noted that all these high biomass yields in the southern EU were obtained under irrigation.

Sweet sorghum biomass yields in the EU are significantly affected by the latitude (Figure 7.12). Thus, total above ground fresh biomass yields ranged from as low as 55t/ha in Belgium (Chapelle, 1996), through 120t/ha in Malaga in southern Spain (Curt *et al.*, 1996) and to 127.3t/ha in northern Italy (Dolciotti *et al.*, 1996), up to 141t/ha in southern Greece (Dalianis *et al.*, 1995).

Total dry matter yields were 12t/ha in Belgium (Chapelle, 1996), 40t/ha in southern Spain and northern Greece, and up to 45t/ha in southern Greece and southern Italy (Dalianis *et al.*, 1995; Cosentino, 1996; Curt *et al.*, 1996).

Several varieties have been tested ('Keller', 'Wray', 'Mn 1500', 'Cowley', 'Dale', 'Sofra', 'Dale', 'Theis' and 'Korall') in various EU regions and under various environmental and soil conditions.

On average, 'Keller' was proven the best, combining wide adaptation, excellent stability (from site to site and from year to year) and very high biomass and sugar yields. However, some other varieties were superior in certain EU regions. Thus, fresh biomass yields of 'Keller' were superior to those of 'Cowley' in Greece (Dalianis *et al.*, 1995); yields of 'Wray' were superior to those of 'Korall' in northern Italy

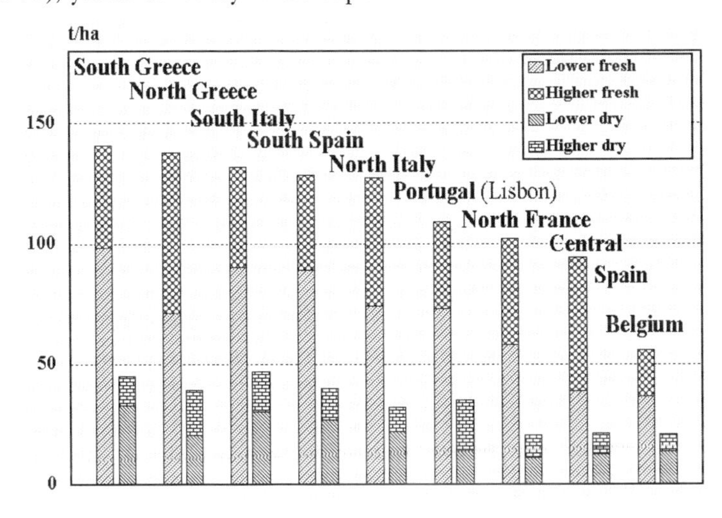

Figure 7.12. Total fresh and dry matter yields of sweet sorghum in various EU regions (ranges within regions are the result of variety, year and/or treatment effects).

(Petrini, 1994); and yields of 'Dale' and 'Theis' were superior to those of 'Wray', 'Keller' and 'Korall' in southern Spain (Curt *et al.*, 1996), all grown under the same conditions at each site.

Processing and utilization

Historically, syrup production was the main use of the sweet sorghum grown in certain tropical and subtropical regions. Attempts to develop a sorghum sugar industry have not been successful because of certain difficulties and limitations that make the sugar more expensive than that from sugarcane or sugarbeet.

Sorghum is gaining attention as a potential alternative crop for energy and industry because of its high yield in biomass and fermentable sugars. As soon as possible after harvesting, the stalks are transported to a processing facility where the stalks are cut and crushed for juice extraction to separate the sugar from the bagasse. Two extraction techniques are being investigated: in one the sorghum is crushed with a mechanical screw or roller press, and the other uses a crushing, diffusion and dewatering system. After extraction the sugar can be concentrated for storage or it can be sent to an energy conversion process. Bagasse can also be processed immediately or it can be dried for long term storage.

Sweet sorghum can be converted into energy through two processes: biochemical and thermochemical. The biochemical process is used to convert the sugar into fuel ethanol. The sugar, which is two-thirds saccharose and one-third monosaccharides, can be processed into ethanol. The thermochemical process, which can consist of carbonization, combustion, gasification, or pyrolysis, is used to convert bagasse into bio-crude oil, gasoline, electricity, pulp for paper, compost and charcoal pellets.

The bagasse is mainly lignocellulose, has 55–62% residual humidity, and contains two-thirds of the plant's energy. A cycle can be created when bagasse is used to produce the energy necessary for ethanol distillation.

The theoretical ethanol yield can be calculated from the formula:

> Total sugar content (%) in fresh matter (FM) × 6.5 (a conversion factor) × 0.85 (the process efficiency) × total biomass (in t/ha of FM).

Sugar percentages are around 9–14% from fresh stem matter and 30–45% from dry stem matter. Based on stem biomass yields and sugar percentages observed in various EU regions, sugar yields have been estimated (Figure 7.13). In southern Spain, estimated sugar yields are as high as 13.8t/ha (Curt *et al.*, 1996) and in Greece more than 12t/ha (Dalianis *et al.*, 1995).

The ethanol potential of sweet sorghum, under Greek conditions in fertile fields and under appropriate cultural management, has been estimated at 6750l/ha (Dalianis *et al.*, 1995); in China, the ethanol potential has been estimated at 6159l/ha (Dajuc and Yonggang, 1996).

It should be mentioned that, beyond this high ethanol potential of sweet sorghum, a significant amount of lignocellulosic material is left as bagasse. The bagasse is sufficient to cover all the energy requirements of the chain from plant production to ethanol distillation; or it may be used for pulp production, or to ferment the cellulosic fraction, thereby increasing ethanol yields by at least 40%. Up to 10,000l/ha of ethanol may be possible.

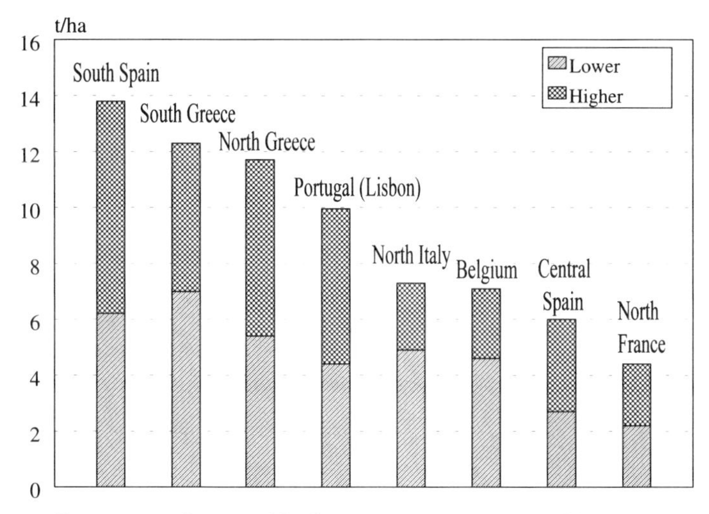

Figure 7.13. Sugar yields of sweet sorghum in various EU regions (ranges within regions are the result of variety, year and/or treatment).

Selected references

Bludau, D. (1987) Landwirtcchftliche Kendaten und Angepasste Produktiontechnik der Zuckerhirse als Energierpflanze. In Der Bundesrepublik Deutschland, Diplomarbeit, Tu-Munchen-Weihenstephan.

Broadhead, D.M. and Freeman, K.C. (1980) Stalk and sugar yield of sweet sorghum as affected by spacing. *Agron. J.* 72: 523–4.

Bryan, W.L., Monroe, G.E., Nichols, R.L. and Bascho, G.J. (1981) *Evaluation of Sweet Sorghum for Fuel Alcohol.* ASAE Paper No. 81–3571, ASAE, St. Joseph, Missouri 49085.

Chapelle, J. (1996) Sorghal, Belgium. Personal communication.

Cosentino, S. (1996) Crop physiology of sweet sorghum (*Sorghum bicolor* [L.] Moench) in relation to water and nitrogen stress. In: *First European Seminar on Sorghum for Energy and Industry*, Institut National de la Recherche Agronomique, France, pp. 30–41.

Curt, M.D., Fernandez, J. and Martinez, M. (1994) Potential of sweet sorghum crop for biomass and sugars production in Madrid. In: Hall *et al.* (eds), *Biomass for Energy and Industry*. Proceedings of the 7th EC Conference, Ponte Press, Bochum, Germany, pp. 632–5.

Curt, M. D., Fernandez, J. and Martinez, M. (1996) Effect of nitrogen fertilization on biomass production and macronutrients extraction of sweet sorghum cv keller. Presented at the 9th EU Conference on Biomass for Energy, Agriculture and Environment, Copenhagen, 24–27 June (in press).

Dajue, L. (1997) Personal communication.

Dajue, L. and Yonggang, L. (1996) Research and production of sweet sorghum in China. In: *First European Seminar on Sorghum for Energy and Industry*, Institut National de la Recherche Agronomique, France, pp. 342–50.

Dalianis, C.D., Sooter, Ch. and Christou, M. (1995) Sweet sorghum [*Sorghum bicolor* (L) Moench] biomass and sugar yields potential in Greece. In: Chartier *et al.* (eds), *Biomass for Energy, Environment, Agriculture and Industry. Proceedings of the 8th EU Biomass Conference*, Pergamon Press, Oxford, pp. 622–8.

Dalianis, C., Alexopoulou, E., Dercas, N. and Sooter, Ch. (1996) Effect of plant density on growth, productivity and sugar yields of sweet sorghum in Greece. Presented at the 9th EU Conference on Biomass for Energy, Agriculture and Environment, Copenhagen, Denmark (in press).

Dercas, N., Dalianis, C., Panoutsou, C. and Sooter, CH. (1995) Sweet sorghum (*Sorghum bicolor* (L) Moench) response to four irrigation and two nitrogen fertilization rates. In: Chartier *et al.* (eds), *Biomass for Energy, Environment, Agriculture and Industry. Proceedings of the 8th EU Biomass Conference*, Pergamon Press, Oxford, vol. 1. pp. 629–39.
Dercas, N., Panoutsou, C. and Dalianis, C. (1996) Water and nitrogen effects on sweet sorghum and productivity. Presented at the 9th EU Conference on Biomass for Energy, Agriculture and Environment, Copenhagen, Denmark (in press).
Dolciotti, I., Mambelli, S., Grandi, S. and Venturi, G. (1996) A comparative analysis of the growth and yield performances of sweet and non sweet sorghum genotypes. In: *First European Seminar on Sorghum for Energy and Industry*, Institut National de la Recherche Agronomique, France, pp. 207–12.
Dogget, H. (1988) *Sorghum*. Longman, UK.
Hayward H.E. and Bernstein, L. (1958) Plant growth relationships on salt affected soils. *Bot. Rev.* 24: 584.
Lipinski, E.S. (1978) *Sugar Crops as a Source of Fuels*, vol. II: *Processing and Conversion*, Research Dept. of Energy. Final Report. Battelle Columbus Labs, OH.
Mastrorilli, M., Katerji, N. and Rana, G. (1996) Evaluation of the risk derived from introducing sweet sorghum in Mediterranean zone. In: *First European Seminar on Sorghum for Energy and Industry*, Institut National de la Recherche Agronomique, France, pp. 323–8.
Martin, J.H. (1941) Climate and sorghum. In: *Climate and Man*, USDA Yearbook, pp. 343–7.
Miller, E.C. (1916) Comparative study of the root systems and leaf areas of corn and sorghums. *J. Agric. Res.* 6: 311.
Miller, E.C. (1938) *Plant Physiology*. McGraw-Hill.
Nam. L. and Ma, J. (1989) Research on sweet sorghum and its synthetic application. In: Grassi *et al.* (eds), *Biomass for Energy and Industry. Proceedings of the 5th EC Conference*, vol. 1. Elsevier Applied Science, London.
Oliviera, P. (1996) UNL, Lisbon. Personal Communication
Quinby, J.R., Kramer, N.W., Stephens, J.C., Lahr, K.A. and Karper, R.E. (1958) Grain sorghum production in Texas. *Bull. Tex. Agric. Exp. Stn.* 912.
Petrini, C., (1994) Response of fibre and sweet sorghum varieties to plant density. In: *8th EC Biomass Conference for Industry Energy and Environment*, Vienna, Austria, 9–11 October.
Shantz, H.L. and Piemeisel, L.N. (1927) The water requirements of plants at Akron, Colo. *J. Agric. Res.* 12: 1093–1191.
Slatyer, R.O. (1955) Studies of the water relations of crop plants grown under natural rainfall in northern Australia. *Aust. J. Agric. Res.* 6: 365–77.
Soileau J.M. and Brandford, B.N. (1985) Biomass and sugar yield response of sweet sorghum to lime and fertilizer. *Agron. J.* 77: 471–5.

SORREL (*Rumex acetosa* L.); (*R. patientia* L. x *R. tianshanious* L.)

Contributed by: V. Petríková

Description

Sorrel was improved in the 1970s in Ukraine as a hybrid of *Rumex patientia* x *Rumex tianshanious*. Sorrel is a perennial plant that can persist on the same stand for more than 10–15 years. This plant is similar in features to the wild growing sorrel, but it is much larger. Sorrel grows as tall as 180–200cm and is a robust, rich flowering plant that creates well ripened fruits. Sorrel is a photophilic plant that grows very quickly. It has a high content of proteins and vitamins: the protein content amounts to as much as 8t/ha (from the whole above ground mass).

Grain yield reaches 7 t/ha. In the most unfavourable conditions the plant provides 2t/ha of grain as a minimum. According to data from Ukraine, the growth of farm animals after feeding with sorrel is two to three times higher than with normal feed. Milk production and milk fat content increased.

Sorrel can also be used as a vegetable. Very early in the spring it develops the large brittle leaves that form the ground level rosette. The leaves are routinely used as the first high quality green feedstuffs of spring. The data mentioned here are for sorrel of Ukrainian origin. For sorrel grown in central European conditions it will be necessary to verify and confirm all these data.

Ecological requirements

Sorrel is a modest plant that is resistant to cold and dry conditions, as well as being resistant to diseases. Sorrel is highly resistant to frost, and hibernates without problems at significantly decreased temperatures. The plant does not demand special treatment, and is covered abundantly with leaves every year. Sorrel is not particularly demanding with regard to nutrition, but it will benefit from rich organic manuring during the establishment of growth.

Propagation

Sorrel is propagated by seed. The seed should be treated by grinding off the small side wings.

Crop management

Sorrel is sown in April in rows 50cm apart at an average of 6kg/ha seeds. The recommended depth of sowing is approximately 1.5–2cm. Because sorrel is a perennial plant, growth can also be established during the summer and autumn if the humidity is suitable for the germination of the seeds.

Its perennial nature means that there is no recommended crop rotation for sorrel. Sorrel can routinely be cultivated on arable soil that may be used to grow the usual agricultural crops after the sorrel has been removed.

Production

Sorrel has been improved as a feeding crop. A large quantity of seed was imported into the Czech Republic in order to test sorrel's as an energy plant for direct combustion at heating plants. Sorrel is a most suitable plant for this purpose. It reaches bumper yields of as much as 40t/ha of dry matter, and has the advantage of prolonged production from the same stand. The energy content of the above ground mass is approximately 18MJ/kg, representing about 800GJ/ha (only the above ground mass, excluding the seed and roots, is included in this assessment).

Processing and utilization

In its second year, sorrel can already be harvested at least twice a year. Leaves for green fodder are harvested in the spring, and the fruits are harvested in the summer. If sorrel is utilized for energy purposes (for direct combustion), then the whole plant is harvested. The plants can be formed into large capacity bales similar to straw bales.

Sorrel seems to be a very promising plant not only as fodder, but also as a source of renewable energy for direct combustion. It can be profitably used on set aside soils in the interests of limiting food production. For these purposes, it would be very useful to undertake detailed research into the production of sorrel under different soil and climatic conditions and to verify the applicability of the Ukrainian data to other parts of the world, with regard to the plant's energy utilization. For energy purposes, sorrel can be regarded as in the same group as *Miscanthus* or other recommended energy species, and may even outperform these plants; the energy content of above ground biomass is approximately 18MJ/kg (800GJ/ha).

Selected references

Magness, J.R., Markle, G.M. and Compton, C.C. (1971) *Food and Feed Crops of the United States*, Interregional Research Project IR-4, IR Bul. 1 (Bul. 828, New Jersey Agr. Exp. Sta.).

Petríková, V., Vána, J. and Ustjak, S. (1996) *Growing and Using of Technical and Energy Crops on Reclaimed Soils*, Metodika UZPI, Praha, no. 17, 24 pp.

Ustjak, S., Honzík, R. and Malírová, J. (1996) Production of biomass as a source of energy, Research Institute of Crop Production, Prague, *Annual Report*, p. 8.

SOYBEAN (*Glycine max* (L.) Merr.)

Description

Soybean is an annual, bush like C_3 plant originating from China. It grows to a height of up to 80cm with all plant parts being hairy. The plant produces a main root with smaller roots branching from it. The smaller roots possess numerous nodules inside of which develop the bacteria *Bradyrhizobium japonicum*. The bacteria is responsible for free nitrogen fixing. Above the cotyledonary nodes two single leaflets develop opposite each other. The plant's other leaves consist usually of three leaflets.

Soybean is for the most part self-fertilizing. Numerous very small purple or white flowers are found in compact racemes in the leaf axils. These racemose inflorescences growing from the leaf axils are responsible for the production of the plant's fruit, which develops in the form of pods. Although 1–5 seeds per pod can be found, the usual number is two or three. The seeds are round to oval, and whitish yellow to brown or black–brown. Their composition is approximately 38–43% protein, 18–25% oil and 24% carbohydrates.

Soybean is a very important crop for the production of oil and protein. Among the largest producers in the world are the USA, Brazil, China and Argentina. Originally, soybean was considered a plant of subtropical areas, growing mainly in North America and East Asia in a band between 35 and 40°N from the equator, but it has since been widely cultivated in tropical zones. Since the 1900s soybeans have been cultivated in Europe.

Ecological requirements

Soybean plants are able to adapt to a wide range of environments, but this adaptability is possible because of the many different cultivars, each having its own characteristics and preferred conditions. Soybean has similar ecological requirements to maize. The plant needs high temperatures in the summer and autumn to facilitate ripening. Loamy soil or loess and black soil with the ability to hold water are recommended. When enough water is present a light sandy soil can also be used. Heavy waterlogged soils are not appropriate. A constantly humid subtropical climate is most favourable for soybean crops. Although 24–25°C is preferred for optimal growth, the 20–25°C range is still excellent for all stages of plant growth. Frost tolerance is better than that of maize.

A sensitivity to photoperiod means that a day length of 12 hours or less leads to premature flowering, which results in diminished yields. In the majority of cultivars, blooming is seen only when the day length is less than 14 hours.

The crop has an approximately 4–5 month growing period. The rainfall required for a good yield in warmer areas is 500–750mm, though less precipitation during ripening is preferred. The pH range of 6–6.5 has been shown to be the most desirable for soybean crops, with certain cultivars performing better at slightly more acidic or basic levels. Approximately 15kg phosphorus and 50kg potassium are used from the soil for every tonne of seeds produced.

Rhizobium symbiosis is an important factor when it comes to crop growth and yield. Soybean seeds need to be inoculated with *Bradyrhizobium japonicum*, and because many cultivars require a specific strain it is recommended that the seeds be inoculated with a composite of many strains.

Propagation

Soybean propagation is performed using seed. Because of the wide variety of characteristics found in soybean plants, it is necessary to choose seed of the plant that is most appropriate for the soil and climate in the area it will be grown. The use or end product of the crop also needs to be taken into consideration. As mentioned previously, an important factor after selecting the seed is that the seeds are inoculated with the appropriate *Bradyrhizobium* strain or composite of strains.

Crop management

Sowing should take place in the middle of April or beginning of May. The ground temperature at a depth of 5cm should be at least 10°C. A sowing depth of 3–4cm and a distance between rows of 25–30cm is recommended. The desired crop density is 40–60 plants per m^2. This corresponds to about 70–90kg/ha of sown seed. Because the N_2 binding potential of soybean's nodules is negatively affected by high soil temperatures, regions in southern Europe and the Mediterranean should use denser crop spacings, thus shielding the soil in hotter climates. Wider spacings usually lead to the production of more branches and seeds.

Because of soybean's nitrogen fixing bacteria, most soils do not need supplementary nitrogen fertilization, especially where the previous crop has been abundantly fertilized.

Weed control is essential during the first few weeks after sowing. This can be performed mechanically, or there are several herbicides that can be used preferentially before seed sowing or before germination. The time from planting to maturity is from 90–180 days, depending on climate and plant variety.

Drought or excessive dryness strongly reduces affects crop yield, especially before flower formation and during granulation. General irrigation may or may not be necessary depending upon the climate, but irrigation is necessary at germination, flowering and during the plant's seed development in semi-arid climates.

Disease damage to soybean crops can be greatly reduced by proper cultivar and seed selection, and by maintaining sanitary practices and appropriate crop rotation. The most destructive diseases are, among others, frogeye leaf spot (*Cercospora sojina* Hara), stem canker (*Diaporthe phaseolorum*), *Pseudomonas glycinea* Coerper, and soybean mosaic virus.

Some of the more damaging pests are the soya bean moth (*Laspeyresia glycinivorella*), found in parts of Asia, the leaf eating beetle (*Plagiodera inclusa* Stål.), the Japanese beetle (*Popillia japonica*) and the soybean cyst nematode (*Heterodera glycines* Ichinohe). If these pests are present, insecticides may be necessary.

Soybean can be used with a large variety of crop rotation plans. Usually, soybean comes between two grain or cereal crops. Possible rotation combinations are: corn/soybean/wheat, soybean/wheat/sorghum, millet/winter wheat/soybean, and corn/soybean/cotton.

Production

In a study of oil crops in the USA by Goering and Daugherty it was determined that the total energy input for a non-irrigated soybean crop was 12,044MJ/ha. This includes the entire range of costs from seed price to oil recovery costs. The energy output was determined to be 20,777MJ/ha, based on a seed yield of 2623kg/ha and an oil content of 20%. The energy output/input ratio was 4.56. This was the highest ratio of the nine different crops tested under conditions of non-irrigation. Rape had the next highest output/input ratio at 4.18.

At present, soybean crops do not have a high enough yield to make affordable use of the vegetable oil as an energy source (diesel fuel substitute). Increased oil yields per hectare appear to be an achievable goal of research and plant breeding. Harvests in Germany resulted in 2.1t/ha of soybeans with an oil content of 18%. This produced a 0.378t/ha oil yield.

Processing and utilization

A combine harvester can be used for harvesting soybean crops. The plants are harvested when the leaves turn yellow and the plant dies. By this time the seeds are fully ripe, but the pods have not yet ruptured. The water content of the seeds should be under 20% at the time of harvest.

The harvested beans are prepared for storage by being cleaned and dried to a moisture content of between 12 and 14%, after which the beans can be stored in elevators or silos. This provides a continuous, year-round supply of beans for further processing.

Soybeans currently have a very wide range of food and non-food uses. The seeds can be used to produce edible products such as soy meal used for cooking and baking, soy milk, soy sauce, cheeses such as tofu, yogurt, and a product similar to meat called TVP (textured vegetable protein). The oil is used to make vegetable oil, margarine, shortening and salad dressings. The unripe seeds and the bean sprouts can be eaten as vegetables. The remaining green plant can be fed as forage to animals.

The seeds can also be used for non-food products. The lecithin in the seeds can be used as an emulsifier for pharmaceuticals, decorating materials, printing inks, pesticides, etc. The protein can be used for the production of synthetic fibres, glues, etc.

For purposes of bioenergy production the oil is the most important element. The oil is usually obtained from the seeds by solvent extraction, and consists of 3–11% linoleic acid. Pure vegetable oil can be obtained through solvent extraction. With the additional step of transesterification, an esterified vegetable oil is obtained that can be used alone or mixed with diesel fuel to produce a biofuel.

In 1991, the market price of a litre of diesel oil in Italy was 327L untaxed and 1,131L taxed, whereas the price of a litre of soybean vegetable oil was 674L, or 830L with esterification. This shows that a subsidy is necessary if soybean oil is to compete with diesel and be used as a biofuel.

Selected references

Bartolelli, V. and Mutinati, G. (1992) Economic convenience of vegetable oils as combustible. In: *Biomass for Energy and Industry*, 7th EC Conference, Florence.

Goering, C.E. and Daugherty, M.J. (1982) Energy accounting for eleven vegetable oil fuels. *Transactions of the ASAE* 25(5): 1209–15.

Rehm, S. and Espig, G. (1991) *The Cultivated Plants of the Tropics and Subtropics*, Verlag Joseph Margraf, Priese GmbH, Berlin.

Schuster, W.H. (1989) Sojabohne. In: Rehm, S. (ed.), *Spezieller Pflanzenbau in den Tropen und Subtropen. Handbuch der Landwirtschaft und Ernährung in den Entwicklungsländern*, 2nd edn, vol. 4, Ulmer, Stuttgart.

SUGARCANE (*Saccharum officinarum* L.)

Description

Sugarcane is a perennial C_4 plant that has many positive qualities, including high sugar content, low fibre content, good juice purity, good ecological adaptability, and disease resistance. It is also a convenient raw material for the production of alcohol and cellulose. Although many cultivars barely flower, the presence of short days can lead to undesired flowering (Rehm and Espig, 1991).

Because of sugarcane's excellent capacity for energy binding with respect to surface area and time, it is considered to be one of the most important economic plants in the world (Husz, 1989).

Ecological requirements

Subtropical and tropical areas have proven to be the most appropriate for growing sugarcane. The areas fall into the latitudes between roughly 37°N and 31°S. Because of significant variations in the conditions of different growing areas, special cultivars are usually grown at breeding stations in the same climatic area as that in which the plant will be grown (Rehm and Espig, 1991).

Although sugarcane can be and is cultivated on a greatly varying range of soils, those preferred for sugarcane cultivation are heavy soils with a high nutrient content and a high water holding capacity, though extended periods of waterlogging are not tolerated. It is recommended that the water table be at or more than 1m below the soil surface (Alvim, 1977).

Soil pH values ranging from 5.5 to 8.5 have been shown to have no negative effects on sugarcane production (Husz, 1989).

With the production of such a large amount of biomass there comes a high nutrient requirement. According to Husz (1972), 1t of fresh cane holds 0.5–1.2kg N, 0.2–0.3kg P, 1.0–2.5kg K, 0.3–0.6kg Ca and 0.2–0.4kg Mg. Here the leaf and top nutrient contents are not included because they are much higher. The presence of an endomycorrhiza to facilitate potassium uptake and free living nitrogen fixing bacteria in the soil are positive conditions.

The most favourable temperature for subtropical and tropical sugarcane cultivation is 25–26°C. Cold tolerance is an attribute of subtropical cultivars, but in the case of tropical cultivars plant growth is hindered at temperatures as low as 21°C, while temperatures around 13°C cause a halt in growth and temperatures below 5°C lead to chlorosis (Rehm and Espig, 1991).

Sugarcane needs a large quantity of rainfall – an average of around 1500–1800mm of rain under most conditions. In hot and dry conditions, as much as 2500mm or more may be needed (Rehm and Espig, 1991).

Propagation

Sugarcane is most commonly propagated using stem cuttings. This form of vegetative propagation is best performed using cuttings that have two nodes and are taken from unripened canes that are 8 or 9 months old. For best results the cuttings should come from stems that have been grown in the area where the cuttings will be planted. Roughly 5 to 6ha can be planted from the cuttings obtained from 1ha of sugarcane. A pre-planting fungicide and insecticide treatment is recommended. To destroy viruses and insects an additional hot water treatment of 1½ hours at 52°C can be performed (Rehm and Espig, 1991).

Crop management

A row spacing of 1.2–1.5m is recommended, but the cultivar and machinery used also need to be taken into consideration. The ground is ploughed to a depth of 30–40cm and the stem cuttings are placed end to end in the furrows, either mechanically or by hand. The furrows are then closed so that the cuttings are 3–5cm beneath the soil surface, or deeper if the soil is sandy or the climate is dry. The root system takes several weeks to form (Rehm and Espig, 1991). About 15,000 to 20,000 stem cuttings will be needed for each hectare to be planted, corresponding to 25 to 35 stem cuttings per 10m (Husz, 1989). Until the leaf canopy can close, it is suggested that a herbicide be used on the plants; but the area between the rows should be weeded mechanically.

According to Husz (1989), the best time to apply fertilizer is at the time of planting, along with an initial watering, and then again before the third month. Because of the greatly varying degree of nutrient requirements among the different cultivars, it is recommended that leaf or internode samples be analysed to determine the most appropriate fertilizer dosages for a particular cultivar (Husz, 1972). The sugarcane nutrient requirements listed earlier can be used as a guideline to help determine fertilizer requirements.

Through resistance breeding, many of the diseases that affected sugarcane such as downy mildew, smut, rust, leaf blight and mosaic disease are no longer major problems. With the planting of resistant cultivars, disinfection of planting materials and good sanitary practices, many fungus and bacterial disease problems can be avoided. Rehm and Espig (1991) draw attention to the eye-spot diseases (*Drechslera sacchari* (Butl.) Subram. et Jain.), pineapple disease (*Ceratocystis paradoxa* (Dade) Mor.), and gumming disease (*Xanthomonas vasculorum* (Cobb.) Dows). Also mentioned is the use of crop rotation and special disinfection steps to combat soilborne diseases of the *Fusarium*, *Pythium* and *Rhizoctonia* species, among others.

Problem pests such as aphids, leafhoppers, mealybugs and stem borers (*Diatraea* spp.) can be properly controlled through the use of appropriately timed insecticides or other insects such as the Cuban fly, *Lixophaga diatraea* (Long and Hensley, 1972).

Because sugarcane is perennial and creates no adverse effects when planted before or after itself, there is no need to develop a crop rotation or use a pause in cultivation.

Production

According to Husz (1989), the amount and quality of sugarcane harvested are affected by the correlation of growing periods with the time of the year. Table 7.60 compares 12 month growth cycles using planting dates in January and May with a theoretical maximum harvest.

Table 7.60. Characteristics and theoretical maximum yield of sugarcane based on planting month.

Characteristic	Month of planting		Theoretical
	January	May	maximum
Dry matter (t/ha)	34	27.5	50
Fresh matter (t/ha)	136	110	200
Fresh leaves (t/ha)	46	49.5	40–60
Cane stalk	90	60.5	140–160
Saccharose (%)	14	8.5	–
Saccharose (t/ha)	12.6	5.14	–
White sugar (t/ha)	10	3.45	14–16

Processing and utilization

Harvesting, using sugarcane harvesters, is most strategically performed after a dry period lasting approximately 2 months. This dry period results in the halt of plant growth and an increase in sugar content (Rehm and Espig, 1991). The harvesting of one year's growth of the crop should be performed between 10 and 13 months after planting, while the harvesting of crops grown for two years should be performed between 19 and 24 months after planting (Husz, 1989).

The harvested cane should be sent, most commonly by truck, to the processing facility within 30 hours of harvest to prevent quality deterioration (Husz, 1989).

The sugarcane is processed by diffusion or pressing to remove the saccharose containing juice from the stalk. The remaining bagasse can be burned to produce heat and electricity or additionally processed to produce building materials, cellulose and paper, among other products. The saccharose juice is purified through crystallization and centrifuging, at which time the 96% pure product is referred to as raw sugar. Additional washing and centrifuging leave the 99.8% pure product, white sugar. Molasses, which is removed during these purifying steps, can be processed to produce alcohol (Husz, 1989).

One tonne of sugarcane can produce approximately 100kg of sugar, 30kg of bagasse and 40kg of molasses (which can produce 10l of alcohol). If processed for alcohol production alone this 1t of sugarcane would produce about 70l alcohol (Husz, 1989).

Selected references

Alvim, P. de T. and Kozlowski, T.T. (eds) (1977) *Ecophysiology of Tropical Crops*, Academic Press, New York.

Husz, G. St. (1972) *Sugar Cane. Cultivation and Fertilization*, Ruhr-Stickstoff, Bochum.

Husz, G. St. (1989) Zuckerrohr. In: Rehm, S. (ed.), *Spezieller Pflanzenbau in den Tropen und Subtropen. Handbuch der Landwirtschaft und Ernährung in den Entwicklungs-ländern*, 2nd edn, vol. 4, Ulmer, Stuttgart.

Long, W.H. and Hensley, S.D. (1972) Insect pests of sugar cane. *Ann. Rev. Entom.* 17: 149–76.

Rehm, S. and Espig, G. (1991) *The Cultivated Plants of the Tropics and Subtropics*, Verlag Joseph Margraf, Priese GmbH, Berlin.

SUNFLOWER (*Helianthus annuus* L.)

Contributed by: C. D. Dalianis

Description

The cultivated species of sunflower is a native of America that was taken to Spain from Central America before the middle of the sixteenth century. It was being grown by Indians for food and for hair oil. Improved varieties had been developed in Europe before 1600. These varieties were introduced in the USA and cultivation of sunflower started.

In recent years, the world's cultivated area has been steadily increasing. This is mainly the result of the breeding of dwarf high yielding hybrids that also facilitate mechanization. Another reason is the emphasis given to polyunsaturated acids for human consumption. Since sunflower oil is rich in polyunsaturated acids, its use for human consumption has greatly increased.

Sunflower belongs to the genus *Helianthus* of the Compositae family. The genus *Helianthus* was named using the Greek *helios* meaning sun, and *anthos*, flower. The inflorescence of the plants of this family are heads in which the fertile flowers are aggregated and bordered by rays, the corollas of sterile flowers.

The genus *Helianthus* includes 67 annual and perennial species. The cultivated sunflower is an annual plant with the scientific name of *Helianthus annuus*. It is an erect, unbranched, coarse annual, with a distinctive large, golden head, the seeds of which are often eaten and are commonly crushed for oil production.

Ecological requirements

Sunflower is a well adapted crop under various climatic and soil conditions. With its well developed root system it is one of the most drought resistant crops and considered suitable for the southern semi-arid countries. However, oil yields are substantially reduced if plants are allowed to become stressed during the main growth period and at flowering. Under moisture stress conditions the number and the size of the leaves are reduced. One of the mechanisms employed by sunflower to resist moisture stress is wilting, since it has been shown in controlled experiments that in limp leaves water loss was reduced to a greater extent than photosynthesis. A satisfactory crop can be produced, without irrigation, even in winter rainfall regions of approximately 300mm.

Sunflower is well adapted to warm southern European regions. Growth is satisfactory when temperatures do not fall below 10°C, but it can resist far lower temperatures. The young plants can withstand considerable freezing until they reach the four to six leaf stage, and the ripening seeds suffer little damage from slight frost. Sunflower requires ample sunlight and is considered insensitive to day length.

Sunflower grows on a variety of soils ranging from sandy soils to clay. In low fertility soils its performance is better in comparison with other crops such as corn, potato and wheat. The best pH range is between 6.5 to 8. It is considered slightly susceptible to salt.

Crop management

The seedbed for sunflowers should be prepared as for corn. Minimum tillage is recommended. However, because of the pericarp of the seed a seedbed with high moisture content is required for satisfactory germination. The seed should be placed in moist soil; rapid drying out of the seedbed should be avoided.

Sunflower seeds are capable of germinating even at 5°C, but in practice temperatures over 10°C are required for satisfactory germination and even higher for satisfactory emergence. Generally, it is possible to sow sunflower fairly early in spring before other summer crops, such as corn, cotton or sorghum. Usually, early plantings lead to higher seed yields and higher oil content of seeds.

Optimum plant population is a key factor for obtaining high yields. This is mainly because each plant produces only one disc, so sunflower has a limited ability to adjust yields when the deviation from the optimum number of plants is large. When deviations are not large thinner populations may produce larger discs, but they usually have more empty seeds in the centre of the disc and consequently oil yields are usually reduced.

In areas with more than 500mm annual rainfall, populations range from 35,000 to 60,000 plants per hectare. Under irrigation, the number of plants per hectare is 25–50% higher, but plant lodging probabilities are also higher. An empirical method for determining the optimum population density is the disc diameter at maturity. If it is smaller than 12cm, plant density is too high, and if it is larger than 25cm, plant density is too low.

Planting is done with a corn planter equipped with special sunflower plates, or with a grain drill with most of the feed cups closed off. Distances of 75cm between rows are very common. Under irrigation, shorter between row spacings of 50 to 70cm and within row spacings of 15–30cm give higher yields. If possible, the rows should be sown with an east–west orientation because most of the discs remain orientated towards east and mechanical harvesting is thus easier. Seed rates range between 5 and 15kg/ha.

In the early growth stages sunflower is sensitive to weed competition. A pre-emergence harrowing, followed by cross-harrowing of the crop in the seedling stage, is recommended. Immediately after plant establishment, cultivation should be undertaken as required. Pre-sowing and/or pre-emergence herbicides are used for the control of certain weeds.

Large areas of sunflower throughout the world are grown without irrigation, because farmers prefer to use irrigation for other crops that are more dependent on irrigation or more profitable. In summer rainfall areas growth and production are satisfactory. In winter rainfall areas, as in most southern European regions, although satisfactory yields are obtained without irrigation in comparison with many other summer crops, sunflower's response to irrigation is very pronounced. In the same field, irrigated sunflower yields have been observed to increase up to 100% or more in comparison to non-irrigated plants. However, water use efficiency depends on fertilization.

A significant yield reduction is observed under water stress conditions, mainly because the proportion of fertile seeds is reduced as a result of poor fertilization, and also because the abortion of flowers shortly after fertilization is greatly increased. As a rule, a favourable soil moisture regime must be maintained throughout the period from the differentiation of floral organs, when the inflorescence bud is about 15mm in diameter, up to harvest.

In non-irrigated fields sunflower's response to fertilizers is very limited, as is the case with many other non-irrigated spring grown crops. In irrigated fields of low fertility, nitrogen fertilization response is positive. Nitrogen fertilization rates from 50 to 80kg/ha are recommended. Nitrogen application must be related to the availability of phosphate and potash, for if these are deficient nitrogen usually depresses yield and seed oil content.

Production

In southern European countries a total of 2.66 million hectares of sunflower are grown each year, mainly for oil production. Annual sunflower seed production is equal to 3.92 million tonnes with an average seed yield of 1473kg/ha. However, there are very large productivity differences among the various southern European countries (Table 7.61). These are related to the soil type, but mainly to climatic conditions in each country as well as to the irrigated versus non-irrigated areas. If we assume an average 50% oil content of the sunflower achenes, oil production ranges from as low as 280kg/ha in Portugal to 1349kg/ha in Italy, with an average overall yield of 738kg/ha.

Recently, in southern European regions, sunflower has been considered as an energy crop for biodiesel production as is rapeseed in northern European regions.

Table 7.61. Acreage, production and yields of seed sunflower in southern EU countries and Germany.

Country	Acreage (ha)	Production (tonnes)	Yields (kg/ha, seed)
France	991,000	2,143,000	2162
Spain	1,455,000	1,363,000	937
Italy	127,000	349,000	2748
Portugal	73,000	42,000	575
Greece	14,000	22,000	1571
Germany	65,000	160,000	2461
Total	2,725,000	4,076,000	1496

However, sunflower biodiesel can be produced only under strong and continuous financial incentives (subsidies and/or tax exemptions). In Italy there are tax exemptions of 125,000t/year for biodiesel either from sunflower or from rapeseed. In France, tax exemptions exist for all liquid biofuels, whereas in Spain, Portugal and Greece there are no incentives for liquid biofuels.

Processing and utilization

Sunflower heads should be fully mature before harvesting. It is estimated that during the last fourteen days of maturation the dry matter of the seeds may increase by 50–100%. However, seed maturation in a head is not uniform. Delaying harvesting until all the seeds are fully mature may cause significant losses due to shattering and losses from birds that eat the seeds. A practical indication of the appropriate stage for harvesting is when the backs of the sunflower heads are yellow and the outer bracts are beginning to turn brown. However, even at this stage a fair proportion of seeds contain up to 50% moisture. This should be taken into account when considering safe storage because the maximum moisture content at storage should not exceed 9%.

Sometimes a pre-harvest desiccant is applied in order to facilitate harvesting. Diquat, applied when the leaves are turning yellow, is very effective and the yield losses are negligible. Dwarf varieties are harvested with a cereal combine machine equipped with a special attachment on the header.

The head of the matured sunflower contains about 50% of the dry matter of the whole plant. Nearly one-half the weight of dried heads is seed.

In oil producing varieties the pericarp ranges from 22 to 28% and the kernel from 72 to 78%. Kernels are rich in oil. The average oil composition of the achenes ranges from 40 to 50%, the protein content from 15 to 20%, and the fibre content from 10 to 15%. In non-oil producing varieties the pericarp ranges from 45 to 50% and the kernel from 50 to 55%. These achenes have a lower oil content (30 to 35%) and higher fibre content (20 to 25%).

It is estimated that 90% of the world's sunflower seed production is crushed for oil extraction. Sunflower oil is characterized by its high content of linoleic acid and medium content of oleic acid. Sunflower oil produced in cool, northern climates usually contains 70% or more linoleic acid, whereas sunflower oil produced in southern warm regions contains 30% linoleic acid. There is usually a negative relationship between linoleic and oleic acid contents. The nutritive value of sunflower oil is equal to that of olive oil. Sunflower oil is mainly used for human consumption.

In growing sunflower for biodiesel production it should be taken into account that, beyond the seeds, large quantities of other sunflower plant parts could be exploited for energy purposes. A sunflower crop produces a significant amount of plant residues, mainly stalks, that remain in the field. On the average, stalk residues are estimated as twice as much the seed production. These residues have a high gross heating value of about 17 to 18MJ/kg of dry matter.

The non-seed part of the heads is also an important biomass source that is almost equal to the seed production. In addition, if dehulling is practised prior to the oil extraction process, a significant quantity of hulls is produced. In oil producing hybrids about 25% of the achene weight is the hull, which is composed of equal proportions of lignin, pentosans and cellulosic material, representing 82 to 86% of the total weight. These materials also have a high heating value. The gross heating value is about 17MJ/kg of dry matter.

The long term potential of sunflower oil for biodiesel is not encouraging for three main agronomic reasons. First, sunflower is a low yielding crop, giving on average less than 1t/ha of oil. Second, as a worldwide cultivated crop of many years' standing it has been considerably improved, so future productivity improvements are expected to be small and difficult. Third, under the same soil and climatic conditions many other biomass crops give much higher yields of raw material that could be exploited for energy purposes. For example, irrigated corn may give, under southern European conditions, approximately 4000l/ha ethanol, whereas irrigated sunflower gives only 1000 to 1200kg/ha of biodiesel.

It is understood that from a strictly agronomic point of view in a future competitive market for liquid biofuels, sunflower biodiesel will be in a very disadvantageous position to compete with other liquid biofuels such as corn bioethanol. Sunflower is in a similar position with regard to other irrigated agricultural crops or wild grown plants (giant reed, sweet and fibre sorghum), and when non-irrigated sunflower is compared with other non-irrigated crops (wheat, etc.). Of course, sunflower biodiesel has the great advantage that the oil extraction plants already exist in many rural areas, so there is usually no need for costly new investments.

It should be also taken into account that sunflower oil for biodiesel would have to compete with the strong market for human consumption that is steadily increasing both worldwide and Europe-wide. Its high quality and value for human consumption means that the bulk of present European sunflower oil production is directed towards this end.

Finally, it is well known that the GATT agreement imposes severe restrictions on the area in Europe that could be given to oil crops for biodiesel production. It is estimated that throughout Europe no more than a million hectares can be given to oil crops, rapeseed and sunflower for biodiesel production.

Selected references

Carter, J.K. (ed.) (1978) *Sunflower Science and Technology*. Agronomy monograph, Amer. Soc. Agron., Madison, USA.

Helia, the newsletter of the FAO/ESCORENA research network on sunflowers.

Liquid Biofuels Newsletter (1995) BLT, Wieselburg, Austria, vol. 2.

Weiss, E.A. (1983) *Oilseed Crops*, Longman, London.

SWEET POTATO (*Ipomoea batatus* (L.) Poir. ex Lam.)

Description

The sweet potato is a herbaceous vine with a creeping growth habit. Its habitat is the part of the Andes from northern South America to Mexico, where it was already in use in pre-Columbian times (Lötschert and Beese, 1984). The plant is hexaploid, 2n=90. It is assumed that its ancestors were tetraploid and diploid species found in this region. Nowadays, it is grown everywhere in the tropics and subtropics (Rehm and Espig, 1991). The sweet potato can even be grown in some parts of the temperate zones with warm summers. The reason for its wide range is the rather short vegetation period. The sweet potato can be grown where the temperature during the growing period is above 18°C and temperatures below 10°C do not occur (Franke *et al.*, 1988). All parts of the plant contain latex tubes with a neutral tasting milky juice, which, in contrast to cassava, is not toxic (Franke, 1985).

Ecological requirements

The sweet potato grows well if there is an average daytime temperature of more than 18°C. At 10–12°C growth comes entirely to a standstill. The slightest frost kills all plant parts above the ground. The sweet potato is very sensitive to temperatures below 1°C for brief periods and below 3°C for longer periods. Sweet potato is a warm season perennial crop that is cultivated as an annual and may be harvested after 80–360 days. A frost free period of 110–170 days is necessary. However, the sweet potato's need for warmth is less than most other root and tuber crops. In the tropics it grows up to 2500m.

Distinct differences between day and night temperatures encourage the formation of tuberous roots. Cultivars with short vegetation times grow well in the subtropics, up to 40° latitude. Long day conditions delay the flowering and reduce the formation of tuberous roots: a photoperiod of less than 11 hours induces flowering; less flowering occurs at 12 hours, and no flowering at all at 13 hours daylight.

Sweet potato cultivation is limited to regions with 750–1250mm rainfall. The plant survives long dry periods, but for high yields and for good tuber quality there must be a water supply (precipitation or irrigation) that is evenly spread throughout the growing period. Dryness is desirable during the ripening period, for then the roots will be firm and keep well. In subtropical climates the roots contain more starch and less sugar than in tropical climates. High yields are only achieved in areas with large amounts of sunshine. Sweet potato needs a well aerated and well drained soil. Where there are heavy soils or there is a danger of waterlogging it is grown on ridges or mounds. Neither pH below 5 nor salt is acceptable (FAO, 1996). Calcium deficiency of the soil makes planting impossible.

Propagation

Propagation is done vegetatively using stem cuttings of 30cm length, with the leaves stripped off before planting. The cuttings are placed into moist warm soil, and after 2–3 days they take root. The sweet potato cannot be planted in dry soils. Another way is to plant roots that were stored during the winter, but this is disadvantageous.

Because the plants are frequently self-sterile, cross-breeding is necessary as a rule. Parent plants are easy to bring to flowering at the same time in any particular climate if they are grafted onto an annual *Ipomoea* species. The *in vitro* technique is used in breeding improved cultivars – for example, to select cold resistant cultivars (Irawati, 1994).

Crop management

The cultivation of the sweet potato can be completely mechanized. Planting distances are 90cm between and 60cm within the rows. Weeding by hoeing or with herbicides (linuron, diphenanid) is necessary in the first 6 weeks; thereafter the leaves suppress the weeds' growth.

The greatest damage is caused by several viral diseases; then less susceptible or new virus free cultivars have to be planted. Losses by fungal diseases are generally slight because resistant cultivars are available in most regions. Infestation is possible by leaf eating beetles whose larvae hollow the roots, and leaf eating caterpillars can cause serious losses. Nematodes attack the roots and tubers. To prevent severe damage, insecticides, harvesting at the right time and crop rotation are useful measures.

The sweet potato has a positive influence on weed control; it can suppress perennial weeds like nutsedge (*Cyperus rotundus* L.). It is self-compatible. Some tribes in New Guinea plant sweet potatoes in monoculture, but mostly the plant is grown in rotation with other field crops.

Production

The sweet potato should be harvested when the leaves begin to yellow because the starch content is highest at that point. Harvesting can be fully mechanized, as for potatoes. In small scale farming the tubers can be dug up with a plough or manually with spade or hoe. For cultivars with a growing time of 5 months a successful yield is 20t/ha, and 40–50t/ha can be achieved with high yielding cultivars. With early ripening cultivars, less improved local cultivars, or under bad cultivation conditions, the yields are much lower. The global average is a yield of 14.6t/ha.

Processing and utilization

The major portion of production is consumed fresh. Drying or starch extraction is possible. The sweet potato, like other starch and sugar producing plants, is a potential energy crop. The carbohydrate content of the edible part is 9–25% starch and 0.5–5% sugar. The concentration of sugar becomes higher the nearer to the equator the plant is grown (Franke, 1985). If the sweet potato is grown as an energy source the starch of the roots is processed to alcohol; the roots are ground and the starch is cracked into sugars by treatment with acids or hydrolytic enzymes. After the conversion of starch into sugars, fermentation takes place to give ethanol. The development of continuous working fermentation techniques and the use of improved varieties of yeast have improved efficiency and lowered costs (FAO, 1995).

One tonne of sweet potatoes yields 125l of ethanol (Wright, 1996). Leaves, foliage, unsaleable tubers and remnants from processing are used as fodder. Wastes and by-products can be used to produce energy – the tubers for ethanol production, and leaves and foliage for biogas production.

Selected references

FAO (1995) *Future Energy Requirements for Africa's Agriculture*, Food and Agriculture Organization of the United Nations, Rome.
FAO (1996) Ecocrop 1 Database, Rome.
Franke, G., Hammer, K., Hanelt, P, Ketz, H.-A., Natho, G. and Reinbothe, H. (1988) *Früchte der Erde*, 3rd edn, Urania Verlag, Leipzig, Jena and Berlin, Germany.
Franke, W. (1985) *Nutzpflanzenkunde*, 3rd edn, Georg Thieme Verlag Stuttgart and New York.
Irawati (1994) Selection for cold hardiness of sweet potatoes from Baliem Valley by *in vitro* culture. *Acta Horticulturae* 369: 220–5.
Lötschert, W. and Beese, G. (1984) *Pflanzen der Tropen*, 2nd edn, BLV Verlagsgesellschaft München, Wien and Zürich.
Rehm, S. and Espig, G. (1991) *The Cultivated Plants of the Tropics and Subtropics*, Verlag Joseph Margraf, Preise GmbH, Berlin.
Wright, R.M. (1996) *Jamaica's Energy: Old Prospects New Resources*, Twin Guinep (Publishers) Ltd / Stephenson's Litho Press Ltd.

SWITCHGRASS (*Panicum virgatum* L.)

Contributed by: D. G. Christian and H. W. Elbersen

Description

Switchgrass is a perennial grass with slowly spreading rhizomes and a clump form of growth. Culms (stems) are erect and 50–250cm tall, depending on variety and growing conditions (Figure 7.14). The stiff, tough stems are capable of withstanding moderate snowloads. The inflorescence is a panicle 15–50cm long. Seeds are hard, smooth and shiny. Seed weights are variable and range between 0.7 and 2.0g per 1000. Dormancy level may be high in newly harvested seed. Storing seed or other dormancy breaking techniques may be required before it can be sown.

The grass is highly polymorphic, outcrossing, and has a basic chromosome number of nine. Most cultivars are tetraploids or hexaploids. Two major ecotypes can be distinguished (Moser and Vogel, 1995). Lowland types are generally found in flood plains and are taller and coarser, while upland types are found in areas that do not flood, and have finer leaves and generally slower growth than lowland types (Sanderson *et al.*, 1996). The geographical distribution of switchgrass is from North America to Central America, but it is also found in South America and Africa.

In the USA, switchgrass is an important warm season native range grass. Warm season grasses, which have the C_4 photosynthesis pathway, produce most of the biomass in hot summers when cool season (C_3) grasses become semi-dormant. The more efficient C_4 carbon assimilation pathway potentially gives warm season grasses much higher yields than cool season grasses. Compared to C_3 grasses C_4 grasses have higher base optimum temperatures; Hsu et al (1985a) found a minimum germination temperature of 10.3°C for switchgrass, while the optimum temperature for seedling growth was more than 30°C (Hsu *et al.*, 1985b). Switchgrass has several characteristics that make it suitable as a biomass crop, including high potential productivity, low nutrient demand, efficient water use and good persistence (Jung *et al.*, 1988; Hope and McElroy, 1990; Jung *et al.*, 1990; Sladden *et al.*, 1991).

Considerable information is available on the use and management of switchgrass as a forage grass. In recent years, switchgrass has been evaluated as a potential biomass feedstock for energy production in the USA. It has been selected as a model herbaceous crop for the Oak Ridge National Laboratory's Biofuel Feedstock Development Program (Parrish *et al.*, 1993; McLaughlin *et al.*, 1996; Sanderson *et al.*, 1996). The suitability of this grass as a biomass crop in Europe has yet to be demonstrated. Therefore this discussion draws heavily on American publications and the reader should bear this in mind.

Ecological requirements

Switchgrass has considerable cold tolerance after winter hardening; switchgrass stands can be maintained up into Canada (Hope and McElroy, 1990). In a study at Rothamsted in southern England, seven varieties have successfully overwintered; however, the adoption of varieties to a range of European conditions has not yet been evaluated. The adaptability of switchgrass means there are cultivars available for a wide range of environmental conditions.

Switchgrass will tolerate acid conditions but establishment and growth will be better if soils are limed to correct the pH to neutrality (Jung *et al.*, 1988). Seedbeds are normally prepared using traditional ploughing and secondary cultivation to produce a firm seedbed with a fine textured surface. The seedbed should be weed free because switchgrass is not competitive during the establishment phase.

Figure 7.14. Switchgrass, a C_4 energy crop, Germany.

Crop management

Switchgrass is established by seed. Sowing is normally carried out when soil temperatures have reached at least 10–15°C. Because of dormancy problems in newly harvested seed, freshly harvested seed can be 95% dormant; year old seed often germinates better than fresh seed. The amount of pure live seed (PLS), not the weight of seed, is used to calculate the seed rate. PLS is calculated by multiplying the purity by the germination. Germination rates of above 75% are desirable.

The seed can be sown using normal small seed drills or cereal drills fitted with fine seed rolls. Direct drilling (no-till planting) may be appropriate in some circumstances (Parrish *et al.*, 1993). Seed depth should not be more than 1cm. The number of plants established can be up to 400 plants per m^2, depending upon weed control strategy after sowing (Vogel, 1987). In the experiment at Rothamsted a common seed rate of 10kg/ha was used to sow seven varieties of switchgrass. The number of plants established ranged from 189 to 301 per m^2 (Christian, 1994).

For optimum yields adequate fertility levels are required, though reasonable yields can be attained under low fertility conditions (Moser and Vogel, 1995). If levels of phosphorus and potassium are low, these nutrients may be added to the seedbed to benefit seedling growth. Nitrogen should not be applied to the seedbed or during the early establishment phase because it may stimulate weed growth to the disadvantage of the slow growing grass. There may also be a risk in the established crop that nitrogen carried over from the previous year and from early nitrogen application will stimulate weed growth at a time when switchgrass is still not competitive because of low temperatures in spring.

Weed competition is a major cause of seedling establishment failure (Moser and Vogel, 1995). It is important to reduce weeds prior to planting. Pre-emergence herbicides are often used to reduce weed competition when switchgrass is not competitive. Broad leaf weeds can be suppressed with herbicides until switchgrass seedlings are large enough. Grassy weeds can be mown to reduce light competition. There are no herbicides approved for use in switchgrass in Europe, but simazine and atrazine will effectively control weeds even in young crops. Isoproturon has been used during the plant's dormant period in winter.

Production

In the UK, shoots from the rhizomes emerge in late spring, anthesis takes place during August or September depending on the cultivar, and senescence starts with falling temperatures in the autumn. Growth may stop sooner than for temperate grasses.

Biomass production in the year of planting is likely to be low but substantially higher in the subsequent years (Table 7.62). The four years of evaluation at Rothamsted provided evidence of wide variation in the yield of different cultivars. The yield of the three best cultivars averaged 11t/ha in 1996.

The yield of individual cultivars has increased each year, except for 'Dacotah' which had the same average yield in 1995 and 1996 (Table 7.62). It is not known if maximum yields have been reached. It is noteworthy that yields have increased despite drought conditions in the summers of 1995 and 1996. In 1993/94 the date of maturity varied from October to January, depending on the cultivar. No differences in maturity date

Table 7.62. The yield (DM t/ha) of seven varieties of switchgrass between 1993 and 1996.

Variety	1993		1994		1995		1996	
	N0	N60	N0	N60	N0	N60	N0	N60
Pathfinder	2.42	2.06	6.89	5.52	8.48	8.22	9.91	9.01
Sunburst	1.34	1.03	6.05	4.16	6.98	8.05	7.36	8.60
Cave-in-rock	1.66	1.66	6.92	6.06	8.11	8.03	10.88	10.11
Nebraska 28	1.55	1.51	5.34	5.25	8.59	7.30	10.56	9.27
Dacotah	1.11	1.35	4.44	4.10	5.28	5.70	5.40	5.37
Kanlow	2.15	1.64	6.77	4.34	7.10	5.34	12.48	10.78
Forestburgh	1.53	1.47	6.27	6.28	7.35	8.72	11.72	9.95
Mean	1.68	1.53	6.10	5.10	7.41	7.34	9.76	9.01
SED, Varieties XN		0.517		0.938		0.529		1.434

Crops were sown in May 1993 and received no fertilizer (N0), or 60 kg/ha/year (N60). Crops were harvested in winter when stems were dead.

Table 7.63. The amount of various minerals present in different varieties of switchgrass at harvest 1995/6.

Cultivar	Fertilizer (N at 0 or 60kg/ha)	Nitrogen (kg/ha)	Nitrogen (kg/t)	Phosphorus (g/ha)	Phosphorus (kg/t)	Potassium (kg/ha)	Potassium (kg/t)	Magnesium (kg/ha)	Magnesium (kg/t)
Pathfinder	N0	43.9	5.2	3.5	0.4	11.8	1.4	7.3	0.9
	N60	43.8	5.3	3.2	0.4	8.2	1.0	5.4	0.7
Sunburst	N0	39.3	5.6	2.9	0.4	11.1	1.6	4.9	0.7
	N60	36.5	4.5	2.5	0.3	9.3	1.2	4.1	0.5
Cave-in-rock	N0	48.1	5.9	4.0	0.5	10.5	1.3	6.6	0.8
	N60	49.2	6.1	3.6	0.4	10.9	1.4	6.4	0.8
Nebraska 28	N0	42.5	4.9	2.8	0.3	9.0	1.0	4.2	0.5
	N60	38.2	5.2	3.4	0.5	9.3	1.3	4.4	0.6
Dacotah	N0	32.7	6.2	2.6	0.5	5.7	1.1	2.2	0.4
	N60	32.6	5.7	2.3	0.4	5.3	0.9	2.4	0.4
Kanlow	N0	39.3	5.5	3.6	0.5	11.1	1.6	10.4	1.5
	N60	44.1	8.2	3.6	0.7	10.1	1.9	9.5	1.8
Forestburgh	N0	35.6	4.8	3.0	0.4	8.8	1.2	4.0	0.5
	N60	40.5	4.6	2.8	0.3	8.4	1.0	4.0	0.5

Mineral content is reported as total offtake (kg/ha) and kg/t of biomass.

were found in 1995, probably as a result of the drought conditions that occurred in the summer. Response to nitrogen has been inconsistent and frequently negative, possibly because in the first year there was adequate N supplied from soil sources whereas in the second season drought may have been influential. Work in USA has shown 50kg/ha N to be about optimal for yield on marginal soils (Parrish *et al.*, 1993).

In the Rothamsted experiment the amount of the major nutrients removed at harvest was found to be similar for all cultivars except Dacotah (Table 7.63). Nutrient content at harvest is lower than in *Miscanthus* grass which was of the same age, grown in the same field and harvested in the same winter. The *Miscanthus* contained 8kg N, 0.6kg P and 8.7kg K per tonne of dry biomass. The ash content of switchgrass and *Miscanthus* was 7.8% and 8.4% respectively. The low mineral content of switchgrass is a desirable characteristic for efficient combustion and low exhaust gas emissions.

Table 7.64. The yield of switchgrass at four locations in the eastern USA (cutting regimes vary).

Region (location)	Nitrogen (kg/ha N)	Dry matter yield (t/ha)		Notes	Reference
North-east (Pennsylvania)		1981–1984		Mean of five varieties cut at heading	Jung *et al.*, 1990
	0	7.5	(6.9–7.8)		
	75	9.9	(9.1–12.3)		
East (Virginia)		1988–90		Various sites on marginal land. Single variety 'Cave-in-rock'. Reduced yield variation with November cut, wide yield variation with site and season.	Parrish *et al.*, 1993
		Cut Sept.	Cut Nov.		
	0	8.4	8.1		
	50	10.7	9.1		
	100	10.6	9.3		
	Residual year, previously –	1991		Cut after stems senesced. Heavier yield where crop previously cut in November.	
	0	5.8	8.2		
	50	8.2	9.8		
	100	9.8	10.2		
Upper south-east (Virginia, Tennessee, West Virginia, Kentucky, North Carolina)	100	One cut system		Seasonal yields from six varieties grown at eight locations. One cut in October or November or two cuts, at heading and in October or November.	Parrish *et al.*, 1997
		13.3	(11.0–19.4)		
		Two cut system			
		15.7	(13.4 21.2)		
South-east (Alabama)	84	1989		Mean of eight varieties. Yield is total of two cuts, anthesis and regrowth. Two varieties, 'Alamo' and 'Kanlow', gave outstanding yields in both years. Other varieties included 'Cave-in-rock' and 'Pathfinder'. Most had a higher yield in the second year.	Sladden *et al.*, 1991
		10.4	(6.8–17.5)		
		1990			
		14.2	(6.8–34.6)		

The yields of switchgrass grown at four locations at different latitudes in the eastern USA are presented in Table 7.64. Although the Rothamsted site has a more northerly location, the yield of some cultivars compares favourably with yields obtained at the US sites. This would indicate that if switchgrass is grown in Europe heavy yields of biomass might be achieved, especially at more southerly locations. In northern Europe, switchgrass may have a lower yield than some alternative biomass crops such as *Miscanthus* grass (Schwarz *et al.*, 1994; Van der Werf *et al.*, 1993) or energy coppice (Cannell, 1988). However, switchgrass has lower production costs compared to *Miscanthus* because it can be established by seed and may require less fertilizer to replace minerals exported in biomass; and, in common with *Miscanthus*, the annual harvest gives a quicker payback than short rotation coppice which has a 3–5 year harvest cycle.

Processing and utilization

There is no technical reason why the crop cannot be cut and harvested using traditional grass harvesting machinery. The thin woody stems of switchgrass allow good dry-down in the winter. At Rothamsted, moisture content at harvest was down to 30%; this may allow the baled crop to be stored for a short period before use without the need for drying. McLaughlin *et al.* (1996) report moisture contents of 13 to 15% at baling in the USA. No records of harvesting methods or storage in the European context have been found.

End uses of switchgrass being considered are ethanol production, combustion and thermal conversion. Yields of 280l of ethanol per tonne of dry switchgrass appear possible, which compares favourably to other species that have been examined (McLaughlin *et al.*, 1996). The main attributes that determine usefulness for combustion are total energy content, moisture content, and fouling and corrosion characteristics. The energy content of switchgrass is about 18.4kJ per kg of dry weight, which is comparable to woody species but about 33% less than that of coal (McLaughlin *et al.*, 1996). Moisture content is variable and depends on time of harvest and storage methods, but compares favourably to woody species. McLaughlin *et al.* (1996) reported the ash content of switchgrass to be variable (2.8–7.6%), and generally lower than that of coal.

There may be some concern about the alkali and silica content of switchgrass, which could lower ash melting points and thus contribute to slagging in boilers. Particularly high potassium and phosphorus can be a problem. Ragland *et al.* (1996) conducted experiments in which switchgrass was co-fired with pulverized coal. They found an ash content of 4.6% for switchgrass, which was about half of that of coal. Also switchgrass had 40% as much nitrogen and 5% as much sulfur as coal. Co-firing switchgrass with coal can thus reduce air pollution emissions per unit of energy produced. The switchgrass ash contained 3.4 times more potassium and 50 times more phosphorous than coal ash.

Appropriate harvest management can reduce the mineral content of switchgrass and thus increase combustion quality. Sanderson and Wolf (1995) concluded that allowing switchgrass to reach maturity would minimize the concentration of inorganic elements in the feedstock. In Europe, combustion, co-firing and gasification are good options for switchgrass utilization in the short run, but other options for its use considered in the USA have so far not been evaluated.

With respect to switchgrass the following conclusions can be reached:

- Evaluation of switchgrass as a biomass crop for European conditions is strictly limited at present.
- The range of cultivars available makes it likely that switchgrass could be grown in many regions of Europe and may suit drier areas.
- Switchgrass is established by seed and therefore has a low cost of establishment and potentially low costs of production. Establishment is the most critical part of switchgrass cultivation.
- Yields so far obtained in Europe have been comparable to yields obtained in similar climatic regions of the USA.

Selected references

Cannell, M.G. (1988) The scientific background in Biomass Forestry in Europe: A strategy for the future In: Hummel, F.C., Palz, W. and Grassi, G. (eds), *Energy from Biomass*, 3, Elsevier Applied Science, pp. 83–140.

Christian, D.G. (1994) Quantifying the yield of perennial grasses grown as a biofuel for energy generation. *Renewable Energy* 5(2): 762–6.

Hope, H.T. and McElroy, A. (1990) Low temperature tolerance of Switchgrass. *Can. J. Plant Sci.* 70: 1091–6.

Hsu, F.H., Nelson, C.J. and Matches, A.G. (1985a) Temperature effects on germination of perennial warm-season forage grasses. *Crop Sci.* 25: 215–20.

Hsu, F.H., Nelson, C.J. and Matches, A.G. (1985b) Temperature effects on seedling development of perennial warm-season forage grasses. *Crop Sci.* 25: 249–55.

Jung, G.A., Shaffer, J.A. and Stout, W.L. (1988) Switchgrass and Big Bluestem responses to amendments on strongly acid soil. *Agron. J.* 80: 669–76.

Jung, G.A., Shaffer, J.A., Stout, W.L. and Panciera, M.T. (1990) Warm-season grass diversity in yield, plant morphology and nitrogen concentration and removal in Northeastern USA. *Agron. J.* 82: 21–6.

McLaughlin, S.B., Samson, R., Bransby, D. and Wiselogel, A. (1996) Evaluating physical, chemical, and energetic properties of perennial grasses as biofuels. In: *Bioenergy '96. Proceedings of the Seventh National Bioenergy Conference*, 15–20 September, Nashville, Tennessee, vol. 1, pp. 1–8.

Moser, L.E. and Vogel, K.P. (1995) Switchgrass, Big Bluestem, and Indiangrass. In: Barnes, R.F., Miller, D.A. and Nelson, C.J. (eds), *An Introduction to Grassland Agriculture. Forages*, 5th edn, vol. I, Iowa State University Press, Ames, pp. 409–20.

Parrish, D.J., Wolf, D.D. and Lee Daniels, W. (1993) *Perennial Species for Optimum Production of Herbaceous Biomass in the Piedmont*. Management study, 1987–1991, ORNL/Sub/85-27413/7. National Technical Information Service, US Department of Commerce, 5285 Port Royal Road, Springfield, VA 22161.

Parrish, D.J., Wolf, D.D. and Daniels, W.L. (1997) *Switchgrass as a Biofuel for the Upper Southeast: Variety Trails and Cultural Improvements*. Final report 1992–1997. Biofuels feedstock development program environmental sciences division, Oak Ridge National Laboratory, Oak Ridge, Tennessee.

Ragland, K.W., Aerts, D.J. and Weiss, C. (1996) Co-firing switchgrass in a 50MW pulverized coal burner. In: *Bioenergy '96. Proceedings of the Seventh National Bioenergy Conference*, 15–20 September, Nashville, Tennessee, vol. 1, pp. 113–20.

Sanderson, M.A. and Wolf, D.D. (1995) Switchgrass biomass composition during morphological development in diverse environments. *Crop Sci.* 35: 1432–8.

Sanderson, M.A., Reed, R.L., McLaughlin, S.B., Wullschleger, S.D., Conger, B.V., Parrish, D.J., Wolf, D.D., Taliaferro, C., Hopkins, A.A., Ocumpaugh, W.R., Hussey, M.A., Read, J.R. and Tischler, C.R. (1996) Switchgrass as a sustainable bioenergy crop. *Bioresource Technol.* 56: 83–93.

Schwarz, K.U., Murphy, D.P.L. and Schnug, E. (1994) Studies on the growth and yield of *Miscanthus* x *giganteus* in Germany. *Aspects of Applied Biology* 40: *Arable Farming under CAP Reform*, vol. 2, pp. 533–40.

Sladden, S.E., Bransby, D.I. and Aiken, G.E. (1991) Biomass yield, composition and production costs for eight switchgrass varieties in Alabama. *Biomass and Bioenergy* 1(2): 119–22.

Van der Werf, H.M.G., Meijer, W.J.M., Mathijssen, W.F.J.M. and Darwinkel, A. (1993) Potential dry matter production of *Miscanthus sinensis* in the Netherlands. *Industrial Crops and Products* 1: 203–10.

Vogel, K.P. (1987) Seeding rates for establishing big bluestem and switchgrass with pre-emergence Atrazine applications. *Agron. J.* 79: 509–12.

TALL FESCUE (REED FESCUE) (*Festuca arundinacea* Schreb.)

Description

Tall fescue is a perennial cool season grass commonly grown for forage, turf, or as a conservation grass. The plant grows to a height of 0.6–1.8m. The leaf blades are grooved and the underside is keeled. The inflorescence is a panicle. The main flowering period is from June to July (Siebert, 1975; Kaltofen and Schrader, 1991)

Most tall fescue is infected (mutualistic symbioses) by the fungal endophyte (*Acremonium coenophialum*). Although this fungus makes the plant toxic for grazing, it is ecologically necessary for the plant's success and has been proven to confer insect resistance on the plant.

Tall fescue plant material has mineral concentrations of approximately 9–10% ash, 2–3% silicon dioxide (SiO_2), 2.5% nitrogen, 100mg/kg iron, 60–70mg/kg manganese and 5–6mg/kg copper (Pahkala *et al.*, 1995).

Ecological requirements

Tall fescue can adapt to a wide range of soils and climates. It responds well to fertile soils, in which it develops open sods. Temperate zones, subtropics and tropical highlands are the areas with the most suitable climates for tall fescue. In mild climates, growth continues throughout winter. Sustained summer growth is usual. Tall fescue grows well even on poor soils. Although tall fescue is not tolerant of wet soils and salt, it has been shown to be tolerant to drought (Rehm and Espig, 1991)

Propagation

Tall fescue is propagated using seed.

Crop management

The amount of seed sown is approximately 20kg/ha. Fertilizer application tests on tall fescue by Moyer *et al.* (1995) in the USA showed that subsurface (knife) application of fertilizer led to a 20% increase in yield when compared with surface broadcast application. In addition, a yield increase of 69% was observed when the nitrogen fertilizer level was increased from 13kg/ha to 168kg/ha.

Common weeds that infest tall fescue crops are, among others, crabgrass (*Digitaria* spp.) and white clover (*Trifolium repens* L.). Herbicide trials on tall fescue crops by Muellerwarrant *et al.* (1995), using pre-emergence herbicides in mid-October, post-emergence herbicides in early December, or both, showed that although herbicides may control weeds effectively, crop damage (yield reduction) does occur. The least damaging herbicide tested was pendimethalin applied pre-emergence with no post-emergence herbicide.

Among the most common fungi known to attack tall fescue plants are rhizoctonia blight, white blight, and rust (*Puccinia graminis* ssp. *graminicola*). Insect pests such as grass grub larvae are known frequently to feed on the roots of tall fescue.

Production

Field tests at four locations in Germany where tall fescue was harvested once a year at the onset of flowering provided yields averaging between 11.4 and 13.1t/ha dry matter. The plant's dry matter content averaged around 30% at the time of harvest. The heating value averaged approximately 17MJ/kg (Feuerstein, 1995).

Processing and utilization

Tall fescue crops are harvested by mowing, starting in the first year they reach maturity. When grown for the production of biomass, the crop may be harvested several times each year beginning in early summer, usually with about 4 weeks between harvests. The harvested crop can be baled and allowed to dry for storage. It is important that the crop matter be sufficiently dried for its economic utilization as a biofuel.

In addition to its traditional uses for hay and pasture, tall fescue has also taken on the roles of turf or conservation grass and has been widely and successfully used to stabilize ditch banks and waterways. Like many other grasses, tall fescue has the potential to be used as a biofuel. This involves burning the dry matter for the direct production of heat or indirect production of electricity.

One of the most heavily researched and promising uses of tall fescue is as pulp for making fine papers. The qualities that are required of short fibres in printing grade papers are: high number of fibres per unit weight, and stiff, short fibres (low hemicellulose content, and low fibre width to cell wall thickness ratio). Tall fescue has short, narrow fibres and produces pulp with a greater number of fibres than most hardwoods.

Selected references

Feuerstein, U. (1995) *Erzeugung standortgerechter zur Ganzpflanzenverbrennung geeigneter Gräser für die Nutzung als nachwachsende Rohstoffe*. GFP-Project F 46/91 NR-90 NR 026, Deutsche Saatveredelung Lippstadt – Bremen GmbH.

Kaltofen, H. and Schrader, A. (1991) *Gräser*, 3rd edn, Deutscher Landwirtschaftsverlag, Berlin.

Moyer, J.L., Sweeney, D.W. and Lamond, R.E. (1995) Response of tall fescue to fertilizer placement at different levels of phosphorus, potassium, and soil pH. *Journal of Plant Nutrition* 18(4): 729–46.

Muellerwarrant, G.W., Young, W.C. and Mellbye, M.E. (1995) Residue removal method and herbicides for tall fescue seed production. 2: Crop tolerance. *Agronomy Journal* 87(3): 558–62.

Pahkala, K.A., Mela, T.J.N. and Laamanen, L. (1994) Mineral composition and pulping characteristics of several field crops cultivated in Finland. In: *Biomass for Energy, Environment, Agriculture and Industry*, 8th EC Conference, Vienna, Pergamon, pp. 395–400.

Rehm, S. and Espig, G. (1991) *The Cultivated Plants of the Tropics and Subtropics*. Verlag Joseph Margraf, Priese GmbH, Berlin.

Siebert, K. (1975) *Kriterien der Futterpflanzen einschließlich Rasengräser und ihre Bewertung zur Sortenidentifizierung*, Bundesverband Deutscher Pflanzenzüchter e.V., Bonn

TALL GRASSES

The 'tall grasses' topic includes brief descriptions of various plant species that lack adequate information. The tall grasses have high biomass yields because of their linear crop growth rates over long periods of 140–196 days, and sometimes even longer. They can yield oven dry biomass of 20–45Mg/ha in colder subtropical or warmer temperate zones, and over 60Mg/ha annually in Florida. Grasses as bioenergy crops can be used as feedstocks for industrial processes or burned directly to produce energy. The juice of sugar and energy cane can be extracted for ethanol production. Cellulose and hemicellulose can be converted to sugars and then to alcohol, or used to make methane (Prine *et al.*, 1997). Table 7.65 shows the cellulose contents of several tall grasses.

Table 7.65. Cellulose composition of some tall grasses.

Tall grass	Hemicellulose (%)	Cellulose (%)
Elephant grass	27.95	39.4
Erianthus	26.64	40.8
Energy cane	27.78	37.6
Sugar cane	23.31	31.8

Grasses are cosmopolitan plants that live from the equator to the Arctic Circle. Grasses have the power to grow again after they have been eaten or cut down almost to the roots. High yielding grasses have been introduced in many parts of the world. First, the aim was to use them as fodder plants. However, their high biomass yields make them interesting subjects for energy production (Burton, 1993). Desirable characteristics for energy feedstocks include efficient conversion of sunlight, efficient water use, capture of sunlight for as much of the growing season as possible, and low external inputs.

Perennial plants do not require annual establishment costs. In northern climates, the sunlight interception of perennial plants is more efficient, because annual plants spend much of the spring developing a canopy (Samson and Omielan, 1994). The plant should remain erect without lodging, and retain most of the biomass produced during the season in the standing plant (Prine and McConnel, 1996).

Grass tribes may be grouped into panicoid grasses and festacoid grasses. The panicoid group contains most of the African grasses and fixes carbon via the C_4 pathway. Their optimum temperature is 30–40°C and optimum light intensity is 50–60klux. Panicoid grasses can fix 30–50g/m^2/day of dry matter whereas festacoid grasses fix up to 20g/m^2/day of dry matter. However, panicoid grasses do not make use of this advantage if temperatures and light intensities are too low (Burton, 1993).

Napier grass, elephant grass, Uganda grass (*Pennisetum purpureum* Schumach.)

P. purpureum is a robust bunch grass that may reach a height of 6m, growing in dense clumps (Figure 7.15) (Burton, 1993). The leaves are 30–90cm long, and up to 3cm broad (Duke, 1983). It grows on a wide range of well drained soils and is drought tolerant. The plant does not tolerate much frost. The above ground biomass may be

Figure 7.15. Tall grass (elephant grass), USA.

killed by frost, but it will thrive again from the rhizomes if the soil is not frozen (Burton, 1993). The plant requires a rich soil for best growth. The tolerated climate ranges from warm temperate dry to wet, through tropical dry to wet forest life zones. *P. purpureum* is reported to tolerate annual precipitation of 200–4000mm, annual temperatures of 13.6–27.3°C and pH of 4.5–8.2. In the African centre of diversity, elephant grass is reported to tolerate drought, fire, waterlogging, sewage sludge, and monsoons.

The grass is usually propagated vegetatively. A good supply of nitrogen is required for high yields (Duke, 1983). The grass is established from stem cuttings or crown divisions. If cut once a year, *P. purpureum* can produce more dry matter per unit area than any other crop that can be grown in the deep south of the USA (Burton, 1993). It is one of the highest yielding tropical forage grasses. The annual productivity ranges from 2 to 85t/ha. Miyagi (1980) yielded 500t/ha wet matter (70t/ha DM), spacing the plants 50cm × 50cm.

Some reported dry matter yields were: 19t/ha/year in Australia, 66t/ha/year in Brazil, 58t/ha/year in Costa Rica, 85t/ha/year in El Salvador, 48t/ha/year in Kenya, 14t/ha/year in Malawi, 64t/ha/year in Pakistan, 84t/ha/year in Puerto Rico, 76t/ha/year in Thailand and 30t/ha/year in Uganda. The stems of elephant grass provide primarily lignocellulose with virtually no juice sugars. Experimental yields in Queensland, Australia, have attained 70t/ha/year DM, of which 50t were stem. Expected farm yields might be 50–55t/ha/year DM.

Yields of other *Pennisetum* species: *P. americanum* is reported to yield 1–22t/ha/year, *P. clandestinum* 2–25 t/ha/year, *P. pedicellatum* 3–8t/ha/year, *P. polystachyum* 3–10t/ha/year (Duke, 1983).

Weeping lovegrass (*Eragrostis curvula* (Schrad.) Nees)

E. curvula is a bunch grass with long, narrow drooping leaves, reaching a height of 0.5–1.5m. It grows best on sandy soils from southern Texas to northern Oklahoma (Burton, 1993).

Buffelgrass, African foxtail (*Cenchrus ciliaris* L.)

C. ciliaris is a drought tolerant perennial bunch grass that produces forage for the livestock industry in Texas. The plant is 1–1.5m high (Burton, 1993). The leaves are green to bluish-green, 2.8–30cm long and 2.2–8.5mm broad. It thrives from sea level to an altitude of 2000m in dry sandy regions, with rainfall from 250–750mm, but it tolerates much higher rainfall. It grows on shallow soils of marginal fertility. The tolerated climate ranges from warm temperate thorn to moist through tropical desert to moist forest life zones. It is reported to tolerate an annual precipitation of (250) 380–2670mm, an annual temperature of 12.5–27.8°C and pH of 5.5–8.2. The annual productivity ranges from 1 to 26t/ha/year, though some strains yield up to 37t/ha/year (Duke, 1983).

Pangola grass (*Digitaria decumbens* Stent)

D. decumbens is a sterile stoloniferous natural hybrid of unknown parentage. It lacks winter hardiness and is restricted to the southern two-thirds of Florida. It grows to a height of 0.4–0.8m (Burton, 1993).

Kleingrass, coloured Guinea grass (*Panicum coloratum* L.)

P. coloratum is a leafy bunch grass that grows well on sandy clay soils from Texas to Oklahoma. It is drought tolerant and grows earlier in the spring and later in the autumn than many warm season grasses. Seed shattering is a serious problem that limits the plant's use. The plant grows to a height of 0.4–1.4m (Burton, 1993).

Guinea grass (*Panicum maximum* Jacq.)

Guinea grass rows to a height of 0.5–4.5m (Burton, 1993).

Bermuda grass (*Cynodon dactylon* (L.) Pers.)

Bermuda grass originated in Africa, but is now common in many parts of the tropics and temperate zones. It may become a weed. Improved cultivars out-yield the common and are suited to other climatic conditions.

Energy cane (*Saccharum officinarum* L.)

Energy cane and sugar cane are identical. 'Energy cane' means that the total biomass (leaves and stems) are harvested together and used as an energy feedstock. The crop can be harvested for several years. *S. officinarum* originated in the South Pacific islands and New Guinea. Nowadays it is found throughout the tropics and subtropics. Sugarcane is cultivated as far north as 36.7° (Spain) and as far as south as 31° (South

Africa). The tolerated climate ranges from warm temperate dry to moist through tropical very dry to wet forest life zones. *S. officinarum* is reported to tolerate an annual precipitation of 470–4290mm, an annual temperature of 16.0–29.9°C, and pH of 4.3–8.4.

Energy cane is propagated by stem cuttings. It is generally grown for many years without rotation or rest (Duke, 1983). Sugarcane is sensitive to cold, the tops being killed by frost. In the continental USA, where frost may occur, it is planted in late summer or early autumn and is harvested one year later. The growing period is 7–8 months. In the tropics it can be planted at any time of year, because there the plant does not have a rest period and grows continuously. The yields of sugar are higher under tropical conditions. The mature stems are 4–12 feet (1.3–4m) high, and have a diameter of 0.75–2 inch (20–50mm) (Magness *et al.*, 1971).

Energy cane occurs gregariously, growing in sunny areas, on soils unsuitable for trees. It needs aeration at the roots and grows in sand, but not in loam. Lime is considered beneficial for the proper development of the sugar content. For good growth it requires a hot, humid climate, alternating with dry periods. It thrives best at low elevations on flat or slightly sloped land. Occasional flooding is tolerated (Duke, 1983).

At harvesting time the stems are cut and the leaves are removed. The cane should be processed without delay to avoid loss of sugar (Magness *et al.*, 1971). The cane is cut as close to the ground as possible, because the root end is richest in sugar. Harvest takes place after 12–20 months. The canes become tough and are turning pale yellow when they are ready for cutting. The rhizomes will regenerate the crop for at last 3–4 years, and sometimes up to 8 or more years.

Reported dry matter yields are from 16 to 73t/ha/year. Yields achieved at some sites are: Hawaii 67.3t/ha/year and Brazil 54t/ha/year (Duke, 1983). During sugar manufacture the residual bagasse is a by-product. The bagasse can be used for energy production. The advantage of biomass energy derived from residues is that no additional land is demanded. Residues are an already available renewable resource (Swisher and Renner, 1996).

Table 7.66 displays the average yields of sugarcane, energy cane, erianthus and elephant grass cultivars on phosphatic clay soils in Polk County, central Florida. The yields were recorded from 1987 to 1990 and 1992 to 1994.

Elephant grass and energy cane are limited to the lower south of the USA (Prine and McConnel, 1996). If appropriate cultivars are chosen, the productivity of switchgrass is high across much of North America. In studies near the Canadian border, winter hardy upland ecotypes have yielded 9.2t/ha in northern North Dakota and 12.5t/ha in northern New York. Table 7.67 shows possible tall grass prairie species for the USA and Canada.

Table 7.66. Average yields of several cultivars of tall grasses on phosphatic clay soils (Prine et al., *1997).*

Tall grass Cultivar	Annual average biomass (t/ha/year)	
	1987–1990	1992–1994
Sugarcane (US 78-1009)	49.7	32.3
Energy cane (US 59-6)	52.2	36.5
Erianthus (IK-7647)	48.8	17.9
Elephant grass (N-51)	45.2	19.0

Table 7.67. Promising warm season grasses for biomass production in the USA and Canada.

Dry prairie	Dry-mesic prairie	Mesic prairie	Wet-mesic prairie	Wet prairie
Sand bluestem, little bluestem, prairie sandreed (*Calamovilfa longifolia*)	Indian grass (*Sorghastrum nutans*), Little bluestem	Big bluestem (*Andropogon geradii*), Indian grass (*Sorghastrum nutans*)	Big bluestem (*Andropogon geradii*), Indian grass (*Sorghastrum nutans*), switchgrass (*Panicum virgatum*), prairie cordgrass (*Spartina pectinata*)	Prairie cordgrass (*Spartina pectinata*)

Some common tall grass prairie species like big bluestem or Indian grass have high levels of productivity. The seeding of all three of the major tall grass prairie species in mixtures rather than switchgrass monoculture may play an important role in reducing potential disease and insect problems.

Switchgrass is classified as a wet-mesic prairie species. Other species are better adapted than switchgrass to drier or wetter prairie conditions. For example, prairie sandreed has out-yielded switchgrass in the dry regions of the northern US Great Plains when 25–35cm of annual rainfall occurred. In the higher rainfall areas of the prairie region, eastern gamagrass (*Tripsacum dactyloides*) has also proven to be very productive. Prairie cordgrass has a more northern native range than switchgrass; it has a greater chill tolerance, which enables earlier canopy development. Nitrogen fertilization of native prairie cordgrass stands in Nova Scotia have produced yields of 7.7t/ha. Small plot biomass studies in England have yielded 8–23t/ha (Samson and Omielan, 1994).

Selected references

Burton, G.W. (1993) African grasses. In: Janick, J. and Simon, J.E. (eds), *New Crops*, Wiley, New York, pp. 294–8, via Internet.

Duke, J.A. (1983) *Handbook of Energy Crops*, published via Internet. http://www.hort.purdue.edu/newcrop/indices/

Prine, G.M. and McConnel, W.V. (1996) Growing tall-grass energy crops on sewage effluent spray field at Tallahassee, FL. In: *Bioenergy '96, Proceedings of the Seventh National Bioenergy Conference*, 15–20 September, Nashville, Tennessee.

Prine, G.M, Stricker, J.A. and McConnel, W.V. (1997) Opportunities for bioenergy development in the lower south USA. In: *Making Business from Biomass, Proceedings of the Third Biomass Conference of the Americas*, Montreal, Canada, 24–29 August.

Magness, J.R., Markle, G.M. and Compton, C.C. (1971) *Food and Feed Crops of the United States*, International Research Project IR-4, IR Bul. 1 (Bul. 828 *New Jersey Agr. Expt. Sta.*), via Internet.

Miyagi, E. (1980) The effect of planting density on yield of napier grass (*Pennisetum purpureum* Schumach). *Sci. Bul. Coll. Agr., U. Ryukus* No. 27: 293–301.

Rahmani, M, Hodges, A.W., Stricker, J.A. and Kiker, C.F. (1997) Economic analysis of biomass crop production in Florida. In: In: *Making Business from Biomass, Proceedings of the Third Biomass Conference of the Americas*, Montreal, Canada, 24–29 August.

Samson, R.A. and Omielan J.A. (1994) Switchgrass: *A Potential Biomass Energy Crop for Ethanol Production. Proceedings of the Thirteenth North American Prairie Conference*, Canada. pp. 253–8.

Swisher, J.N. and Renner, F.P. (1996) Carbon offsets from biomass energy projects. In: *Bioenergy '96, Proceedings of the Seventh National Bioenergy Conference*, 15–20 September, Nashville, Tennessee.

TIMOTHY (*Phleum pratense* L.)

Description

Timothy is a short lived perennial plant most commonly found in meadows, pastures, ditches and along trails. It has moderately leafy stems that reach a height of about 1m. The leaves are ungrooved with a matte underside, and are light green to mid-green. The plant's leaves grow to a length of approximately 18–30cm and a width of about 3–9mm. The inflorescence is a dense cylindrical false ear which is usually about 6–7mm in diameter and 10cm in length. Flowering usually starts at the beginning of June (Kaltofen and Schrader, 1991; Siebert, 1975).

Plant material from timothy crops has mineral concentrations of approximately 5% ash, 1% silicon dioxide (SiO_2), 1.0–1.5% nitrogen, 50mg/kg iron, 30–40mg/kg manganese and 4–5mg/kg copper (Pahkala *et al.*, 1995).

Ecological requirements

Timothy readily adapts to most types of soil, especially to fertile soils where it responds by producing a high yield. Nutrient rich soils in locations ranging from fresh to damp are most appropriate for timothy crop cultivation. The plant is unable to tolerate long periods of dry weather, though it can tolerate winters and high levels of precipitation (Kaltofen and Schrader, 1991).

Propagation

Timothy is propagated by seed.

Crop management

Timothy is usually sown in the beginning to middle of summer (June to August). The crop has a high demand for fertilizers, especially nitrogen. Table 7.68 summarizes the fertilizer rates obtained from field tests at sites in Germany. A large percentage of the fertilizer was supplied in the form of liquid manure – in some cases as much as 100%. The fertilizer rates vary with soil composition.

Because timothy is a perennial plant that has most commonly been used for fodder purposes, timothy–legume mixtures in hay and pasture seedings have been used in

Table 7.68. Fertilizer rates used for timothy at several locations in Germany (Kusterer and Wurth, 1992, 1995; Wurth, 1994).

Fertilizer	Amount (kg/ha)
Nitrogen (N)	214–411
Phosphorus (P_2O_5)	42–130
Potassium (K_2O)	166–544
Magnesium (MgO)	35–55

crop rotations in the USA. Crop rotation with respect to timothy as a source of biomass for biofuels has not received as much attention.

Production

In Germany, field tests using timothy, harvested five times a year, produced in 1994 (3 years after sowing) fresh matter yields averaging 66.32t/ha and dry matter yields averaging 12.28t/ha. The variety 'Tiller' produced the highest fresh matter yield, and the variety 'Barnee' produced the highest dry matter yield (Kusterer and Wurth, 1995).

Mediavilla *et al.* (1993, 1994, 1995) provide the dry matter yields (t/ha) from timothy field tests that were conducted using liquid manure fertilizer at two locations in Switzerland, Reckenholz and Anwil. The yields in 1993, 1994 and 1995 were, respectively, 15.7, 12.5 and 14.5t/ha at Reckenholz, and 18.5, 17.9 and 11.2t/ha at Anwil. The crop was harvested three times in 1993 and twice times in each of 1994 and 1995. In Reckenholz between 70 and 90kg/ha/year NH_4-N liquid manure was used, while at Anwil the rate was up to 180kg/ha/year NH_4-N.

Processing and utilization

Timothy is harvested by mowing, with the number of harvests per year varying from one to five for mature crops. Harvests of five times per year involve harvesting approximately every 3–4 weeks beginning in the middle of May through until the middle of October. The harvested material then needs to be dried relatively quickly if it is to be used for fodder or stored. If it is to be used as biomass for conversion into a biofuel the quality of the harvested material is not important, but drying is necessary for its economic utilization.

Traditionally, timothy has been one of the most important hay grasses for cooler temperate humid regions. Through the application of present and developing technologies timothy will be used as biomass for conversion into a biofuel. This most commonly involves burning the crop matter for the direct production of heat and indirect production of electricity.

Selected references

Kaltofen, H. and Schrader, A. (1991) *Gräser*, 3rd edn, Deutscher Landwirtschaftsverlag, Berlin.

Kusterer, G. and Wurth, W. (1992) *Ergebnisse der Landessortenversuche mit ausdauernden Gräsern 1985–1991. Information für die Pflanzenproduktion*, 1, Landesanstalt für Pflanzenbau Forchheim.

Kusterer, B. and Wurth, W. (1995) *Ergebnisse der Landessortenversuche mit ausdauernden Gräsern 1994. Information für die Pflanzenproduktion*, 3, Landesanstalt für Pflanzenbau Forchheim.

Mediavilla, V., Lehmann, J. and Meister, E. (1993) *Energiegras/Feldholz – Teilprojekt A: Energiegras*, Jahresbericht 1993, Bundesamt für Energiewirtschaft, Bern.

Mediavilla, V., Lehmann, J., Meister, E., Stünzi, H. and Serafin, F. (1994) *Energiegras/ Feldholz – Energiegras*, Jahresbericht 1994, Bundesamt für Energiewirtschaft, Bern.

Mediavilla, V., Lehmann, J., Meister, E., Stünzi, H. and Serafin, F. (1995) *Energiegras/ Feldholz – Energiegras*, Jahresbericht 1995, Bundesamt für Energiewirtschaft, Bern.

Pahkala, K.A., Mela, T.J.N. and Laamanen, L. (1995) Mineral composition and pulping characteristics of several field crops cultivated in Finland. In: *Biomass for Energy, Environment, Agriculture and Industry, 8th EC Conference*, Vienna, Pergamon, pp. 395–400.
Siebert, K. (1975) *Kriterien der Futterpflanzen einschließlich Rasengräser und ihre Bewertung zur Sortenidentifizierung*, Bundesverband Deutscher Pflanzenzüchter e.V. Bonn.
Wurth, W. (1994) *Ergebnisse der Landessortenversuche mit ausdauernden Gräsern 1993. Information für die Pflanzenproduktion*, 2, Landesanstalt für Pflanzenbau Forchheim.

TOPINAMBUR (JERUSALEM ARTICHOKE) (*Helianthus tuberosus* L.)

Description

Topinambur is a short day perennial C_3 plant that originated in North America. A relative of the sunflower, topinambur grows to a height varying from 1 to 4m. Opposing one another on the stalk, grow rough, heart shaped to lancet shaped leaves. Flowering begins in late summer with yellow flowers of 4–8cm diameter (Franke, 1985). In central Europe, the plant's flowers scarcely produce seeds (Hoffmann, 1993).

Underground stolons form irregularly shaped tubers that are similar to potatoes except that topinambur tubers have a higher water content and produce adventitious roots. Depending on the variety, the tuber has a yellow, brown or red peel with white flesh. The tubers consist of 75–79% water, 2–3% protein and 15–16% carbohydrates, of which the fructose polymer inulin can constitute up to 7–8% or more (Franke, 1985). By the end of the growing cycle the carbohydrates, which are initially in high levels in the stems, have been transferred to the tubers. Topinambur can be placed in the group of plants producing the highest total dry matter yield per hectare (Dambroth, 1984).

Ecological requirements

Topinambur does not make special demands on the environment. Most soils have been shown to be suitable for topinambur cultivation, though the soil should be siftable so excessively clayey soils should be avoided. In addition, the soil should have a low number of stones to facilitate tuber harvesting (Hoffmann, 1993). The pH of the soil should be between 5.5 and 7.0 (Bramm and Bätz, 1988). The plant is very sensitive to frost, but the tubers can tolerate a ground temperature as low as –30°C.

Because of topinambur's efficient ability to hold water and nutrients, areas with low precipitation or periods of low precipitation can be endured by the crop and may even have little effect on the yield, though drought results in a halt in growth (Hoffmann, 1993).

Propagation

According to Hoffmann (1993), topinambur can be propagated using seed, tuber sprouts or rhizome pieces.

Crop management

Topinambur planting begins in the middle of April using a potato planter to minimize field preparation. The rows should be 75cm apart and the plants should be 33cm apart in their row. The desired crop density is 38,000 to 40,000 plants to the hectare (Bramm and Bätz, 1988).

Correct fertilizer amounts are important for the success of the crop, but organic fertilizers should not be used during tuber formation. Table 7.69 shows suggested fertilizer levels, as given by Bramm and Bätz (1988).

Table 7.69. Recommended fertilizer levels for topinambur.

Fertilizer	Amount (kg/ha)
Nitrogen (N)	40–80
Potassium (K_2O)	240–300
Phosphorus (P_2O_5)	90–140
Magnesium (MgO)	60–90

The crop can be mechanically weeded, but usually topinambur's rapid growth and leaf formation quickly crowd out most weeds (Dambroth, 1984). Diseases that are most likely to attack topinambur crops include the stem rot *Sclerotinia sclerotiarum*, the sunflower rust *Puccinia helianthi* and the mildew *Erysiphe cichoracearum*. The crop is also susceptible to damage by rabbits, deer, wild boar and mice.

The crop planted before topinambur is not of importance in establishing a crop rotation, but appropriate crops to follow it are maize and summer wheat. Because topinambur is susceptible to *Sclerotinia*, a break in cultivation of 4 to 5 years is recommended (Bramm and Bätz, 1988). In addition, topinambur crops have a higher yield when planted as an annual crop than when planted as a perennial crop (Dambroth, 1984). It should be noted that because topinambur is a perennial plant, it may also reappear in subsequent crops.

Production

The lack of a market for energy crops means that it is difficult to determine their value, but there are estimates for topinambur, based on the selling price for the crop (in Italy) that produces a gross margin equal to that from a traditional crop. Taking into account labour costs, topinambur's variable costs, production and yield, marketable co-products and traditional crops' gross margin, the farmer would need to receive from 18 to 47Ecu per tonne of tubers from a topinambur biomass crop to gain an income comparable to that from a traditional crop (Bartolelli *et al.*, 1991).

The topinambur clone 'Violet de Rennes' has been shown to be very productive in Spain. Dry matter yields for tubers have been approximately 16t/ha/year (Fernandez *et al.*, 1991). Portugal has shown tuber yields of up to 30-43t/ha for some clones tested on both sandy and clay soils (Rosa *et al.*, 1991).

Breeding research in the Netherlands using selected clones of 'Bianka', 'Yellow Perfect', 'Columbia' and 'Précoce' on light sandy soil produced in 1989 fresh matter tuber yields of 92–105t/ha, inulin percentages of 15–18%, and inulin yields of approximately 16t/ha. The following year produced tuber yields of 50–57t/ha, inulin percentages of 13–15%, and inulin yields of approximately 7t/ha. The reductions in 1990 are the result of an extensive period of low precipitation during the tuber filling stage (van Soest *et al.*, 1993). Table 7.70 shows research results from Germany (Schittenhelm, 1987).

Table 7.70. Characteristics of three topinambur varieties.

Characteristic	'Bianka'	'Waldspindel'	'Medius'
Growth height	low	very high	very high
Ripeness	very early	middle	late
Tuber size	very large	large	middle
Dry matter (%)	19.27	26.59	24.33
Fructose (%)	10.67	17.90	16.24
Total sugar (%)	13.71	20.80	19.50
Tuber yield (t/ha)	49.667	61.417	62.333
Dry matter (t/ha)	9.571	16.331	15.166
Fructose (t/ha)	5.299	10.994	10.123
Ethanol (l/ha)	3969.85	7447.67	7086.33

Processing and utilization

Tuber harvesting can begin in November. The tubers' winter hardiness means that the harvest can be extended into the early part of the following year, which has the advantage of permitting a more continuous harvest based on demand. Topinambur harvesting techniques are quite similar to those of potato harvesting, so the tubers are most easily harvested using a potato harvester. Because of their poor storage ability – worse than that of potato – topinambur tubers should be processed within 14 days of being harvested.

Inulin is the product of greatest value obtained from the tubers. This D-fructose polymer can be used to produce ethanol or, through hydrolysis separation and isomerization, fructose and glucose (Delmas and Gaset, 1991). The green mass of the plant, which has a heating value of 17.6MJ/kg dry matter, can also be processed as paper pulp or used as fodder.

The tubers are crushed and the pulp pressed to obtain an extract. The inulin can then be obtained from the extract through physical, chemical or microbiological processes. Physically, tangential ultrafiltration can be used to recover a high sugar yield. Chemically, ethanol precipitation can be used to recover the sugars, but this produces a significantly lower sugar yield. Biologically, yeast fermentation can be used to obtain both sugar and ethanol (Fontana *et al.*, 1992). Approximately 8–10l of ethanol can be produced from 100kg of tubers (Franke, 1985). Topinambur stalks also show potential for industrial uses after processing. Using catalytic dehydration, 5-hydroxymethyl-2-furfural (HMF) can be obtained from fructans. HMF can be used to produce agrochemicals, detergents, pharmaceuticals, solvents, etc.

Selected references

Bartolelli, V., Mutinati, G. and Pisani, F. (1991) Microeconomic aspects of energy crops cultivation. In: *Biomass for Energy, Industry and Environment, 6th EC Conference*, Athens, Elsevier Applied Science. pp. 233–7.

Bramm, A. and Bätz, W. (eds) (1988) Topinambur. In: *Industriepflanzenbau: Produktions- und Verwendungsalternativen*, Bundesminister für Ernährung, Landwirtschaft und Forsten, Bonn, pp. 28–9.
Dambroth, M., 1984. Topinambur-eine Konkurrenz für den Industriekartoffelanbau? *Der Kartoffelbau* 35(11): 450–3.
Delmas, M. and Gaset, A. (1991) A new industrial crop network for the production of animal food, sugars and derivatives, and paper pulps. In: *Biomass for Energy, Industry and Environment, 6th EC Conference*, Athens, Elsevier Applied Science. pp. 1262–68.
Fernandez, J., Curt, M.D. and Martinez, M. (1991) Water use efficiency of *Helianthus tuberosus* L. 'Violet de Rennes', grown in drainage lysimeter. In: *Biomass for Energy, Industry and Environment, 6th EC Conference*, Athens, Elsevier Applied Science. pp. 297–301.
Fontana, A., Schorr-Galindo, S. and Guiraud, J.P. (1992) Inulin-containing crops: improvement of fermentation economics through co-products valorization. In: *Biomass for Energy and Industry, 7th EC Conference*, Florence, Ponte Press, pp. 1060–65.
Franke, W. (1985) Inulin liefernde Pflanzen. In: *Nutzpflanzenkunde*, 3rd edn, George Thieme Verlag, Stuttgart, pp. 109–11.
Hoffmann, G. (1993) Der Anbau von Topinambur: Alternative oder botanische 'Kuriosität'? *Neue Landwirtschaft* 12/93.
Rosa, M.F., Bartolomeu, M.L., Novias, J.M., Sá-Correia, I., Barradas, M.C., Romano, M.C., Antunes, M.P. and Sampaio, T.M. (1991) The Portuguese experience on the direct ethanolic fermentation of Jerusalem artichoke tubers. In: *Biomass for Energy, Industry and Environment, 6th EC Conference*, Athens, Elsevier Applied Science. pp. 546–55.
Schittenhelm, S. (1987) Topinambur – eine Pflanze mit Zukunft. *Lohnunternehmer Jahrbuch*, pp. 169–74.
van Soest, L.J.M., Mastebroek, H.D. and de Meijer, E.P.M. (1993) Genetic resources and breeding: a necessity for the success of industrial crops. *Industrial Crops and Products* 1: 283–8.

WATER HYACINTH (*Eichhornia crassipes* (Mart.) Solms)

Description

The water hyacinth is a perennial aquatic herb. The rhizomes and stems are normally floating. The plant is native to Brazil, but now it is growing in most tropical and subtropical countries. Its habitat is estimated to range from tropical desert to rain forest, through subtropical or warm temperate desert to rain forest life zones. The leaves are killed by frost, and plants cannot tolerate water temperatures greater than 34°C. Often it grows as a weed. Rafts of water hyacinth can block channels and rivers.

Production

Harvesting appears to be more critical than cultivation. Rafts of floating plants may be harvested with specially equipped dredgers and rakes. A floating mat of medium sized plants may contain 2 million plants per hectare and weigh 270–400t (wet). The annual productivity is estimated from 15–30t/ha to 88t/ha. In Florida, compared with other plant species representing many life forms, the water hyacinth is very productive. Once harvested and dried, the dry matter of the water hyacinth is roughly equivalent to those of other species in term of energy (Table 7.71).

Table 7.71. Biomass production of various plant species in Florida.

Plant	Dry matter (t/ha/year)	Plant	Dry matter (t/ha/year)
Azolla	10	*Hydrocotyle umbellata*	20–58
Beta	4.4–11.7	*Ipomoea batatus*	7–23
Brassica	3–10	*Lemna*	12
Casuarina equisetifolia	8.3	*Melaleuca quinquenervia*	28.5
Cichorium intybus	5.5–7.9	*Paspalum notatum*	22.4
Colocasi esculenta	9–19	*Pennisetum* spp.	57.3
Cynodon dactylon	23.5–24.6	*Pinus clausa*	9.0
Daucus carota	2.5–5.5	*Pinus elliottii*	9.4
Eichhornia crassipes	30–88	*Saccharum*	32–54
Elodea	3	*Sorghum*	16–37
Eucalyptus	5.6–20	*Sorghum* 'Sordan'	22.4
Helianthus tuberosus	2.2–9.5	*Typha* spp.	20–40
Hydrilia	15		

Processing and utilization

There are several ways to use the weed for energy production. Bengali farmers dry water hyacinths for fuel. Estimations of biogas production vary: 1kg of dry matter yields 0.2–0.37m^3 biogas. In India 1t of dried water hyacinth yields about 50l of ethanol and 200kg of residual fibre. Bacterial fermentation of 1t yields about 740m^3 of gas with 51.6% methane, 25.4% hydrogen, 22.1% CO_2, and 1.2% oxygen. Gasification of 1t by air and steam at high temperatures (800°C) gives about 1100m^3 gas containing 16.6% hydrogen, 4.8% methane, 21.7% CO, 4.1% CO_2 and 52.8% N_2.

Selected reference

Duke, J.A. (1984) *Handbook of Energy Crops*, published via Internet. http://www.hort.purdue.edu/newcrop/duke_energy/Eichornia_crassipes

WHITE FOAM (*Limnanthes alba* Hartw.)

Description

Limnanthes alba is an annual herb. The stems are erect or ascending and 10–30cm tall. The leaves are up to 10cm long. The plant is native to the Sierra Nevada foothills and the adjacent rolling plains from Sacramento to Chico, California, USA. The seeds contain 25–30% oil with 1.56% volatile isocyanates, and 20% protein. The high concentration of C-20 fatty acids is unique. More than 90% of the fatty acids have a chain length greater than C-18. Table 7.72 shows various characteristics of white foam seeds.

Ecological requirements

The range extends across warm temperate moist through subtropical dry to moist forest life zones. White foam is reported to tolerate slopes and waterlogging. The plant is also reported to tolerated an annual precipitation of 700–1100mm and an

Table 7.72. Characteristics of white foam seeds (Kleiman, 1990).

Seed characteristic	Content
Oil (%)	17–29
$20{:}1^{5}$ (%)	50–65
$22{:}1^{5}$ and $22{:}1^{13}$ (%)	10–29
$22{:}2^{5,13}$ (%)	15–30
Protein (%)	11–28
Glucosinolates (%)	3–10
Seed weight (g/1000)	4.2–9.8

annual temperature of 12–19°C (Duke, 1983). The killing temperature is 0°C. The pH value should be 6–6.5 (FAO, 1996).

White foam is commonly found on banks and the gravelly bars of small intermittent streams in the Sierra Nevada foothills. It grows on porous, quick drying soils. Essentially, it is a xerophyte, flowering and setting seed on the last seasonal soil and stem moisture. It has about the same water requirement as dry farmed winter grains, and seems to require less moisture than other species of this genus.

No pests or diseases have been reported for this species.

Propagation

Limnanthes alba is propagated by seed. Normally, it is sown in the autumn and harvested in early summer (FAO, 1996). In experiments, seeds germinated at 4.5°C, gave poor germination at 21°C, and went dormant at 26.5°C. Growth and seedling periods are closely match those of oats and barley. Good weed control is essential for good yields (achieved using propachlor and diclofog). As small streams dry in late spring, growth is terminated and plants rapidly mature their seeds. Seed dates range from 5–30 May. Seed shattering may cause 16–54% of seed loss in *Limnanthes alba*, and 71–93% in *Limnanthes douglasii*. With improved cultivars and earlier harvests a seed recovery greater than 95% has been achieved with direct combine harvesting or windrowing and then combining.

Production

Experimental plantings with a seed rate of 2.9kg/ha yielded 1650kg/ha; 24.7kg/ha yielded about 2000kg/ha. With respect to the two seed rates, applying 48–50kg/ha N following a non-legume crop in early March gave seed yields of 790kg/ha and 700kg/ha respectively; without any fertilizer, yields of 530kg/ha were obtained. Yields of 400kg/ha are reported for *Limnanthes bakeri* and 900kg/ha for *Limnanthes alba*, with 1900kg/ha for *Limnanthes douglasii*, the most promising species of this genus.

Researchers hope to increase the yields, but for now, oil yields of 500kg/ha are difficult to obtain. Oregon could already produce 1100kg/ha seed, yielding 275kg oil. At Cornwallis, seed yields have ranged from 900–1800kg/ha. White foam is grown as a winter annual, principally in the Willamette valley, Oregon, where yields over 1.1t/ha are routine.

Processing and utilization

The oil composition of white foam is unique in several ways. Over 95% of the fatty acids are longer than C-18, and about 90% of these fatty acids have double bonds in delta-5 position. The $22{:}2\Delta^{5,13}$ fatty acids react essentially like a monoenoic fatty acid in terms of oxidative stability. The oil should be oxidatively stable because of the absence of polyunsaturated fatty acids and the long chain nature, and because the delta-5 double bond is more stable than olefins with the double bond in the centre of the molecule.

The oil can be used as a fuel source and the stems offer various possibilities as solid biofuels for combustion, gasification or pyrolysis.

Selected references

FAO (1996) Ecocrop 1 Database, Rome.

Duke, J.A.(1983) *Handbook of Energy Crops*, published via Internet. http://www.hort.purdue.edu/newcrop/duke_energy/limnanthes_alba

Kleiman, R. (1990) Chemistry of new industrial oilseed crops. In: Janick, J. and Simon, J.E. (eds), *Advances in New Crops*, Timber Press, Portland, OR, pp. 196–203; via Internet.

WILLOW *(Salix spp.)*

Contributed by: W. Bacher

Description

Willow is grown mainly in continental climate zones including Europe, Western Siberia and Central Asia, as well as the Caucasus, Northern Persia and Asia Minor (Schütt *et al.*, 1994), but it is most successful in northern Europe. Some deposits of willow plantations can be found at altitudes as high as 2000m in mountainous regions, but willow is a typical plant of plains and hilly areas. Most *Salix* spp. are compact shrubs with numerous thick branches and grow to a height between 0.3 and 4m. Others are true trees. *S. fragilis* is able to reach a height of 15m and *S. alba* a hieght of 20m (Schmeil and Fitschen, 1982). Some species of *Salix* build leaves before blossoms in spring (Schütt *et al.*, 1994). Normally, the shrubs build an extensive root system. In Sweden, willow is specially used as a short rotation coppice crop (SRC). Willow species such as *Salix viminalis* (L.) and *Salix dasyclados* (Wimm.) are cultivated on biomass plantations for use as energy sources.

Ecological requirements

Willow can be grown on a wide range of soil types, on light soils as well as on loamy soils, with a pH range from 6.0 to 7.5 and an optimum pH of 6.5. *Salix* can tolerate salinity up to 4dS/m (FAO, 1996). *Salix* species are more dependent on water than other agricultural crops so dry locations should be avoided (Johansson *et al.*, 1992). Willow's water consumption can reach 4.8mm/m^2 per day in June and July. More than 500l water is needed to produce 1kg of dry matter (Pohjonen, 1980). Irrigation is necessary if the yearly rainfall is less than 600l/m^2 (Parfitt and Royle, 1996). Most of the types selected for SRC could not produce economically viable yields on very dry or alkaline soils. To a certain degree they are tolerant of waterlogging. The optimal growing temperature of *Salix* spp. ranges between 15°C and 26°C, with a minimum growing temperature for the different *Salix* species of between 5°C and 10°C and a maximum growing temperature between 30°C and 40°C. Some species are frost tolerant, but not below –30°C.

Propagation

Willow cuttings from one year old shoots can be planted in quite coarsely prepared soil. For ease of planting, successful rooting and subsequent management, the site should be subsoiled if necessary, deep ploughed (25–30cm) in autumn and power harrowed to produce a level and uncompacted tilth (Parfitt and Royle, 1996). Planting is usually delayed until the spring, when about 18,000 cuttings 20cm long and more than 0.8 cm thick are planted per hectare, after storage at –4°C during the winter and subsequent watering for a couple of days (Johansson *et al.*, 1992).

The *Salix* cuttings are placed in double rows with a row spacing of 75cm, and 125cm between the double rows; the plant spacing in the row is 55cm (Johansson *et al.*, 1992). In the first year only a few shoots on each cutting will grow, which will be cut back to the ground level in the first winter and can be used for making cuttings to establish new plantations. The regrowth of cut back SRC-*Salix* is vigorous and dense, reaching up to 4m high in the next vegetation period.

Several laboratories have developed micropropagation techniques, and this is now a satisfactory propagation method.

Crop management

During the establishment phase, weeds are the worst enemy of willow. Weed control is necessary during the first year of establishment and the fields must be kept free from perennial weed rhizomes before planting. After the first year of establishing, willows are able to suppress weeds.

An average application of 60–80kg N, 10kg P and 35kg K is suitable from the second year onwards (Johansson *et al.*, 1992), because in the planting year (first year) the application of fertilizers could enhance weed growth. The fertilizer regime should be adapted to the soil type and natural mineralization, and to the nitrogen entry and deposition via rainwater. In relation to its mass production, willow can be considered a low input plant (El Bassam, 1996).

Willow may be attacked by a wide range of leaf consuming, stem sucking and wood boring insects (Hunter *et al.*, 1988). Nonc of these insects has so far been more than an occasional problem. A *Salix* plantation provides a good habitat for many different insects, birds and animals. Diseases are more serious, and numerous pathogens are described, especially different species of rust. Different clones have varying susceptibilities to fungi and insects and can be selected for better resistance. Generally, there is little reason to introduce inputs for insects and disease control. Considerable damage may be caused by mice and deer during the establishment phase of a *Salix* plantation.

Figure 7.16. Traditional willow coppice (DTI).

Production

Willow can produce its first yield three years after planting. Current expectations are that there are eight or more cycles of harvests during the *Salix* plantation life period of three years, each as SRC (Figure 7.16). Estimated annual yields per hectare, in oven dried tonnes (ODT), are 5 (Ireland), 8–10 (Sweden), 8–20 (UK) and 15–20 (Italy). Increased yields are expected if the full benefits from improved breeding materials have been obtained and when agricultural practices have been optimized.

Processing and utilization

Harvesting of stems take places in winter when the leaves have fallen off. There are two different approaches to harvesting: direct chipping and whole shoot harvesting. Direct chipping is practicable if chips can be combusted at the harvest moisture content (50%) in large heating plants. Chips with a high moisture content will quickly deteriorate through microbial activity, so must be burned immediately or ventilated for storage. Drier chips are needed in smaller plants and stationary ovens on farms. In this case shoots are harvested whole and dried in piles during summer. This material reaches a dry matter content of about 70%.

Dry wood chips for energy production need no further processing and can be burned or gasified (Parfitt and Royle, 1996). Willow chips can be utilized in two ways. Direct combustion in special automated boilers will give low grade space heating of 50 to 400kWh. High pressure steam techniques can be used to drive steam turbines of up to 10MW installed capacity (Graef, 1997). Another possibility is the gasification (pyrolysis) of wood chips to drive an engine to power a generator and also produce heat, a system known as combined heat and power (CHP). A growth rate of 15ODT/ha/year is equivalent to 7000l of oil.

Selected references

DTI (Department of Trade and Industry) (1994) *Short Rotation Coppice Production and the Environment*. Technology Status Report 014.
El Bassam, N. (1997) Unpublished data.
FAO (1996) Ecocrop 1, Crop Environmental Requirements Database, Software Library, September.
Graef, M. (1997) Personal communication.
Hunter, T., Royle, D.J. and Stott, K.G. (1988). Diseases of Salix biomass plantations: results of an international survey in 1987. *Proceedings of the International Energy Agency/Bioenergy Agreement Task II*. Workshop Biotechnology Development, Uppsala, Sweden, pp. 37–48.
Johansson, H., Ledin, S. and Forsse, L.S. (1992) *Practical Energy Forestry in Sweden: A Commercial Alternative for Farmers*. In: Hall, D. O. *et al.* (eds), Biomass for Energy and Industry, Ponte Press Bochum, pp. 117–26.
Parfitt, R.J. and Royle, D.J. (1996) Willow, Poplar. In: El Bassam, N. (ed.), *Renewable Energy – Potential energy crops for Europe and the Mediterranean region*, REU Technical Series, 46, FAO.
Pohjonen, V. (1980) Energiaviljely sitoo auringon energgiaa. *Tyotehoseuran Metsatiedotus* 3.
Schmeil, O. and Fitschen, J. (1982) *Flora von Deutschland und seinen angrenzenden Gebieten*. 87. Auflage, Quelle & Mayer, Heidelberg.
Schütt, P., Schuck, H.J., Aas, G. and Lang, U.M., (eds) (1994) *Enzyklopädie der Holzgewächse. Handbuch und Atlas der Dendrologie*, Ecomed Verlagsges Ischaft Landsberg am Lech, III-3.

CEREALS

Barley (Hordeum vulgare *L.*), ***Maize*** (Zea mays *L. ssp.* mays), ***Oats*** (Avena sativa *L.*), ***Rye*** (Secale cereale *L.*), ***Triticale*** (x Triticosecale), ***Wheat*** (Triticum aestivum *L.*)

Grains from cereals should continue to serve as a major source for food and feed. Cereal straw combined with rice straw and maize husks constitutes one of the principal biomass sources arising from present agricultural activities in Europe. This represents a large potential resource for utilization as energy feedstocks. Although there are some plans to use cereals as an energy source, the main justifications are overproduction of cereals and advanced technology in seed production, tillage, sowing, harvesting, baling and storage. Cereal crops that are whole crop harvested may be used under certain circumstances as energy crops.

In recent decades, plant breeding activities have made good progress in increasing grain yields and improving the genetic make up of crops to increase nutrient utilization efficiency and to improve their adaptability for wider environmental conditions. This has been achieved by improving the harvest index (the ration of grain to straw) to almost 0.5 or even higher. The total biomass has been influenced very little by breeding activities. Cereals as energy crops on set aside areas might be grown according to the following guidelines:

- Suitable genotypes will have a low harvest index – that is, they should produce considerably more straw than grain. Land races and wild types are most suitable.
- The nitrogen fertilization rate could be 50% of the normal rate for grain cereals because most of the nitrogen needed by cereals is for grain production and to

improve the protein content of the grain. This is not needed in whole crop harvesting for energy feedstocks.

- The whole crop harvesting procedure will involve cutting and collecting the unfractionated crop from the field.
- Harvesting should take place earlier than for grain cereals.

This system will prevent conflicts with the production of food cereal grains. Breeders should try to produce cultivars with higher straw proportion as energy cereal crops. This would minimize the inputs and decrease environmental damage. There are already attempts to breed cereals, especially barley, with higher enzyme contents that enable efficient fermentation and ethanol production from the grain.

Worldwide average cereal production rates (t/ha of grain) in 1994 were: barley 2.18, maize 4.33, oats 1.70, rye 2.05, sorghum 1.39, wheat 2.22 (FAO, 1995). The average total biomass to be expected by following the above (environmentally consistent) guidelines might be calculated as follows: wheat 10t/ha, barley 6t/ha, rye 6t/ha, oats 4t/ha, triticale 12t/ha, sorghum 12t/ha and maize 12t/ha.

Significantly less total biomass should be expected from non-European Mediterranean countries. From the ecological point of view, rye and triticale possess some advantages because these two types can be considered as relatively low input genotypes. No attempt has been made here to describe any of these cereal crops – they are not new crops, and breeding and cultivation procedures are already well established.

Selected reference

FAO (1995) *Production Year Book, 1994*, FAO, Rome.

PSEUDOCEREALS

Amaranthus (Amaranthus *spp.*), ***Buckwheat*** (Fagopyrum esculentum *Moench*), ***Quinoa*** (Chenopodium quinoa *Willd.*)

'Pseudocereals' is an expression used to refer to plant species cultivated for their starch containing seeds. They have a very low harvest index and low cereal yields as a result of extensive breeding activities, though they have been cultivated for a very long time. Some of them have high biomass potential: amaranthus (*Amaranthus* spp.), quinoa (*Chenopodium quinoa* Willd.) and buckwheat (*Fagopyrum esculentum* Moench). These plant species can be grown in cool regions, and are the focus of research programmes in different parts of Europe.

Amaranthus is a herbaceous annual plant species with simple leaves. Only a few of the many amaranthus species are cultivated. Most cultivated species are hybrids. Current investigations show that the total biomass achieved from amaranthus field trials was nearly 11t/ha dry matter, with a seed production of 2.4t/ha. Some genotypes have delivered dry matter total biomass yields of up to 20t/ha.

Quinoa possesses high genetic variability and ecological adaptability characteristics, and can be grown in different environments. There has been some breeding activity in Austria and Denmark and new varieties could be released in the near future.

Quinoa has been grown in Northern Europe as an alternative crop for 10 years. Its total biomass yield lies between 4 and 10t/ha dry matter with a seed yield of between 1 and 2t/ha. Others have achieved higher figures – 10–13t/ha dry matter and 2–3t/ha grain (Aufhammer *et al.*, 1995).

Buckwheat is an annual plant species with branched stems growing from about 50 to 150cm high. The total biomass yield could reach 8.5t/ha dry matter and 3–4t/ha. But under common cultivation practices the grain yield lies between 1 and 2t/ha and the dry matter total biomass is approximately 5.5t/ha.

Selected reference

Aufhammer, W., Lee, J.H., Kübler, E., Kuhn, M. and Wagner, S. (1995) Anbau und Nutzung der Pseudocerealien Buchweizen (*Fagopyrum esculentum* Moench), Reismelde (*Chenopodium quinoa* Willd.) und Amarant (*Amaranthus* ssp. L.) als Körnerfruchtarten, pp. 125–139. *Mitteilungen der Gesellschaft für Pflanzenbauwissenschaften*, Band 8, Wissenschaftlicher Verlag Giessen.

MICROALGAE (*Oleaginous* spp.)

Contributed by: M. Satin

Description

Microalgae comprise a vast group of photosynthetic, heterotrophic organisms which have an extraordinary potential for cultivation as energy crops. They can be cultivated under difficult agroclimatic conditions and are able to produce a wide range of commercially interesting by-products such as fats, oils, sugars and functional bioactive compounds. As a group, they are of particular interest in the development of future renewable energy scenarios. Certain microalgae are effective in the production of hydrogen and oxygen through the process of biophotolysis, while others naturally manufacture hydrocarbons suitable for direct use as high-energy liquid fuels. It is this last class that forms the subject here.

Once algae species have been grown, their harvesting and transportation costs are lower than with conventional crops, and their small size allows for a range of cost-effective processing options. They are easily studied under laboratory conditions and can effectively incorporate stable isotopes into their biomass, thus allowing effective genetic and metabolic research to be carried out over a much shorter period than with conventional plants.

Ecological requirements

Microalgae represent an immense range of genetic diversity and can exist as unicells, colonies and extended filaments. They are ubiquitously distributed throughout the biosphere and grow under the widest possible variety of conditions. Microalgae can be cultivated under aqueous conditions ranging from freshwater to situations of extreme salinity. They live in moist, black earth, in the desert sands and in all the conditions in between. Microalgae have been found living in clouds and have long been known to be essential components of coral reefs. This wide span of ecological requirements plays a significant role in determining the range of metabolic products they produce.

Propagation

Microalgae can be grown both in open culture systems such as ponds, lakes and raceways, and in highly controlled closed culture systems similar to those used in commercial fermentation processes. Certain microalgae are very suitable for open system culture where the environmental conditions are very specific, such as high salt or high alkaline ponds, lakes or lagoons. The extreme nature of these environments severely limits the growth of competitive species, though other types of organisms may contaminate the culture. The advantages of such systems are that they are generally require low investment, and are very cost-effective and easy to manage. Closed culture systems, on the other hand, require significantly higher investments and operating costs, but are independent of all variations in agroclimatic conditions and are very closely controlled for optimal performance and quality.

Open culture systems take advantage of natural sunlight and are totally subject to the vagaries of weather unless some form of shading system is utilized. Highly controlled closed systems use photobioreactors for phototrophic culture and conventional fermenters for heterotrophic growth.

The range of sophistication available for the two systems is very great, as is the associated investment. As an example, photobioreactors can vary from simple, externally illuminated glass jars to highly engineered fermenters saturated with light transmitting fibre optic filaments to ensure even lighting to all cells and infused with specific gas mixtures to control metabolism and growth rates. Certain new photobioreactors incorporate an α-type tubular design for greater cost-effectiveness and commercial efficiency (Lee *et al.*, 1995).

Energy production

In the production of energy from microalgal biomass, two basic approaches are employed depending upon the particular organism and the hydrocarbons that they produce. The first is simply the biological conversion of nutrients into lipids or hydrocarbons. The second procedure entails the thermochemical liquefaction of algal biomass into usable hydrocarbons.

Anabolic production of lipids and hydrocarbons by microalgae

Lipids and hydrocarbons can normally be found throughout the microalgal cell mass. They occur as storage product inclusions in the cytoplasm and as functional components of various membranes. In some cases, they are excreted extracellularly into the microalgal colony matrix as almost pure oleaginous globules. In certain cases, the lipid composition can be regulated through the addition or restriction of certain components in the diet. For example, nitrogen or silicon starvation and other stress provocateurs may increase total lipid production.

The type and level of hydrocarbons produced is often affected by environmental factors such as light, temperature, ion concentration and pH. While it is not uncommon to find levels of 20–40% lipids on a dry basis, on occasion the quantities of lipids found in microalgae can be extraordinarily high. For example, in one particular genus, *Botryococcus*, the concentration of hydrocarbons in the dry matter may exceed 90%, under certain conditions (Largeau, 1980).

Thermochemical liquefaction of microalgae

The convenience of microalgal harvesting and handling makes it equally suitable for thermochemical processing such as liquefaction. Following a process reminiscent of the origin of petroleum products, microalgae are converted into oily substances under the influence of high temperature and high pressure. Yields in the 30–40% range of heavy-type oil can be obtained in this manner (Kishimoto, in press). Because of the high levels of proteinatious materials in the system, nitrogen levels leading to NO_x formation have to be carefully controlled. Yields close to 50% of liquid hydrocarbon have been obtained using a very high temperature, high pressure, catalysed hydrogenation process (Chin, 1979). Table 7.73 provides some examples of the lipid contents of various microalgae.

Table 7.73. Lipid contents of different algae.

Strain	% Lipid (on a dry basis)
Scenedesmus spp.	12–40
Chlamydomonas spp.	21
Clorella spp.	14–22
Spirogyra spp.	11–21
Dunaliella spp.	6–8
Euglena spp.	14–20
Prymnesium spp.	22–38
Porphyridium spp.	9–14
Synechoccus spp.	11

Processing and handling

In cases where the hydrocarbons are produced anabolically by the microalgae, direct extraction is the simplest and most effective way of obtaining products. This can be effected through the employment of solvents, through the direct expression of the liquid lipids, or a combination of both methods. The thermochemical liquefaction process often results in a heavy oily or tarry material which is then separated into different fractions by conventional catalytic cracking. As with hydrocarbons derived from other forms of renewable biomass, microalgal lipids can be converted into suitable gasoline and diesel fuels through transesterification.

Future prospects

The future of microalgal fuel production will be dependent upon their economies compared to other sources. Economies of production are steadily coming closer to fossil fuels and will continue to do so in the future. This phenomenon will accelerate as the extent of environmental degradation is factored into the cost of conventional fossil fuels.

Microalgae have the potential to achieve a greater level of photosynthetic efficiency than most other forms of plant life and are very amenable to genetic engineering of their photosynthetic apparatus for increased efficiency. If laboratory production can be effectively scaled up to commercial quantities, levels of up to 200t/ha/year may be obtained. Considering that the carbon source for microalgal hydrocarbon production is CO_2, the net production of CO_2 from fuel utilization will be nil. Whether the production focus is on hydrocarbons, hydrogen or useful energy products from waste conversion, the functional microbial characteristics of microalgae make them important components of all future renewable energy programmes.

Selected references

Chin, L. (1979) Fuels by hydrogenation, Ph.D. Thesis, Pennsylvania State University.

Kishimoto, M. (in press) Oil production. In: Miyamoto, K. (ed.), *Renewable Biological Systems for Alternative Sustainable Energy Production*, FAO Agricultural Services Bulletin.

Largeau, C., *et al.* (1980) *Phytochemistry* 19: 1043–51.

Lee, Y.K., Ding, S.Y., Low, C.S., Chang, Y.C., Forday, W.L. and Chew, P.C. (1995) Design of an α-type Tubular Photobioreactor for Mass Cultivation of Microalgae. *Journal of Applied Phycology* 7(1): 47–51.

Additional Potential Plant Species

Table 7.74. Various potential oil and wax producing plant species (Diercke, 1981; Franke, 1985; Rehm and Espig, 1991).

Botanical name	Common name	Notes/region
Arecoideae		
Euterpe edulis Mart.	Juçara, jiçara, assai	Oil in pulp. Brazil
Jessenia bataua (Mart.) Burret	ungurahui, seje, Patauá, kumbu	Oil in pulp, good tasting, similar to olive oil. The palm is planted near the dwellings of native people. South America
J. polycarpa Karst.	Seje grande, coroba, milpesos, jagua	Oil in pulp, cooking oil of the Chocó Indians. South America
J. repanda Engel	Aricaguá, aricacúa	Oil in pulp. The oil content of the mesocarp and the taste of the oil are similar to seje. Central America
Manicaria saccifera Gaertn.	Sleeve palm, monkey cap palm, temiche, ubussu, guagara, bassu	Oil in seeds, locally used as food. Northern South America
Oenocarpus bacaba Mart.	Turu palm, bacaba	Oil in pulp, good tasting edible oil. The palm is planted near Indian settlements. Northern South America
Betulaceae		
Corylus avellana L.	Haselnuß, hazel, noisetier, coudrier	Nuts contain 60–68% fat

Table 7.74. (cont.)

Botanical name	Common name	Notes/region
Caryocaraceae		
Caryocar villosum (Aubl.) Pers.	Piquiá, pequiá	Mesocarp and seeds contain 70% fat. Tropics
Chrysobalanaceae		
Licania rigida Benth.	Oiticia	Oil contains 73–83% licanic acid, uses similar to tung oil; production of 50,000t seeds (wild grown). Tropics
Cocoidae		
Acrocomia aculeata (Jacj.) Lodd. ex Mart.	Paraguay palm, macaya, micauba, gru-gru	Oil in seeds and pulp, 50–55% oil in the kernels; fruit oil for soap, kernel oil is edible. Tropics
A. mexicana Karw.	Coyol, babosa, Mexican gru-gru palm, wine palm	40% fat in the seeds, sold in local markets. Mexico to Guatemala
A. totai Mart.	Mbocayá palm, totai palm, gru-gru palm	Oil in seeds and pulp; industrial extraction in Paraguay, small amount exported. Latin America
Astrocaryum jauary Mart.	Awarra palm, jauary palm	Edible oil in pulp and seeds. South America
A. murumuru Mart.	Murumuru	43% oil in the seeds, high melting point (33–36°C). Amazon region
A. tucuma Mart.	Tucuma palm, tucum	Oil in pulp and seeds, pulp oil for soap; 30–50% fat in kernels; also exported. Tropical America
A. vulgare Mart.	Awarra palm, cumara palm, aoura palm	Oil in pulp and seeds; grows well on sandy ground. South America
Attalea funifera Mart. ex Spreng	Bahia piasavva, coquilla nut	Oil in seeds, used for margarine and chocolate industry; main product: piassava fibre. South America
Bactris gasipaës H. B. K. (Guilielma gasipaës)	Peribaye, peach palm, paréton piritu, gacipaes, masato, uvito, chonta ruru, pupunha, amana	Oil in seeds, important food plant because of its high-starch fruit flesh; seeds provide macanilla fat. Central and South America
Maximiliana maripa Drude	Maripa, naxáribo	Oil in seeds (maripa fat), fruit flesh tastes of apricot. Guyana
M. regia Mart.	Cocorite palm, cucurita palm, inajá palm jaguá palm	Oil in seeds, similar to *M. maripa*. South America
Orbygnia cohune (Mart.) Dahlgr. ex Standl	Cohune palm, manaca	Oil in seeds; the oil is used as salad oil or for technical purposes. Central America
Scheelea martiana Burret (*Attalea excelsa* Mart. ex Spreng)	Ouricoury, urucuri	Oil in seeds; the kernels are pressed; the underside of the leaves produces wax. Latin America
Syagrus coronata (Mart.) Becc.	Ouricoury, licuri	70% oil in seeds; the leaves provide ouricury wax; the fruit flesh is edible. South America

Table 7.74. (cont.)

Botanical name	Common name	Notes/region
Chrysobalanaceae		
Licania rigida	Oiticicabaum, oiticica	Evergreen tree, occurs mostly wild; seeds contain 55–63% fat, like tung oil. South America
Compositae		
Guizotia abyssinica (L. f.) Cass.	Nigerseed, niger, tilangi, ram-till, nook, nug, guja	Valuable vegetable oil. Tropics, tropical highlands and subtropics
Madia sativa Mol.	Chilean tarweed, coast tarweed, pitchweed, madi, melosa	Fruit contains 35–40% oil, undemanding plant; ancient oil plant of the Indians. Subtropics
Coryphoideae		
Erythea salvadorensis (H. Wendl. ex Becc.) H. E. Moore	Palma	Oil in seeds, used locally. Central America
Cruciferae		
Berteroa incana (L.) DC.	Grey cress	Oil in seeds, used locally
Brassica carinata A. Braun	Abyssinian mustard, Ethiopian cabbage	Oil and vegetable plant. Tropical highlands
B. junkea Czern. et Coss	Chinese mustard, Indian brown mustard, raya, rai	Seeds are used for oil extraction or seasoning. Tropics, subtropics
B. napus L. emend. Metzger var. *napus*	Rapeseed, colza, colsat, navette	Oil for nutrition and technical purposes. Subtropics, temperate zones
B. rapa L. emend. Metzger (*B. campestris*)	Various varieties	Oil of all varieties edible or for technical purposes
var. *dichotoma* Watt	Brown sarson, Indian rape	Temperate zones, subtropics, tropics
var. *sarson* Prain	Indian colza, yellow sarson, sarisa	Temperate zones, subtropics, tropics
var. *silvestris* (Lam) Briggs (var. *oleifera* DC)	Turnip rape, Chinese colza	Cultivation limited to Europe. Temperate zones, subtropics
var. *toria* Duthie & Fuller	Indian rape, toria	Mostly grown with irrigation. Temperate zones, subtropics, tropics
Camelina sativa (L.) Crantz	Leindotter, false flax, gold of pleasure	30–35% drying oil in seeds; originally a weed in flax fields. Temperate regions
Crambe abyssinica	Abyssinian kale, crambe, colewort	The oil is only utilized technically. Tropics, subtropics, tropical highlands
Eruca vesicaria (L.) Cav. ssp. *sativa*	Ölrauke, rocket	26–33% oil in seeds, for nutrition and as fuel. Russia, India
Raphanus sativus L. var. *oleiformis* Pers.	Chinesischer Ölrettich, oil-seed radish, oil radish, fodder radish	Oil in seeds, for nutrition and as fuel. East Asia
Raphanus raphanistrum L.	Hederich	Oil in seeds, used locally

Table 7.74. (cont.)

Botanical name	Common name	Notes/region
Cucurbitaceae		
Telfairia pedata (Sm. ex Sims) Hook.	Kweme, oyster nut tree	Perennial, climbing on trees; fruits, 36% oil in seeds, 60% in shelled seeds; cultivated plants productive for 30 years
Cyperaceae		
Cyperus esculentus	Earth almond, yellow nutsedge, chufa	20–25% oil in tubers, mostly ester of oleic acid, yield 0.5–1.2t/ha. Tropics of East Africa
Dipterocarpaceae		
Shorea stenoptera Burck	Borneo tallow, pontianak kernels, illipe, engkabang	62% saturated fatty acids, 43% stearic acid, 37% oleic acid; several species, but only wild trees, are used. Tropics
Euphorbiaceae		
Aleurites fordii Hemsl., *A. montanus, A. cordatus*	Tung oil tree, alévrite	Related to castor, up to 12m high, seed in capsules; 42–53% drying oil in seed, oil purgative and emetic. Tropics
Sapium sebiferum (L.) Roxb.	Chinese tallow tree, tallow berry, acid; páu de sebo, chii an shu	Mesocarp 62% palmitic acid, 27% oleic kernels 53% linoleic acid, 30% linolenic acid. Subtropics, tropical highlands
Tetracarpidium conophorum Hutchin. et Dalz (syn. *Plukenetia conophora*)	Conophor, awusa nut, African walnut	Seeds contain 54–60% drying oil. Western Africa
Fagaceae		
Fagus sylvatica L.	European beech, hêtre	Seeds contain up to 46% oil, mostly glycerides of oleic and linoleic acid; trees only bloom and form fruit every 5–10 years
Gramineae		
Zea mays L.	Mais, maize, corn, maïs	Embryos, divided during starch production from the endosperm, contain 33–36% oil, by-product of starch production
Labiatae		
Hyptis spicigera Lam.	Moño, nino, kindi, andoka, hard simsim	Seeds eaten or pressed, 24–37% of drying oil; cultivation similar to sesame. Tropics
Perilla frutescens (L.) Britt. (*P. ocymoides* L.)	Perilla, tzu-su, hsiang-sui	Seeds contain 30–51% oil, 63–70% linolenic acid, for cooking oil and technical purposes. Subtropics, tropical highlands

Table 7.74. (cont.)

Botanical name	Common name	Notes/region
Lauraceae		
Persea americana	Avocado	Pulp contains up to 30% oil used for oil production locally in Guatemala
Lecythidaceae		
Bertholletia excelsa Humb. et Bonpl.	Brazil nut tree, paranut tree	Seeds contain about 66% oil, rich in oleic and linoleic acid, used for nutrition
Lepidocaryoorideae		
Mauritia flexuosa L. f.	Burity do brejo, aguaje, bâche, miriti, ita, moriche	Oil in seeds. Northern South America
Linaceae		
Linum usitatissimum L.	Linseed, flax, lin, tétard, lino	Main use for painting materials, biofuels; yields 800–2200kg/ha, linseeds 765–1300kg/ha. Temperate zones, subtropics, tropical highlands
Malvaceae		
Gossypium hirsutum L.	Baumwolle, cotton, coton	16–24% oil in seeds, by-product of fibre production; oil and press cake contain poisonous phenolic dialdehyde; press cake contains 23–44% protein, used as fodder
Myristicaceae		
Virola surinamensis	Ucuhuba	Seeds contain 65% fat, 63–72% is myristic acid; collected from wild trees. Tropics
Papaveraceae		
Papaver somniferum L.	Opium poppy, pavot somnifère, dormideira	Vegetable oil, also used technically. Temperate zones, subtropics, tropical highlands
Pedaliaceae		
Sesamum indicum L.	Sesam, oriental sesame, sésame	Annual herb; seed in capsules, old cultivars with opening capsules; short day plant; non-drying oil, 42–70% protein in the press cake
Polygalaceae		
Polygala butyracea Heckel	Black beniseed, malukang, ankalaki, cheyi	Seeds contain 30% fat (55–60% palmitic acid, 30% oleic acid). Tropics
Madhuca longifolia (J. G. Koenig) Macbr. var. *longifolia*, var. *latifolia* Roxb.) Cheval. (*M. indica* J. F. Gmel.)	Illipé, mahua, moa tree, butter tree, mahawa, mowara	Seeds contain 55–60% fat (23% palmitic acid, 19% stearic acid, 43% oleic acid, 13% linoleic acid), mostly used for technical purposes. Tropics

Table 7.74. (cont.)

Botanical name	Common name	Notes/region
Theaceae		
Camellia sasanqua Thunb.	Teaseed, sasanqua camellia	Seeds contain 40% oil with 72–87% oleic acid, mostly used for nutritional purposes. Tropics
Anacardiaceae		
Toxicodendron succedaneum (L.) O. Kuntze (*Rhus succedanea* L.)	Japan tallow, wax tree, sumac cirier	Fatty oil in kernel; fruit mesocarp contains 65% wax, used for polishes, candles, matches. Subtropics
Euphorbiaceae		
Euphorbia antisyphilitica Zucc.	Candelilla	Wax covers shoots, obtained by boiling, gathered from wild plants, used for polishes, painting materials, waxing of fruits. Tropics
Pedilanthus pavonis Boiss.	Candelilla	Other *Euphorbiaceae* species are used similarly. Tropics
Myriaceae		
Myrica cerifera L.	Wax myrtle, bayberry, arrayán, cera vegetal	Obtained from wild plants; several other *Myrica* species also produce bayberry wax. Subtropics, tropical highlands
Palmae		
Ceroxylon alpinum Bonpl. ex DC. (*C. andicola* Humb. et Bonpl.)	South American wax palm, palma cera	Stem is covered by wax, used locally for production of candles and matches. Tropical highlands
Copernicia alba Morong (*C. australis* Becc.)	Caranday, carandá	Large stands in Mato Grosso, as yet little used. Tropical highlands
C. prunifera (Mill.) H. E. Moore (*C. cerifera* (Arr. da Cam. ex Koster) Mart.)	Carnauba wax palm	Large stands in north-east Brazil, 12,000t annual production; wax covers leaves, high melting point (83–86°C). Tropical highlands
Syagrus coronata (Mart.) Becc.	Ouricury palm, ouricuru, uricuri, licuri, nicuri	Mainly from Bahia, on dry sites; wax from leaves, obtained like carnauba wax. Tropical highlands

Table 7.75. Various sugar and starch producing plant species with potential for ethanol production (Diercke, 1981; El Bassam, 1996; Franke, 1985; Rehm and Espig, 1991).

Botanical name	Common name	Notes/region
Chenopodiaceae		
Beta vulgaris var. *altissima* Döll	Sugar beet	Under favourable conditions: yield 57.4t/ha; sugar content 16% of fresh matter; sugar yield 9.18t/ha; ethanol production 5600 litres/ha. Temperate climates

Table 7.75. (cont.)

Botanical name	Common name	Notes/region
B. vulgaris var. rapacea W.D.J. Koch	Fodder beet, mangel-wurzel	Good yields approach: yield 98.5t/ha; starch content 8.2% of fresh matter; starch yield 8.08t/ha; ethanol production 4923 litres/ha. Temperate climates
Basellaceae		
Ullucus tuberosus Lozano	Ulluco, melloco, papa lisa, kipa uljuco, chugua	Production in Peru about 30,000t/a. Tropical highlands
Cruciferae		
Lepidum meyennii Walp.	Maca, chijura	Cultivated at heights of 3800–4200m. Tropical highlands
Labiatae		
Plectranthus edulis (Vatke) Agnew (*Coleus edulis* Vatke)	Gala dinich, oromo diniche	Tropical highlands
P. esculentus N. E. Br. (*Coleus esculentus* (N. E. Br) G. Tayl.)	Kafir potato, dazo, ndazu, rizga	Widely cultivated in the drier parts of tropical Africa
Solenostemon rotundifolius (Poir.) J. K. Morton (*Coleus rotundifolius* (Poir.) A. Chev. et Perr., *C. parviflorus* Benth.)	Hausa potato, Chinese potato, county potato, fra-fra-salaga, koorkan	Cultivars, generally small tubers; vegetation period 3–4 months. Tropics
Leguminosae		
Flemingia vestita Benth. ex Barker (*Maughania vestita*)	Sòh-phlong	Yields up to 10t/ha, 7 month vegetation period. Tropical highlands
Pachyrhizus ahipa (Wedd.) Parodi	Yam bean, ahipa, jíquima, jícama, ajipa	Old cultivated plant of the Indians. Tropics
P. erosus L. Urb.	Potato bean, yam bean, patate cochon, jícama, sin kama, fan-ko	Young tubers are sweet; tubers over 1 year old are mainly used for starch production; yields up to 95t/ha, propagated by seed. Tropics
P. tuberosus (Lam.) Spreng.	Yam bean, jícama, fejião yacatupé	Similar to *P. erosus*, but the tubers are larger. Tropics
Sphenostylis stenocarpa (Hochst.) Harms	African yam bean, kutonoso, roya,	Growing period 8 months, mostly propagated by seed. Tropics
Nyctaginaceae		
Mirabilis expansa Ruiz et Pav.	Marvel of Peru, mauka	Cultivated in a limited region for its edible roots. Tropical highlands
Oxalidaceae		
Oxalis tuberosa Mol.	Oca, oka, ibia, apillia	Tubers with high sugar and starch content, yields up to 20t/ha; only edible when cooked. Tropical

Table 7.75. (cont.)

Botanical name	Common name	Notes/region
Tropaeolaceae		
Tropeaolum tuberosum	Añu, cubio, isañu, mashua	Yields of 20–30t/ha. Tropical Ruiz et Pav. highlands
Alismataceae		
Sagittaria sagittifolia L. ssp. *leucopetala* (Miq.) Hartog ex van Steenis	Chinese arrowhead, muya muya, ubi keladi	Cultivated in ponds; not edible raw, used for fodder and starch production. Subtropics, tropics
Araceae		
Amorphophallus konjak K. Koch	Elephant yam, konjaku (Japan)	Flour used for the production of vermicelli and cakes. Subtropics, tropics
A. paeoniifolius (Dennst.) Nicolson (*A. campanulatus* Bl. ex Deene.)	Oroy, elephant foot, whitespot arum, telinga potato, suran, pongapong	Old tubers weigh up to 25kg; some forms contain large amounts of oxalate. Tropics
Colocasia esculenta (L.) Schott	Dasheen, taro, 'old' cocoyam, eddo	About 1000 cultivars; single large tuber, cultivated in water (dasheen) or many small daughter tubers, cultivated on dry land (eddoes); 6–15 month growing period, withstands high soil salt contents. Tropics and subtropics
Cyrtosperma chamissonis (Schott) Merr.	Giant swamp taro, maota (Tahiti), gallan, opeves, palauan	Tubers weigh up to 50kg after several years; grows well on coral islands, also on swampy saline soil. Tropics
Xanthosoma sagittifolium (L.) Schott	Yautia, 'new' cocoyam, tannia, tarias, Chinese taro, taye, ocumo, mangarito, taioba	Many cultivated forms. Grown world wide in the tropics
Cannaceae		
Canna edulis Ker-Gawl.	Edible canna, gruya, achira	Yields up to 50t/ha, important fodder. Subtropics, tropics
Cyperaceae		
Eleocharis dulcis (Burm. f.) Trin. ex Henschel (*E. tuberosa* Schult.)	Chinese water chestnut, waternut, pi-tsi, teker, matei (China)	Yields of 20–40t/ha, cultivated in water. Tropics
Dioscorea		
D. alata L.	Water yam, greater yam, winged yam, white yam	Important species; large tubers, 8–10 month growing period
D. bulbifera L.	Aerial yam, potato yam, igname bulbifère, batata de aire	Mostly only aerial tubers edible, tubers store well

Table 7.75. (cont.)

Botanical name	Common name	Notes/region
D. cayenensis L.	Yellow Guinea yam, igname jaune, ñame amarillo	10 month growing period
D. dumetorum (Kunth) Pax	Cluster yam, African bitter yam, three leaved yam	Growing period 8–10 months; both poisonous and non-poisonous forms cultivated
D. esculenta (Lour.) Burk	Asiatic yam, lesser yam, potato yam	Only known in cultivation, no bitter forms; growing period 7–10 months
D. hispida Dennst.	Asiatic bitter yam, intoxicating yam	Mostly gathered wild, all forms poisonous
D. japonica Thunb.	Japanese yam, shan-yu-tsai (China)	Food plant in Japan, medicinal plant in China
D. nummularia Lam.	Kerung (Java), ubing basol (Philippines)	Tubers large, deep lying; usually harvested 2–3 years after planting
D. opposita Thunb. (*D. batatas* Decne.)	Chinese yam, cinnamon yam, igname de Chine	Growing period 6 months; tolerant of cold
D. pentaphyllia	Sand yam, buck yam, ubi passir (Java)	Tubers slightly poisonous
D. rotundata	White Guinea yam, eboe yam, ñame blanco	Growing period 6–10 months; only known in cultivation; large tubers
D. trifida	Cush-cush yam, couche-couche, yampi, mapuey, aja	Growing period 9–10 months
Marantaceae		
Calathea allouia (Aubl.) Lindl.	Guinea arrowroot, sweet corn root, llerén, allouya	Small tubers, similar to potato. Tropics
Maranta arundinacea L.	St Vincent arrowroot, herb aux flèches, araruta	Only cultivated for starch production. Tropics
Solanaceae		
Solanum tuberosum	Potato	Cultivation in the tropics increasing; may yield up to 20t/ha, often below 10t/ha
Zingiberaceae		
Curcuma angustifolia Roxb.	East Indian arrowroot, tikhur, tavakhira	Provides East Indian arrowroot starch. Tropics
Andropogonoideae		
Coix lacryma-jobi	Adlay, capim rosarie, larmes de Job, lágrima de san pedro, Job's tears	Perennial; thin shelled forms are used as a cereal, thick shelled as pearls. Tropics

Table 7.75. (cont.)

Botanical name	Common name	Notes/region
Panicoideae		
Digitaria cruciata (Nees) A. Camus	Raishan	Only cultivated locally in India (Khasi hills)
Echinochloa frumentacea (Roxb.) Link	Barnyard millet, billion dollar grass, sawa, sanwa	Quick ripening, salt tolerant. Cultivated in India, South-east Asia
E. utilis Ohwi et Yabuno	Japanese millet, Japanese barnyard millet	Properties as for *Echinochloa frumentacex* but suitable for cooler climates. Cultivated in Japan, China
Panicum miliaceum L.	Common millet, true millet, mijo común, proso	Beer production, unleavened breads, bird seed. Cultivated in Mediterranean countries to East Asia
P. sumatrense Roth ex Roem. et Schult. (*P. miliare* Lam)	Little millet, samo, samai, kutki (India)	Very undemanding; drought tolerant, also tolerant of wetness. Cultivated in India, Sri Lanka
Paspalum scrobiculatum L.	Bastard millet, ditch millet, koda (India)	Undemanding, drought tolerant; fodder plant. South-east Asia
Pennisetum americanum (L.) Leeke	Pearl millet, bulrush millet, millet à chandelle, mijo perla, dochan, kala sat, bajra (India)	Salt tolerant, drought tolerant. In the Sahel, cultivars are grown with a vegetation period of 60 days. Africa, India
Phalaris canariensis L.	Canary grass, alpiste	Used as bird seed, also green fodder. Mediterranean countries (Morocco), Argentina
Setaria italica (L.) Beauv.	Italian millet, foxtail millet, millet des oiseaux, moha, painço	Less need for warmth than other millets. Mediterranean countries to Japan, particularly China, Central Asia, India
Eragrostoideae		
Eleusine coracan (L.) Gaertn.	Finger millet, dagusa (Ethiopia), ragi (India), kurakkan	Stores well, used for beer production, flat breads. Cultivated in Central and Eastern Africa from India to Japan
Eragrostis tef (Zuccagni), Trotter (*E. abyssinica* (Jacq.) Link)	Tef (Amhari), tafi (Galinya), paturin de Abessinie	Beer production, flat breads, fodder. Ethiopia

Table 7.76. Potential energy plant species in arid and semi-arid regions (S.F. Khalifa).

Botanical name	Common name	Notes
Amaranthaceae		
Aerva		Undershrubs
A. javanica	Eigaab, ghell, toorf, shagaret el-ghazal	Perennial plant, propagated by seed, grows on sandy places; erect, densely stellate, hairy undershrub
A. lanata	Eigaab gebeli	Perennial plant, propagated by seed; grows on sandy and stony places, especially lower parts of mountains; erect woolly undershrub, small leaves

Table 7.76. (cont.)

Botanical name	Common name	Notes
Chenopodiaceae		
Hammada		Shrub with jointed stem
H. elegans	Rimth, remeh, balbal, belbel	Perennial plant, propagated by seed; grows on sandy and stony ground; stout shrub with thick glaucous branches, leaves rudimentary
Kochia		Annual herbs with flat leaves
K. indica	Amm shoor, kokia	Annual plant, propagated by seed, grows in moist conditions; highly branched herb, under favourable conditions reaching a height of 2m
Compositae		
Conyza		Mostly shrubby plants
C. aegyptiaca	Nashash ed-dibbaan	Annual plant, propagated by seed; grows on along canals; densely villous herb,
C. bovei	Qessaniya, belleikh	Perennial plant, propagated by seed; grows on sandy and stony ground; glabrous shrub
C. dioscoridis	Barnoof	Perennial plant, propagated by seed; grows on loamy and sandy places; highly branched, hairy shrub, often 2–3m high
C. linifolia	Hasheesh el-gebl, ain el-katkoot	Annual plant, propagated by seed; grows on canal banks and waste places; densely, pubescent herb, numerous leaves
C. triloba	Heleba, habaq atshaan	Perennial plant, propagated by seed; grows on sandy and stony ground; shrublet
Cyperaceae		
Cyperus		Sometimes with creeping rhizome
C. alopecuroides	Samaar helw, mat sedge	Perennial plant, propagated by rhizomes; grows on moist soils; 1m or more high, up to 2cm broad leaves, nut plano-convex
C. articulatus	Boot, dees medawar	Perennial plant, propagated by rhizomes; grows on marshy places and along canals; stout culms, 1–2m high
C. capitatus	Se'd	Perennial plant, propagated by rhizomes; grows on maritime sand; 10–15 cm long, stout pale green stems
C. conglomeratus	Se'di, oshob	Propagated by rhizomes; grows on coastal sand dunes; 10–30cm, tufted
C. laevigatus	Barbeit, okreish	Perennial plant, propagated by rhizomes; grows on salty lands; stem leafless or with short fleshy leaves
C. michelianus ssp. *pygmaeu*	Se'ed, sa'ad	Annual plant, propagated by rhizomes; grows on muddy and dry clay soils; small, densely tufted, mostly 15cm high
C. mundtii	Qateefa, qateef	Perennial plant, propagated by rhizomes; grows on moist places; stems creeping below, leafing to the middle with short, broad, flat leaves
C. papyrus	Fafeer, bordi, bardi, papyrus, paper reed	Perennial plant, propagated by rhizomes; grows around lakes; leafless, giant plant, usually over 1.5m high, could be cultivated

Table 7.76. (cont.)

Botanical name	Common name	Notes
C. rotundus	Sa'd el-homar, nutsedge, purple nutsedge, purple nutgrass	Perennial plant, propagated by rhizomes; grows on sandy places as well as on moist ground; stem only leafy at the ground, stolons sometimes swollen into tubers that are collected as a drug
Gramineae		
Agrostis		Perennial grasses
A. stolonifera	Creeping bentgrass	Perennial plant, propagated by seed; grows in moist places; procumbent, branched weed
Arundo		Robust perennial reeds
A. donax	Ghaab baladi, ghaab farisi, giant reed, Spanish reed	Perennial plant, propagated by rhizomes; mesophytic, robust reed
Avena		Annuals
A. barbata	Zommeyr, slender oat, barbed oat	Annual plant, propagated by seed; grows on sandy fields; glabrous culm, up to 1m high
A. fatua	Zommeyr, wild oat	Annual plant, propagated by seed; grows on clay and sandy fields; common weed, associated with winter cereals
A. sativa	Zommeyr, oat, oats	Annual herb, propagated by seed; grows on clay and sandy fields
A. sterilis	Zommeyr, sterile oat, winter wild oat, wild red oat	Annual herb, propagated by seed; grows on clay and sandy fields
Bromus		Annual grasses
B. adoensis	Yadaab, yadaab gabal	Annual grass, propagated by seed; grows in sandy places; fine, slender grass, culms glabrous, 30cm tall
B. aegyptiacus	Abufakhour, abu keneitla	Annual grass, propagated by seed; grows on clay and sandy ground; 25–50cm high
B. diandrus	Abufakhour, abu keneitla	Annual grass, propagated by seed; grows on clay and sandy ground; up to 80cm high, softly villous leaves
B. fasciculatus	Sabal abu el-hossein	Annual grass, propagated by seed; grows in sandy places; only 5–10cm high
B. fasciculatus var. *alexandrinus*	Sabal abu el-hossein	Annual grass, propagated by seed; grows on clay and sandy ground; culm upwards, panicle branches, softly pubescent
B. japonicus	Safsoof	Annual grass, propagated by seed; grows in sandy places; culm up to 10cm, hairy leaves
B. lanceolatus	Khafoor, abufakhour	Annual grass, propagated by seed; grows on clay ground; culm 10–15cm
B. madritensis	Khafoor, abufakhour	Annual grass, propagated by seed; grows on sandy and stony ground; medium sized grass, erect contracted panicle, culm glabrous below the panicle
B. rigidus	Khafoor, abufakhour	Annual grass, propagated by seed; grows in sandy places, close in characteristics to the preceding
B. rubens	Imendi, deil et-ta'lab	Annual grass, propagated by seed; grows in sandy fields; small, tufted, softly pubescent grass
B. scoparius	Sabal el-faar, sabat	Annual grass, propagated by seed; grows on clay and sandy ground; culm glabrous, 20–50cm high

Table 7.76. (cont.)

Botanical name	Common name	Notes
B. sinaicus	Sabal abu el-hossein, sasoof	Annual grass, propagated by seed; grows in desert conditions
B. tectorum	Sabal abu el-hossein, sasoof, downy brome, drooping brome	Annual grass, propagated by seed; grows in sandy places; fine, slender grass, culm up to 60cm high
B. unioloides	Khafoor, abu fakhour, rescue grass, prairie grass	Annual grass, propagated by seed; suitable for several habitats; tall grass
Cymbopogon		Perennial, aromatic grasses
C. schoenanthus ssp. *proximus*	Halfa barr., camel grass	Perennial plant, propagated by seed; grows in dry places; stout densely tufted grass with narrow leaves
C. schoenanthus ssp. *schoenanthus*	Hashma, camel grass	Perennial plant, propagated by seed; grows in dry and stony places; stout, densely tufted grass, very slight fragrance
Cynodon		Creeping rhizomes or stolons
C. dactylon	Nigeel baladi, Bermuda grass, Bahama grass, couch grass, dhubgrass	Perennial plant, propagated by creeping rhizome; grows on clay and sandy ground; herb with extensively creeping rhizomes, producing rows of leafy culms
Desmostachya		Perennial rigid grasses
D. bipinnata	Halfa, halfa'a	Perennial plant, propagated by rhizomes; grows on canal banks and waste ground; rigid grass with rosetted very long leaves,
Hordeum		Annual plants
H. leporinum	Abu shtirt, reesh (abu) el hossein, hare barley	Annual plant, propagated by seed; grows on rocky places and dry calcareous sandy or loamy ground; flat leaf blades, small culm
H. marinum	Reesh abu) el hossein (el-hosny)	Annual plant, propagated by seed; grows on sandy soil, Nile alluvium and borders of canals; similar to preceding, but smaller
H. spontaneum	Bahma, bohma, shaeeryia	Annual plant, propagated by seed; grows on desert pastures; tall desert grass
Imperata		Medium sized grasses
I. cylindrica	Halfa, halfaa, alang-alang, blady grass, cogon grass	Perennial plant, propagated by rhizomes; grows in dry or moist sandy places along channels and on waste land; long leaves in a dense basilar rosette
Panicum		Annuals or perennials
P. turgidum	Abu rokba, turgid panic grass, desert grass	Perennial plant, propagated by rhizomes; grows on sandy and stony deserts; desert grass, growing in dense bushes, up to 1m high or more
Pennisetum		Annuals or perennials
P. divisum	Thommaam	Perennial plant, propagated by rootstocks; grows in desert conditions; bushy desert grass, growing in large thickets
P. orientale	Sabat, sasoof	Perennial plant, propagated by rootstocks; grows on stony and rocky ground; usually tufted slender grass with flat leaves
Phragmites		Perennial robust reeds

Table 7.76. (cont.)

Botanical name	Common name	Notes
P. australis	Hagna, ghaab hagna, common reed grass	Perennial reed, propagated by rhizomes; grows in dry or moist places and along channels; robust plant, culm erect and tall, in undisturbed marsh rising some 5m above the water, much shorter in dry areas
P. mauritianus	Hagna, ghaab hagna, heesh-meddaad	Perennial plant, propagated by rhizomes; grows in dry or moist places and along channels; usually tall and stout
Polypogon		Annual grasses
P. monspliensis	Deil el-qott	Annual grass, propagated by seed; grows in moist sandy places, on canal banks and in sandy fields
Saccharum		Tall perennial grasses
S. spontaneum var. *aegyptiacum*	Heesh, boos, boos boos, farisi, wild cane, wild sugarcane, kans grass	Perennial plant, propagated by seed; grows on canal banks, sandy islets, in waste places and also rock fissures; up to 3–5m long, grows in dense clumps, very long leaves
Sorghum		Tall annuals or rarely perennials
S. halepense	Gierraaoo, djarraaoo, Johnson grass	Perennial plant, propagated by rhizomes; grows in sandy and clay places; culm from stout scaly rhizome, up to 150cm tall
S. sudanense	Garawa, Sudangrass	Annual plant, propagated by rhizomes; grows in sandy and clay places; culm 1.5–3m high
S. verticilliflorum	Garawa	Perennial and annual plant, propagated by rhizomes; grows in sandy and clay places; culm 3.5cm high
S. virgatum	Garawa, hasheesh el-faras	Perennial and annual plant, propagated by rhizomes; grows along channels and in fields
Juncaceae		
Juncus		Marsh herbs
J. acutus	Sammar morr, asal	Perennial plant, propagated by rhizomes; grows on salty land, in moist places, and dry sandy localities; stout tufted culm 1m long
J. littoralis	Sammaar	Perennial plant, propagated by rhizomes; grows on salty land, in moist places, and dry sandy localities; tall, pungent, like preceding
J. punctorius var. *exallatus*	Shemoor, dees	Perennial plant, propagated by rhizomes; grows in moist places and in stagnant water; up to 2m high, stem rather thick
J. rigidus	Sammaar morr, sammar hoser el-gibn, sea hard-rush	Perennial plant, propagated by rhizomes; grows on salt marshes and salty land; culms up to 1 m long
Leguminosae		
Alhagi		Spiny shrublet
A. maurorum	Aqool	Perennial plant, propagated by rhizomes; grows on canal banks and in waste places; richly branched, spiny shrublet

Table 7.76. (cont.)

Botanical name	Common name	Notes
Liliaceae		
Asparagus		Perennial plants
A. africanus	Aqool gabal, shoak	Perennial plant, propagated by rootstocks; grows in sandy places; tall, woody, prickly shrub
A. aphyllus	Aqool gabal, shoak	Perennial plant, propagated by rootstocks; grows in sandy places; intricately branched plant, 50–100cm high
A. stipularis	Aqool gabal, shoak, aqool berri	Perennial plant, propagated by rootstocks; grows on sandy and stony ground; stem climbing or trailing over the ground
Asphodelus		Annual or perennial herbs
A. fistulosus var. *tenuifolius*	Baroos, basal ibles, basal esh-sheitaan, fistulate asphodel	Perennial plant, propagated by tubers; grows on sandy and stony ground; small or medium sized herb
A. microcarpus	Basal el-onsol	Perennial plant, propagated by tubers; grows on sandy or stony ground; stout robust herb with root of several spindle shaped tubers, 1–2 cm broad leaves, scape richly branched over 1m high
Pontederiaceae		
Eichhornia		Water weeds
E. crassipes	Yasint el-mayya, ward en-nil, water hyacinth	Perennial plant, propagated by rhizomes and stolons; grows along watery canals; free-floating plant with rosetted glabrous leaves, producing richly branched roots at the nodes; native in Florida, infesting waterways in Egypt
Tiliaceae		
Corchorus		Annual, sometimes perennial, herbs
C. depressus		Perennial plant, propagated by seed; grows in sandy and stony places; mat shaped plant with thick tortuous branches
C. olitorius	Melokhia, long-fruited jute	Annual plant, propagated by seed; grows on loamy and sandy ground; plant glabrous except for petioles
C. trilocularis	Melokhia sheitaani	Annual plant, propagated by seed; grows on loamy and sandy ground; herb with pubescent stem
Typhaceae		
Typha		Tall marsh herbs
T. domingensis	Bordi, bardi, birdi, berdi, dees	Perennial plant, propagated by rhizomes; grows in ditches and marshy places; tall marsh herb, often more than 1.5m high
T. elephantina	Bordi, bardi, birdi, berdi, dees, timeyn, elephant grass	Perennial plant, propagated by rhizomes; grows in ditches and marshy places; taller and more robust than the preceding

Table 7.76. (cont.)

Botanical name	Common name	Notes
Umbelliferae		
Ammi		Annual weeds
A. majus	Khilla sheitaani, greater ammi, toothpick ammi, bishop's weed	Annual plant, propagated by seed; grows in loamy clay places; slender herb
A. visnaga	Khilla, khella, toothpick ammi, lesser bishop's weed	Annual plant, propagated by seed; grows on loamy clay ground; stout tall plant with thick stem

All plants listed in this table are harvested by mowing the aerial vegetative shoots of the plant and combining them into piles for storage.

Table 7.77. Fast growing trees and shrubs for fuelwood production (Duke, 1984; Brewbaker, 1995).

Scientific name	Common name	Description/region	Yields
Mimosaceae			
Acacia seyal	Acacia	Easy to regenerate, main energy species in Sudan; occurs on cracking clay soils	Principal species providing energy, gum and timber in Sudan
Acacia nilotica		Easy to regenerate, can withstand flooding	
Acacia senegal		High coppicing power; produces gum	
Acacia auriculiformis A. Cunn.	Darwin black wattle	Native to the savannas of New Guinea, the islands of the Torres Strait, and northern Australia; widely introduced, e.g. to Africa, India, Fiji, Indonesia, Philippines; from subtropical moist to wet through tropical dry to wet forest life zones. Can produce good fuelwood on poor soils, even where there are extended dry seasons. Tolerates high and low pH, poor soil, sand dunes and savanna.	Yields reported to run higher than 20m^3/ha/year on a 10–20 year rotation; dropping to 8–12m^3/ha/year on poor soils. Density of wood 0.6–0.75g/cm^3, heating value 4800–4900kcal/kg. Wood yields excellent charcoal. Litter, branches and leaves may also be used as fuel. Litter amounts to 4.5–6t/ha/year

Table 7.77. (cont.)

Scientific name	Common name	Description/region	Yields
Acacia mangium Willd.	Forest mangrove, mange	Up to 30m high, native to Australia, with small stands in New Guinea and the Moluccas. Probably capable of growing in tropical very dry to moist through subtropical dry to wet forest life zones	Fast growing, attaining 15m in 5 years, 23 in 9 years. Reported yields 30m³/ha/year, or 20m³/ha/year on poor sites. Heating value 4800–4900kcal/kg; Untended 9 year old stands have yielded 415m³/ha timber
Albizia lebbek (L.) Benth.	East Indian walnut, siris tree, kokko	Up to 30m high tree, native to tropical Africa, Asia, and northern Australia, widely planted and naturalized. Ranging from tropical thorn to tropical wet through subtropical thorn to wet forest life zones	Trees coppice well. Annual wood yields of 5m³/ha and 18–28m³/ha reported. Ripe fruits contain 15% moisture and 17% reducing sugar. Possible fruit yields of 10t/ha suggest an ethanol yield of 1700 l/ha/year
Leucaena leucocep hala (Lam.) de Wit	Wild tamarind, lead tree, jumbie bean, hediondilla	Fast growing tree, up to 20m high, forming dense stands. Native throughout the West Indies, from Bahamas and Cuba and from southern Mexico to northern South America. Requires long, warm, wet growing seasons, doing best in full sun. Trees, propagated by seed or cuttings, coppice well	On 3–8 year old trees, annual wood increment varies from 24 to 100m³/ha, averaging 30–40m³/ha. Dry leucaena wood has 39% of the heating value of fuel oil. Leucaenas are multipurpose trees that withstand almost any type/frequency of pruning or coppicing. Used for fodder, green manure and food.
Leucaena diversifolia		Better than *L. leucocephala* for rainfall of 500–2000m and above	
Albizia falcatarina (L.) Fosberg	Molucca albizia	Native to the eastern islands of the Indonesian archipelago and New Guinea; up to 30m high. Reported to tolerate poor soils. Trees can be closely spaced at 1000–2000 plants per ha	Harvested in the Philippines after 7–8 years, then every 8 years from coppice. Wood soft, with a density of 0.30–0.35g/cm³. In a 9 year old stand the above ground biomass was 102t/ha. Annual net productivity was about 20t/ha

Table 7.77. *(cont.)*

Scientific name	Common name	Description/region	Yields
Prosopis juliflora DC	Velvet mesquite	Small tree, up to 12m high, originated from Central or South America, now pantropically introduced; tolerates drought, grazing, heavy soils and sand as well as saline dry flats; probably ranging from tropical thorn to dry through subtropical thorn to dry forest life zones	Yields of a 15 year rotation 75–100t/ha, of a 10 year rotation 50–60t/ha. The specific gravity of the wood is 0.70 g/cm^3 or higher, the wood burns slowly and evenly, holding the heat well
Samanea saman (Jacq.) Merr.	Rain tree	Large tree, up to 60m high, 80m crown diameter, native from Yucatan and Guatemala to Peru, Bolivia and Brazil, widely introduced elsewhere. Ranging from subtropical very dry to moist through tropical dry to moist forest life zones. Easy to propagate from seed and cuttings	A 15 year old tree yields about 200–275kg pods per season; 10t pods can yield 1150l of ethanol. Branches and stem may be used as fuelwood
Fabaceae			
Dalbergia sissoo Roxb. ex DC	Sisu, sissoo, Indian rosewood	Indigenous to India, Nepal, and Pakistan, now widely grown in the tropics; from sea level to over 1500m. Ranging from subtropical thorn to moist through tropical dry to moist forest life zones. Can stand temperatures from below freezing to nearly 50°C. Propagation may be by direct seeding, seedling transplantation, or with root suckers	Young trees coppice vigorously and reproduce from suckers. A height of 7m has been reported in 20 months. Yields (m^3/ha): 20 year old stand 100, 30 year stand 210, 40 year stand 280, 50 year stand 370, 60 year stand 460. Heating values: sapwood 4908kcal/kg, heartwood 5191kcal/kg. The wood is an excellent fuel and gives good charcoal

Table 7.77. (cont.)

Scientific name	Common name	Description/region	Yields
Sesbania grandiflora (L.) Pers.	Agati, corkwood tree, West Indian pea	Propagated readily by seeding or cuttings, cultivated for fuelwood, fodder, and reforestation. After harvest, shoots resprout with vigour. Rapid growth rate, particularly during first 3–4 years. The wood is white, soft, with a density of only 0.42g/cm³. The tree is frost sensitive, ranging from tropical dry through tropical moist forest life zones; native to many Asian countries	If planted along field edges, 3m³/ha of stacked fuelwood in a 2 year rotation are yielded. In Indonesia it is grown in 5 year rotation and yields 20–25m³/ha
Sesbania bispinosa (Jacq.) W. F. Wight	Canicha, danchi, dunchi fibre	Provides a strong fibre, grown as green manure and for fuelwood production. Annual subshrub, up to 7m tall, native to northern India, China, and Sri Lanka; adapted to wet areas and heavy soils. Ranging from subtropical moist through tropical dry to moist forest life zones	The wood has a density of 0.3g/cm³. In Italy 15t/ha dry matter yields are reported; in the tropics yield may be higher
Myrtaceae			
Melaleuca quinquenervia (Cav.) S. T. Blake	Cajeput	Ranging from subtropical dry to wet through tropical dry forest life zones. Evergreen tree, up to 30m high, native from eastern Australia through Malaysia and Burma, now widely introduced. Forms brackish swamp forests immediately behind mangroves. Propagated by seed or cuttings of immature wood. Source of tea tree oil	Annual wood production 10–16m³/ha air dried wood. Provides an excellent fuel; average heating values (kJ/kg): wood 18,422, bark 25,791, terminal branches 19,301, foliage 20,139. Seven year rotations were suggested in Malaysia. The heat of combustion of the bark is unique. Moderately heavy wood: 0.740–0.785g/cm³

Table 7.77. (cont.)

Scientific name	Common name	Description/region	Yields
Casuarinaceae			
Casuarina cunninghamiana Miq.	River she-oak	Ranging from warm temperate dry to moist through tropical thorn to dry forest life zones; native to eastern and northern Australia, medium sized tree, 15–20m high; has survived 8°C frost	*Casuarina* spp. has dense wood of 0.8–1.2g/cm^3, a heating value of about 5000kcal/kg. The wood burns slowly with little smoke or ash, can be burned green, and is good for charcoal, losing only two-thirds of its weight
Casuarina glauca Sieber	Swamp she-oak	Native to eastern Australia. Grows on estuarine plains that are flooded with brackish tidal water; thrives on seaside dunes, often in the path of ocean spray. In fuelwood plantations trees rapidly regenerate from root sprouts	In Israel reaching 20m in 12–14 years, even on saline water tables. From the fourth year trees shed annually about 4t cones that make a good pellet sized fuel
Casuarina equisetifolia J. R. & G. Forst.	Common ironwood, beefwood, Australian pine, Polynesian ironwood	Evergreen tree, up to 30m high, indigenous from Indonesia and Malaysia to India and Sri Lanka, and north-east Australia; ranging from subtropical woodland to wet through tropical thorn to wet forest life zones. Does not coppice readily, works under a clear felling system, with a rotation of 7–35 years	With spacing of 2m and 7–10 year rotation, trees may yield 75–200t/ha wood, i.e. 10–20t/ha/year. In China litterfall is 4t/ha/year. Wood burns with immense heat, even when green, heating a good charcoal; calorific values: wood 4956kcal/kg, charcoal 7181kcal/kg
Casuarina junghuhniana Miq.	Jemara	Up to 35m high, native to highlands of eastern Indonesia to East Java, Bali, lesser Sunda Islands; pioneers natural revegetation of deforested grassland. There are commercial plantings in salt marsh areas, sometimes inundated with saline water. Propagated by cuttings, or by coppice or root sprouts	As other *Casuarina* spp.

Selected references

Brewbaker, J.L. (1995) *Leucaena*, New Crop Fact Sheet, via Internet. http://www.hort.purdue.edu/newcrop/crops/cropfactsheets/leucaena.html

Diercke (1981) *Weltwirtschaftsatlas* 1. Deutscher Taschenbuchverlag/Westermann, Germany

Duke, J.A. (1984) *Handbook of Energy Crops*, published via Internet. http:/www.hort.purdue.edu/newcrop/indices/

El Bassam, N. (1996) *Renewable Energy – Potential Energy Crops for Europe and the Mediterranean Region*, REU Technical Series 46, FAO.

Franke, W. (1985) *Nutzpflanzenkunde*, 3rd edn, Georg Thieme Verlag, Stuttgart and New York.

Khalifa, S.F. (nd) *Report on Potential Energy Crops of the Mediterranean Region in Egypt*, Ain Shams University, Abbassia, Cairo, Egypt.

Rehm, S. and Espig, G. (1991) *The Cultivated Plants of the Tropics and Subtropics*, Verlag Joseph Margraf, Weikersheim, Germany.

Appendices

1. Addresses

Author

Dr N. El Bassam
Institute of Crop Science
Federal Agricultural Research Centre (FAL)
Bundesallee 50
38116 Braunschweig
Germany
Tel: +49 531 596 605
Fax: +49 531 596 365
Email: ElBassam@pf.fal.de

Contributors

Dr Stephen Gaya Agong
Department of Horticulture
Jomo Kenyatta University of Agriculture and Technology (JKUAT)
PO Box 62000
Nairobi
Kenya
Tel: +254-151-22646-9 Ext. 4430
Fax: +254-151-21764
Email: jku_lib@nbnet.co.ke

Dr Wolfgang Bacher
Institute of Crop Science
Federal Agricultural Research Centre (FAL)
Bundesallee 50
38116 Braunschweig
Germany
Tel: +49 531 596 633
Fax: +49 531 596 365
Email: Bacher@pf.fal.de

Ing. Carlo Baldelli
AGRICOLA ARNA
Via Archimede 161
00197 Rome
Italy
Tel: +39 6 8070620
Fax: +39 6 8083617
Email: bald003@it.net

Dr Dudley G. Christian
Institute of Arable Crops Research (IACR)
Rothamsted, Harpenden,
Hertfordshire AL5 2JQ
UK
Tel: +44 1582 763133
Fax: +44 1582 760 981
Email: dudley.christian@bbsrc.ac.uk

Prof. Dr Li Dajue
Beijing Botanical Garden
Institute of Botany
Chinese Academy of Sciences
Beijing 100093
China
Tel: +86 10 62591431
Fax: +86 10 62592686
Email: Lidj@ns.ibcas.ac.cn

Prof. Constantine Dalianis
Centre for Renewable Energy Resources (CRES)
19th Km. Marathonos Ave.
190 09 Pikermi
Greece
Tel: +30 1 6039900
Fax: +30 1 6039905
Email: Dalianis@cresdb.cress.ariadne-t.gr

Dr ir. Wolter Elbersen
Agrotechnological Research Institute
(ATO-DLO)
Bornsesteeg 59
PO Box 17
6700M Wageningen
The Netherlands
Tel: +31 317 475000
Fax: +31 317 475347
Email: elbersen@tref.nl

Prof. Jesús Fernández González
Botánica Agrícola
E.T.S. de Ingenieros Agrónomos
Universidad Politécnica de Madrid
Av. Complutense
28040 Madrid
Spain
Tel: +34 1 5492470 / 5492692
Fax: +34 1 5498482

Dr Ashok Kumar Gupta
Department of Agronomy
Rajasthan Agricultural University
S.K.N. College of Agriculture
At & PO Jobner – 303 329
Dist. Jaipur
India
Tel: +91 1425 4736

Dr Katrin Jakob
Institute of Crop Science
Federal Agricultural Research Centre (FAL)
Bundesallee 50
38116 Braunschweig
Germany
Tel: +49 531 596 609
Fax: +49 531 596 365
Email: Jakob@pf.fal.de

Prof. Dr Sayed Farag Khalifa
Director of Flora and Phytotaxonomy Dept
Department of Botany,
Faculty of Science
Ain Shams University
Abbassia, Cairo
Egypt
Tel: +20 2 821633
Fax: +20 2 822284

Dr Vlasta Petriková
Research Institute of Crop Production
Drnovská 507
16106 Prague
Czech Republic
Tel: +42 2 360851
Fax: +42 2 365228

Dr Arthur Riedacker
Institut National de la Recherche
Agronomique (INRA)
65, boulevard de Brandebourg
94205 Ivry Cedex
France
Tel: +33 1 49 59 69 89
Fax: +33 1 46 70 41 13
Email: riedacke@ivry.inra.fr

Dr Shika Roy
University of Rajasthan
Department of Botany
Jaipur Rajasthan
India

Mr Morton Satin
Agriculture Department
Food and Agriculture Organization of the
United Nations (FAO)
Via delle Terme di Caracalla
I-00100 Rome
Italy
Tel: +39 6 5225 3334
Fax: +39 6 5225 4960
Email: Morton.Satin@fao.org

Prof. Thomas Silou
PRANCongo Equipe Pluridisciplinaire de
Rechercheen Alimentation et en
Nutrition
Equipe mixte DGRST-UMNG
BP 389, Brazzaville
Congo

H. K. Were
Crop Science Department
University of Nairobi
PO Box 29053
Nairobi
Kenya

2. Acronyms

ABE acetone butanol ethanol (fermentation)
BIG biomass integrated gasifier
CAM Crassulacean acid metabolism
CAP Common Agriculture Policy (of the European Union)
CC combined cycle
CHP combined heat and power
CEST condensing extraction steam turbine
DM dry matter
Ecu European Currency Unit
EPA Environmental Protection Agency
ETBE ethyl tertio butyl ether
EU European Union
FCV fuel cell vehicle
GEF Global Environment Facility of the World Bank
GT gas turbine
GTCC gas turbine combined cycle
ICEV internal combustion engine vehicle
MSW municipal solid waste
MTBE methyl tertio butyl ether
NPP net primary productivity (t/ha/yr)
NUE nutrient use efficient
ODT oven dried tonne
PAR photosynthetically active radiation
PEM proton exchange membrane
PET potential evapotranspiration
P/PET precipitation/PET
PV photovoltaics
RIGES Renewable Intensive Global Energy Scenario
RME rape seed methyl ester
RVP Reid vapour pressure
SME sunflower methyl ester
SRC short rotation coppice
STIG steam injected gas turbine
ISTIG intercooled STIG
TER total energy radiation
WUE water use efficiency

3. Units and conversion factors

International units

J = Joule
W = Watt
Wh = Watt hour
1Btu (British Thermal Unit) = 1055J = 0.2931Wh
1kWh = 3600J
1W = $1Js^{-1}$
1 calorie (cal) = 4.19J
1 metric tonne (t) = 1000kg

Unit prefixes

n = nano = 10^{-9}
μ = micro = 10^{-6}
m = milli = 10^{-3}
c = centi = 10^{-2}
d = deci = 10^{-1}
da = deka = 10
h = hecto = 10^{2}
k = kilo = 10^{3}
M = mega = 10^{6}
G = giga = 10^{9}
T = tera = 10^{12}
P = peta = 10^{15}
E = exa = 10^{18}

Energy contents (in lower heating values, LHV)

1t wood = approx. 15GJ (air dry, 20% moisture); 20GJ (0% moisture)
1t oil equivalent = 41.868GJ
1t coal = 29.308GJ
1t charcoal = 28GJ (approx.)
1 cubic metre (cm^3) gas = 31.736kJ
1 bbl = 1 barrel oil = 42 US gallons = 159 litres = 1/7t (approx.)

Chemicals

C = carbon
N = nitrogen
P = phosphorous
K = potassium

CO = carbon monoxide
CO_2 = carbon dioxide
SO_x = sulphuric oxides
NO_x = nitrogen oxides
N_2O = nitrous oxide
CH_2OH = methanol
C_2H_4OH = ethanol

Metric conversions

LENGTH
1 inch (in) = 2.540 centimetres (cm)
1 foot (ft) = 0.3048 metres (m)
1 yard (yd) = 0.9144 metres (m)
1 mile = 1.609 kilometres (km)

AREA
1 square inch = 6.452 square centimetres (cm^2)
1 square foot = 0.09290 square metres (m^2)
1 square yard = 0.8361 square metres (m^2)
1 acre = 0.4047 hectares (ha)
1 square mile = 2.590 square kilometres (km^2)

VOLUME
1 cubic inch = 16.387 cubic centimetres (cm^3)
1 cubic inch = 0.016387 litres (l)
1 cubic foot = 0.02832 cubic metres (m^3)
1 US quart (liquid) (qt)= 0.9464 litres (l)
1 US gallon (liquid) (gal)= 3.785 litres (l)

MASS
1 ounce (dry) = 0.02835 kilograms (kg)
1 pound (lb) = 0.4536 kilograms (kg)
1 ton (2240 lb) = 1016.05 kilograms (kg)
1 ton (2240 lb) = 1.01605 tonne (t)

TEMPERATURE
Celsius (C), Fahrenheit (F), Kelvin (K)
$C = (F - 32) / 1.8$
$F = (1.8 \times C) + 32$
$K = C + 273.15 = 273.15 + (F - 32) / 1.8$

Conversion factors

	kJ	kcal	kWh	kg coal	kg oil	m^3 gas
1 kilojoule (kJ)	–	0.2388	0.000278	0.000034	0.000024	0.000032
1 kilocalorie (kcal)	4.1868	–	0.001163	0.000143	0.0001	0.00013
1 kilowatt hour (kWh)	3600	860	–	0.123	0.086	0.113
1kg coal	29 308	7000	8.14	–	0.7	0.923
1kg crude oil	41 868	10 000	11.63	1.486	–	1.319
1m³ natural gas	31 736	7580	8.816	1.083	0.758	–

Index

1 General terms

2 Energy plant species by common names

3 Energy plant species by scientific names